Dudley's Universe

A celebration of Light, Time, Gravity and The Universal Consciousness

D.A.O. Basson

First Published in South Africa 2010
by
New Voices Publishing
Cape Town, South Africa
www.newvoices.co.za

First Edition November 2009 in paperback
re-print October 2010
re-print October 2012

ISBN-13: 978-1-920094-93-5

CONTENTS

for Colette
a very special daughter

INTRODUCTION

Yet another book on physics for popular reading? Yes, the case is easily made. We take you on two scary roller coaster rides: one through time from the start of the Big Bang to the present and the other through the scale of the Universe from the unimaginably small to the unimaginably large and a good look an the interaction of human consciousness along the way. It has been a lot of fun writing this book and it is hoped that readers will be able to share in the excitement.

Our appreciation of the Universe will take us on a quest of perception embracing not only our natural faculties with or without the aid of instruments, but also the astonishing discoveries by the great names of theoretical physics and mathematics. Our quest will culminate in the perception of a multi-dimensional Universe through the ultimate mode of perception – absolute consciousness.

We must, of course, refer to the giants of science history, Einstein, Planck, Maxwell, Newton and many more going all the way back to Anaxagoras, but the broad canvas attempted here will cause us to refer to some unexpected bedfellows who will pop up from time to time as delicious surprises. We will also look into the work of a sadly neglected modern Leonardo da Vinci, Nikola Tesla, whose ideas might well have mind boggling implications for the people of this planet. The book is interspersed with brief biographical vignettes of the great and famous with their triumphs, tragedies and all too human frailties.

In dealing with aspects of physics it becomes necessary to mention some very large and small numbers. A string of leading or trailing zeroes right across the page is of no use to anyone. We make use of the SI prefixes for the extreme numbers. Everyone will be familiar with milli, kilo and mega, but if you are not sure about femtosecond or zeptometre please consult the full list of SI prefixes at the start of the appendix and examples of how to deal with extreme quantities. It is really quite easy. The appendix is there with some very juicy material to be read with the rest of the book. It has only been removed to the end so as not to break the speed of the roller coaster ride.

The subject matter is presented without a mathematical rodomontade in what is essentially an intensely mathematical subject. The term 'billion' where used will refer to the 'Short Scale' *échelle court* definition (10^9) as used in the USA and was officially accepted by the UK in 1974.

ACKNOWLEDGEMENTS

The list of indebtedness extends to several very special people as well as to some inanimate publications. First and foremost to Heidi Schmutz and Paul van der Merwe. In the early nineties, we would often have tea break discussions on cosmological and ecological matters usually centred on articles in the *Scientific American* magazine and it was these which kept my interest in physics alive. Heidi was a young pretty mathematician who usually wore a bow of broad white ribbon in her hair and was the only person that I ever knew who kept an Impala antelope as a house pet. I often wondered how the *bokkie* managed to find a purchase for his hooves on the smooth floors of the farmhouse. Heidi was on my staff at the time as a mainframe computer systems programmer and had also of necessity become expert at fielding remarks about her surname from people who could understand German. Paul was a man of many and varied talents and the three of us were working for a manufacturer that employed engineers, physicists and mathematicians. These were good times.

The origins of this book go back to 1964 when I was staying in student lodgings. There was a small rectangular thatched room in the garden occupied by a post graduate engineering student Leon Malherbe. There were no 'bad weeds' about Leon despite his amusing surname. One evening Leon came home with a large green covered book called *Occult Chemistry* by C.W. Leadbeater which he had taken out at the local Theosophical library. I had never heard of Leadbeater or the Theosophical society. Leon was troubled. He had taken a great interest in the writings of Leadbeater but *Occult Chemistry* presented a huge problem. It made no sense at all. Anyone who had taken science as a school subject had a good idea of what atoms were supposed to look like but this book showed atoms as complex structures within structures and containing no solid material at all. Why would anyone go to the huge effort of writing a large book only to be exposed to ridicule for writing nonsense? I could not help and thought no more about it but the drawing of the hydrogen nucleus imprinted on my memory. Leon was in poor health at the time and sadly died a few years later without ever realising what rich treasures lay between the green boards of that strange book. The next stage of this story came as a shock when my May 1985 issue of the *National Geographic* magazine arrived in the post. My attention was immediately drawn to an article titled 'Worlds Within The Atom'. There for all to see was a drawing of a proton containing three quarks linked by forces in a triangle. This was exactly what I had seen in *Occult*

Chemistry except that there the quarks were drawn in more detail showing each quark to contain three stringy objects coiled up to an unimaginable degree. The heady exhilaration of this discovery was short lived. It was impossible to speak to anyone about it. People knowledgeable in particle physics would react as if I mentioned that I had a notifiable disease at the slightest mention of clairvoyance and people sympathetic to clairvoyant observation had not the slightest clue about physics. When *Occult Chemistry* was written the atomic electrons presented a major problem. At the time electrons were thought to whiz around the nucleus in orbits like a tiny solar system but Leadbeater could show no such thing – instead the atoms were drawn with inexplicable envelopes. A major bonus in the magazine article was to find the atomic electrons shown as orbitals – clouds as superpositions of probability density functions where the electrons could appear anywhere at any time and looking almost identical to the clairvoyantly observed electrons. When C. Jinarajadasa who wrote the introduction to *Occult Chemistry* was asked what could be done to reconcile the book to modern science he replied, "nothing – leave the scientists to catch up". It has taken one hundred years for this to happen! There remains yet another tantalising morsel of a problem. In *Occult Chemistry* the quarks are also shown with envelopes – could the quarks be enclosed in orbitals of gamma wavelength leptons?

A special word of thanks must go to Mr. Gabriel Korvink for proofreading and comments. A good friend since high school days. Time will tell whether I should have heeded his advice to leave out the science and concentrate on the spooky stuff. Thanks also to Monsieur Pierre Marais, a retired chemical engineer, for his much appreciated comments and encouragement and especially his help with geological matters. Thanks also to Dr. Ian Gaigher for his help with genetic matters and Prof. Alko Meijer for suggestions about the great mathematicians. A final word of thanks to Mr. Derek Woodburn, my Dickensian friend Steerforth since high school days, for critical comment, moderator services and generous tolerance of some outrageous ideas despite his Calvinistic background.

SOMETHING COMES FROM NOTHING

We begin our quest of perception of the universe with fundamentals and must first establish the basic definitions. The method employed, to use computer parlance, is to 'reboot' a hypothetical universe.

We start with nothing. Really nothing. Not even time or space. This degree of nothingness places us in deep philosophical waters where few can swim so we move quickly on to deal with fundamentals. We start our universe with a dimensionless point. We now add more points equally spaced in a line. This gives us a one dimensional line of points. The spacing of the points represents the ultimate digital granularity of space so there cannot be other points between them – the spaces remain undefined. We now replicate the points sideways to produce a two dimensional plane of points and then one step further, move the plane sideways again to obtain a three dimensional matrix of points. It would be tempting to now flesh-out some of the points so that we can have a 'thing' which is not 'nothing'. The choice of material is simple – there is only one – energy, or if you prefer 'Light'. Energy however cannot be made to perch on dimensionless points no matter how little we try to use and energy cannot exist without time and movement. Our model is still totally inadequate.

Let us now add time. We take our model as before and move it sideways again with the same digital spacing. Most people will be hard pressed to imagine a time vector at right angles to all three orthogonal coordinates of a three dimensional matrix, but mathematically this is no problem at all. We now have for our three dimensional model two digital cosmological instants of time. There is yet another problem with our model. There is no way for energy quanta to pass from one point to another as they cannot occupy undefined space on the way. Movement is of course from one time instant to another. A single time instant is completely static with no possibility of change. How can the energy quanta get from one point in one time instant to a point in the next? This is not so simple. We have to add seven extra dimensions to the familiar four that we have already. These extra dimensions, called

Kaluza-Klein dimensions, are curled up to an unimaginable degree and provide both a perch for the energy quanta as well as a means of skipping to the next time frame. The concept of dimensions beyond the familiar three is not new. As early as 1919 German physicist Theodor Franz Eduard Kaluza (1885–1954) presented a paper to Einstein proposing an extra curled up dimension in addition to the four space-time dimensions. This dimension was circular with a length of the smallest possible natural length, the Planck Length. When viewed while taking the time dimension into account, this dimension becomes a spiral. The purpose of Kaluza's fifth dimension was to provide a means of unifying Einstein's theory of gravitation with Maxwell's electromagnetic theory of light. Despite its early promise, Kaluza's theory did not survive the introduction of Max Planck's quantum theory which described electromagnetic energy as discrete packets of energy called quanta. Swedish physicist Oskar Klein (1894–1977) applied Kaluza's theory in 1926 in an attempt to accommodate quantum theory. Kaluza-Klein theory would later become an essential part of modern string theory. Matter energy particles contained in the matrix are configured in a very particular way and are called superstrings – these are the ultimate 'particles' of all matter – strings of dimensionless points which will be described more fully in chapter nine. We now have an eleven dimensional universe: one time dimension, the familiar three spatial dimensions and seven Kaluza-Klein dimensions. Why seven curled up dimensions? There are two ways of arriving at this number. There is a highly complex mathematical method which is too advanced to be given here or the number can simply be determined by counting! This is not a facetious answer – it really can be done and will be taken up again in chapter nine. It is the Kaluza-Klein dimensionality that determines the structure of the superstring so that matter energy cannot exist in any other form. This is not to say that there is only one configuration for the superstring. The Kaluza Klein dimensionality also allows for the smaller and simplified string of the electron. There is at present no evidence to support the idea of a string structure for the photon. We will deal with the detailed format and dimensionality of superstrings, known mathematically as manifolds, in chapter nine.

At this stage, we should also mention that there are theories of space dimensionality that propose different numbers of dimensions and some that maintain that the entire universe is so tightly curled up that objects separated by light years are actually only separated by atomic distances in higher dimensions. A universe rolled up tightly to the size of a football may be too much for most people to swallow but mathematically this is not absurd.

The eleven dimensional universe that we have arrived at here refers to the physical objective universe. In chapter nine, we will take a look at a hierarchy of dimensionality of which the physical objective universe forms only a subset.

Our universe model is now ready to be switched on. The time dimension can now start replicating with the same digital spacing. If some particles could be completely at rest, they would retain the same position in each successive set of coordinates. For an object to move it must first be accelerated and a force must be applied to cause occasional acceleration flips to adjacent points in subsequent time frames. An object in constant motion will not require sideways flips. A moving object or 'observer' has its own coordinates and its own perception of the speed of light. The accelerating force works exactly the same way as gravity and is indistinguishable. Mass, gravity and acceleration are inseparable and will be dealt with later. For successive time frames the energy particles can make only one flip – in the time direction or one of the three space directions. Objects at rest will therefore move at the velocity of light in the time direction. Flipping only in a space direction would result in the maximum speed possible but this can only be done by photon energy quanta which cannot move along the time vector and can have no mass. Objects in motion will appear to lose some time and seem to function more slowly to outside observers but they do not get left behind by the time vector. Past time slices cannot support movement or any interference from the present at all.

At this point, it would be appropriate to take a look at William Blake's famous painting *The Ancient of Days* depicting the creation of the Universe. We could now describe it as The Grand Geometrician of the Universe placing the point of the compasses on the Centre and about to start the Big Bang.

What of the three dimensional frames left in the wake of the switched on time vector? Do they disappear? No, they cannot go away and not a single particle can be changed. The energy is of course carried from the present instant to the next but the past frames are not left blank. There is quite a close analogy to time on a reel of IMAX film. The frame right at the end is the present instant where more frames are being added and the rest of the reel is the past. A closer analogy can be found in an old fashioned fairground peep show box typically titled 'What the butler saw'. Only the more senior amongst us will be able to remember these. The box contains a wad of paper with pictures similar to movie frames which can be flipped through by cranking a handle to see the 'movie'. In our space-time model the thickness of the paper and the grain of the pictures represent the space-time grid spacing.

Shortly after writing this paragraph, I visited a magnificent exhibition of Meccano models and there amongst the dockside cranes, vehicles and beam engines was a working model of a peep show titled 'What the butler saw'.

What has happened to the past? It is still there but you can't send in the paparazzi. To search and replay the past, also known as the Akashic Record, would require consciousness outside the space-time continuum – a daunting task but more of this later.

A quote from Albert Einstein:

The distinction between past, present and future is only a stubbornly persistent illusion.

Thus far our model is still very much over-simplified: matter, after all, is not constituted from individual superstrings. The next stage of particle is the quark – a tightly bound and very stable grouping of three superstrings. A detailed description of the grouping of strings within the quark is given in chapter nine. The strings carry the property of electric charge which is positive or negative depending on rotational direction (chirality). Quarks can have charges of 1 or 2 positive or negative string charges. There are many possibilities for quark configurations but the only two that we will consider are u-quarks and d-quarks which are the only ones needed for the world that we are familiar with. U-quarks contain three strings which add up to two positive string charges. D-quarks contain three strings which add up to one negative charge. We now come to the familiar building blocks of matter – protons and neutrons.

Protons consist of two u-quarks and one d-quark giving them a nett charge of three positive string charges. Neutrons contain two d-quarks and one u-quark resulting in no charge. We need at this stage to mention the electron - also a superstring 'particle' with mass and a total of three negative charges. The electron is one of the most fundamental particles of matter - one would intuitively feel that it should have a single fundamental unit of electric charge but this leaves us with the problem of fractional charges for the quarks. At the present time there is no established structure for the electron available. It is not sufficient to assume that the electron is a group of three negative strings as this does not account for the very small mass of the electron. The electron is a highly complex phenomenon with a large 'presence', remarkable wave properties and a very intimate relationship with massless photons.

The stability of superstrings, quarks and nucleons is phenomenal; most of these particles which exist today have survived since the Big Bang. Curiously, atomic nuclei are only stable if they contain a correct excess of neutrons over protons, and neutrons are only stable when contained within an atomic nucleus. Before we leave the structures of the basic particles, it is necessary to mention forces which are essential to their existence. An atomic nucleus is made up of protons and neutrons in approximately equal numbers. There is a tremendous force of electric repulsion here which should make the protons fly apart at nearly the speed of light – this does not happen – the nucleus is held together with phenomenal stability by what is termed the 'strong force'. There is a similar scenario to be found within the protons and neutrons. These nucleons contain quarks of like and unlike charges which should cause part of the nucleon to collapse. Again there are strong forces at work to provide great stability. The scene is again repeated within the quarks. The quarks contain three strings which always add to one or two string charges so that there will always be an odd member in the quark. Again the quark strings are bound together with great stability. We will return to the 'strong force' in a later chapter.

Now that we have established nucleons and electrons, the periodic table and the world of chemistry unfold. This is the heavy engineering side of particle physics. We take leave of chemistry here as this vast subject is beyond the scope of our present concern.

Thus far, we have described the fundamentals of space and time but the playing field is nothing without the players – in this case the Big Bang. In 1927, Msgr. Georges Henri Joseph Eduard Lemaître (1894–1966) proposed that the universe was continuously expanding after an initial creative explosion, and which would later become popularly known as the 'Big Bang'. This idea had previously been suggested by Alexander Alexandrovich Friedland in 1922 but met with skepticism from Einstein. Georges Lemaître, a most astonishing person, was a Belgian priest, professor of physics and an astronomer. He received support for his expanding universe theory from astrophysicist Sir Arthur Eddington. The theory finally received universal acceptance when Edwin Powell Hubble (1889–1953) and Milton L. Humanson, in 1929, discovered a rough proportionality of galaxy distances and their redshifts. The redshifts of distant galaxies had been noticed by other astronomers without them appreciating the great significance. Hubble deduced that there was only one explanation – the entire universe is expanding and the red shift is due to the effect of the galaxies moving away with increasing velocity the further they were away. No particular part of the universe is the centre – the centre is anywhere so wherever you happen to be the galaxies will be moving away from you and away from each other. The expansion of the universe is by no means a simple matter. It has been found that the expansion applies only to space between galaxies but not to space within the galaxies and that the expansion is taking place at a greater rate between extremely distant galaxies. The Big Bang is an awesome proposition to contemplate – infinite temperature and energy density – ideas far beyond the scope of human comprehension. No one can hope to have a feeling for the entire universe contained within the space of a single atom. Can you take a bar of steel or a cup of water and compress it to half its volume? – no, of course not. There is no physical means at our disposal to do this, not even with high explosives yet cosmologists have been able to backtrack the Big Bang to the size of an atom! Modern science can give us a description of the progress of the Big Bang in steps of Planck Time but this unit is so small that it is little more than incomprehensible numbers printed on paper. The Planck Time, $5,39 \times 10^{-44}$ second, is the time taken by light to travel the distance of one Planck length – a distance far smaller than the smallest subatomic particle.

The following is the latest idea of the progress of the Big Bang. It starts with an instantaneous burst of energy. The ancient author who described the Creation with "Let there be Light" wrote with an uncanny foresight of modern physics. We skip the initial stages as this is a matter of theoretical physics taken to extremes. After about a femtosecond we reach the stage of quarks forming protons, neutrons and mesons. After one second protons and neutrons start forming the nuclei of hydrogen and also some of the very light elements. It is only after about 300 000 years that electrons bind with the nuclei to form the first atoms. About one million years after the Big Bang stars and galaxies begin to form – but how?

We now have a universe exploding at the speed of light and containing hydrogen and helium. This is primordial chaos and totally inhospitable to life of any description. But we are here and have cars, computers, books and DVD players. How did all this emerge from infinite chaos? We can take a look at the laws of thermodynamics and in particular the concept of entropy. Entropy is a measure of the availability of energy. Without an input of energy the availability of energy can only become less – never more. Simply put – if you take quantities of hot and cold water and mix them you will get water with a temperature somewhere in between and no amount of additional mixing will ever result in the hot and cold water that you started with.

The laws of thermodynamics treat fluids as statistical populations of vast numbers of molecules in a closed system. They cannot account for the chaotic nature of particles in an expanding universe. Scientists have had to look elsewhere, and fortunately have found two new branches of science waiting in the wings: Chaos Theory and Fractals facilitated by the number-crunching capabilities of modern supercomputers.

Chaos theory found prominence in meteorological calculations where it was found that very slight differences in initial data values had a profound effect on final results. This is exactly what we need to describe the origins of star and galaxy formation from a vast expanse of hydrogen and helium. In order for objects to form due to gravitational attraction it is necessary for there to be some unevenness in the distribution of the gases.

In the early part of the twentieth century, our own Milky Way galaxy with its millions of stars was thought to be the entire universe. A shocking awakening was to come. In 1924, Edwin Hubble discovered that the nebula in the Andromeda constellation, a fuzzy patch of light with approximately the apparent size of the Moon, was not a cloud of dust illuminated by starlight as previously thought, but in fact a galaxy of millions of stars much like our own Milky Way. From 1758 to 1782 Charles Messier (1730–1817) made a study of celestial nebulae without realising that these were in fact galaxies. The Messier catalogue of 110 galaxies is still in use to this day although this represents only a tiny part of the millions of galaxies in the universe. The universe was now many millions of times larger than originally thought. We now have a problem of measurement – how can we possibly know how large and far away the galaxies are? With nearby stars in our own galaxy the distance can be computed by taking the orbit of the Earth as a base line and measuring the parallax angle of the star. This has resulted in the measurement unit called the parsec (parallax arc second – 3,262 light years). Distances between galaxies are commonly given in megaparsecs. A second of arc is an extremely small angle – smaller than can be measured by ordinary land survey instruments. A spread of less than 5 mm at a distance of one kilometre may seem the limit but the new VLTs (Very Large telescopes) with their ten metre diameter mirrors and interferometric methods are able to resolve points down to the milliarcsecond range. Points 5 mm apart at a distance of one thousand kilometres would be a hopeless aspiration for terrestrial telescopes were it not for computer controlled adaptive optics which can compensate for star twinkling caused by currents of variable density air in the atmosphere. Adaptive optics is discussed in more detail in chapter six. There is no possibility of taking parallax measurements on galaxies – enter the cepheid. Cepheids are large stars with a most remarkable property. They are variable stars which go bright and dim with precise regularity. These are stars which have exhausted their hydrogen and have started consuming helium. They are so bright that they can be seen in distant galaxies but the most remarkable property of all is that their brightness is closely related to their periodicity. By measuring the period of the brightness cycle, it is possible to compute the actual brightness of the star, and by measuring the observed brightness it becomes possible to compute the distance from the Earth. What could possibly be more convenient. There are actually two types of cepheids. More detail of cepheids is given in chapter five. A new technique has been developed for measuring the distances of extremely distant galaxies using red-shift measurements on supernovas.

A groundbreaking experiment was launched in 1981 to probe for unevenness in the microwave background radiation of the universe. This short wavelength radiation is a remnant of the Big Bang which has cooled to the extremely low temperature of about three degrees above absolute zero. This experiment called COBE (COsmic Background Explorer) was run for four years suspended from a high altitude balloon in the stratosphere. This experiment was superseded by a satellite version launched in November 1989. The experiment was developed by NASA's Goddard Space Flight Centre. Needless to say the measurements were extremely delicate almost at the threshold of observation and needing exceptional protection from unwanted signals. The experiment detected an unevenness of about one part in 100 000. This result met with jubilation in academic circles – at last the trigger for the formation of stars had been found. This primordial unevenness supplied the possibility for the hydrogen and helium of the Big Bang to clump together under the force of gravity to eventually form stars, galaxies and even clusters of galaxies. The COBE experiment actually contained three instruments each of which provided results of major scientific value. The instruments were DIRBE the Diffuse Infrared Background Experiment, DMR the Differential Microwave Radiometer and FIRAS the Far Infrared Absolute Spectrophotometer. At this point it must be mentioned that gravity is not sufficiently strong to clump hydrogen and helium into stars or even to prevent galaxies from flying apart – it is only by the presence of 'dark matter' that the clumping and containment can occur. There is about 5.5 times as much dark matter as ordinary physical matter in the universe. Dark matter is extremely difficult to detect as it does not consist of ordinary atoms and molecules. So far dark matter has only been detected by its gravitational effects. In 2006, Dr John Mather and professor George Smoot were awarded the the Nobel Prize in Physics for their work on the background radiation. Professor Per Carlson of the Royal Swedish Academy of Sciences concluded his award presentation address in December 2006 with:

Dr Mather, Professor Smoot:

Cosmology has become a precision science and your ground-breaking research has laid the foundation for that. With your carefully calibrated instruments you have shown the cosmic microwave background radiation to follow very precisely a blackbody form. The in-depth analysis of the radiometer data has shown the presence of the long sought small temperature anisotropies, the seed of the structures in the universe that we observe today. In your successful experiments you have used space-based instruments on board the COBE satellite. We are now all together at sea level in Stockholm, and on behalf of the Royal Swedish Academy of Sciences it is my privilege and pleasure to congratulate you for your outstanding work and I now ask you to step forward to receive your Nobel Prizes from the hands of His Majesty the King.

A successor satellite to the COBE and MIDEX probes was launched by NASA in 2001 and placed in a lissajous orbit at the L2 Lagrangian point. This is the WMAP (Wilkinson microwave anisotropy probe) which had 33 times the resolution of its predecessors. Significant results were:

Age of Universe: 13,73 billion years

Proportions of mass in Universe:

  72,1% dark energy

  23,3% non baryonic

  4,6% baryonic matter – atoms

The dark energy is a complex matter in theoretical physics which is used to explain an increasing rate of the expansion of the universe.

The WMAP mission also yielded a number of high resolution maps of the background radiation in several wavelengths. Since 2000, the three most highly cited papers in all of physics and astronomy were WMAP scientific papers.

A further successor to the COBE was launched on 14 May 2009 by the European Space Agency (ESA). Two observatories, the Herschel Space Mission and The Planck Mission were launched together, then separated and placed in separate halo orbits at the L2 Lagrangian point of the Earth's orbit. These space missions are described in more detail in chapter six.

This is certainly not the end to a discussion of the development of the physical universe; there is far more to the universe than hydrogen fuelled stars. Stars have complex life cycles depending on their size. When a large star has exhausted its hydrogen it will start consuming helium and then progressively create heavier elements, swell up to become a red giant and eventually a supernova. Supernovas are crucial to our existence. It is only in a supernova that the light elements can fuse to form the heavier elements that we and our planet are made of. The supernova is something like a scaled down Big Bang. The explosion of a single star amongst the thousands of millions in a galaxy may seem an almost insignificant incident but it is nevertheless a cataclysmic event of unimaginable proportions. At the supernova stage the star can actually radiate more light than an entire galaxy. Unlike the Big Bang ejecting hydrogen and helium, the various types of supernova can eject atoms of every element in the periodic table. Although this heavier material has much better clumping potential than hydrogen it does not clump into ready made planets. It is currently thought that the entire Solar System was formed from ejecta from a single supernova. The life cycle of stars is described in more detail in chapter five.

The formation of elements in the cosmos has until recently been understood as a two step process but new insights have been gained into a fundamental nuclear reaction, the triple-alpha process, which accounts for the large amounts of carbon throughout the universe. Carbon is the fourth most common element in the universe and of this, carbon-12 is the most common isotope. The occurrence of carbon in the solar system is unusually low.

At this point, we should take a look at Saturn and its rings. This is a natural shape for aggregated rotating material in the universe. The central sphere is simply drawn together by gravitation but further out centrifugal force is sufficient to keep the material in orbit. Material spread along the axis of rotation will gravitate towards the disc giving the typical Saturn configuration. Even galaxies have a tendency to this configuration. A simple example of the sphere and disc configuration can be found in the production of window glass in the early days of glassmaking except that here cohesion and surface tension take the place of gravity. The sheet glass was made by taking a gob of glass heated to plasticity on the end of a stick and then rotating it. The glass would spread out into a thin disc with a lump remaining in the middle. The glass disc would be cut into window panes and even the central part would be sold as lower grade glass rather that be re-melted. In the accretion disc of the Solar System, further clumping then took take place to form orbiting planets around the star. Evidence of this process can be seen in the rotation of the parts of the Solar System. Nearly everything in the Solar System rotates in the same direction. There are a few exceptions, notably Venus, but this is probably due to a very violent event involving a collision with a large body from outside the Solar System. In 1950, a highly controversial theory about the rotation of Venus was proposed by Immanuel Velikovsky in his book 'Worlds in Collision' but this has never enjoyed general acceptance. The accretion disk was by no means uniform in its composition. We are extremely fortunate to have such a good mix of materials in our part of the disc. The larger outer planets are largely composed of gas and the Kuiper belt and Oort cloud where the comets come from contain vast quantities of frozen water. The comets are sometimes disturbed by gravitational attraction of the large planets which can cause them to take elongated elliptical orbits around the Sun. It is thought that comets colliding with the Earth contributed to the abundant water on our planet. The Solar System is discussed in more detail in chapter five.

Can the planets clump themselves together in deep space and then be captured whole by a star? This is extremely unlikely. If the planets were rigid objects they could never be captured into orbit – only diverted if not collided with. For capture some braking mechanism is required. Tidal friction within the planet could cause considerable loss of kinetic energy but the best albeit extremely unlikely chance for capture would be for other planets of the star to gravitationally slingshot the new arrival into orbit. One result of tidal depletion is easily observed. The Moon always presents the same side to the Earth – any additional rotation that it once had has long since been stopped by tidal friction. Something somewhat similar to the tidal braking of planets can be demonstrated by means of a two egg experiment. Take two eggs – one raw and the other hard boiled. Spin the eggs on a hard table and it will be seen that the raw egg will slow down more quickly than the hard boiled one. The viscosity of the contents of the raw egg will reduce the kinetic energy of the egg and slow it down. The supernova is a stupendous event which can eject material for a huge number of stars and planets. Our own Solar System has managed to produce quite a large assortment of planets and there is no reason to suppose that this is an unusual occurrence – there may well be millions of stars with planets in our own galaxy. Many planets orbiting stars have been detected by the huge modern telescopes and the list is growing steadily. There are also cases of two stars orbiting each other (binary stars) and even a planet orbiting two stars.

The SETI Institute was founded in 1984. (Search for ExtraTerrestrial Intelligence) This is a project working at very long odds. No intelligent communication message has at yet been received from deep space but when this happens this will be surely the most sensational news in the history of journalism. The Arecibo big dish is an ideal instrument for searching for extraterrestrial radio messages but the effort of analysing radio signals received is enormous. A novel way of processing the vast quantities of data is to offer data to the general public to process in a program running as a screen saver. The program and data can be downloaded and run putting idle computer time to good use. The Arecibo big dish radio observatory is described in more detail in chapter six. There are many other radio observatories also running SETI projects. The idea that there might be life on other planets has been speculated on since ancient times.

Roman poet Lucretius (99–55 BC) wrote:

> Granted then, that empty space extends without limit in every direction and that seeds innumerable are rushing on countless courses through an unfathomable universe..., it is in the highest degree unlikely that this earth and sky is the only one to have been created. So we must realise that there are other worlds in other parts of the universe, with races of different men and different animals.

These words have acquired new meaning. It is now widely suggested that the basic organic chemicals required for the generation of life are created in space and were brought to the Earth by comets. If this is the case then life might be commonplace throughout the universe.

Metrodorus of Chios (4th cent. BC) put it this way:

> To consider the Earth as the only populated world in infinite space is as absurd as to assert that in an entire field of millet only one grain will grow.

Epicurus (341–270 BC) had this to say on the matter:

> ... there are infinite worlds both like and unlike this world of ours... we must

believe that in all worlds there are living creatures and plants and other things that we see in this world.

Christiaan Huygens (1629–1695) physicist, astronomer and mathematician in his book 'Cosmotheros' described a universe brimming with life.

The search for extraterrestrial life was given a huge boost when the NASA Kepler Mission was launched in March 2009. On 4 April 2012 the mission life was extended to 2016. This instrument is specifically designed to detect the presence of planets orbiting stars. It is quite different to space telescopes such as the Hubble as this is not a telescope but a multiple channel photometer. The instrument is designed to observe a much larger area of sky than a telescope and is not intended to produce images. The photometer will lock on to 100 000 main-sequence dwarf stars, in the Cygnus-Lyra region, and continuously measure their brightness in order to detect the extremely slight reduction in brightness when a planet transits in front of the star. Measurements of such sensitivity would be hopeless from an Earth based instrument due to the presence of the atmosphere. Only a small percentage (about one in 200) of stars will have their planets orbiting in a plane permitting transits to be observed but with such a large sample of stars there should nevertheless be many candidate solar systems. The mission will concentrate on habitable zone planets – those that are able to maintain liquid water on their surfaces. This will include planets both larger and smaller than the Earth. The data gathered will permit the calculation of the size, orbit and length of year of the planets. The Kepler Mission spacecraft was placed in an Earth trailing heliocentric orbit so that it will be able to permanently lock on to target without any interference from a solar system body. If this mission succeeds in all of its objectives the results should be sensational.

As at June 2012, KEPLER had detected 2321 candidate planets orbiting 1790 stars, of which 74 planets have been confirmed.

Current estimates show that 5,4% of all stars host Earth-size planets and that 17% of all stars have multiple planets. It was further estimated that there are at least 30 000 habitable planets within a thousand light years of the Earth and that there are at least 50 billion planets in the Milky Way.

This means that there are about seven times as many planets in our Milky Way galaxy as there are people living on planet Earth. Can there be any doubt that millions of these planets will be inhabited by living creatures?

Data from the Kepler space telescope has been made available on website

Planethunters.org so that volunteers from the public can take part in the hunt for new planets. Since December 2010 more than 170 000 members of the public have taken part in this project. One of the most sensational successes was the discovery of a Neptune size gas giant planet orbiting a pair of binary stars which are in turn orbited by two other stars. This is the only case so far where a planet illuminated by four stars has been found. A five body orbiting system is analytically orders of magnitude more complex than the classical three body problem.

There are several theories regarding the abundance of planets with living organisms as well as those with intelligent life. One of these, possibly the least optimistic, is the Rare Earth Hypothesis which considers several criteria for the existence of life. The following is a brief list of criteria of this most complex hypothesis:

- Distance from galactic centre. This influences the star metallicity which would be too low at the outer reaches of the galaxy. X-ray and gamma ray radiation from the black hole at the galactic centre and a high density of stars, supernovae and neutron stars rules out stars close to the galactic centre. A lower star density reduces the risk of major bolide impacts. The star would also need a nearly circular orbit about the galactic centre so as not to stray into inhospitable regions. In cosmology, 'metals' refers to all elements other that hydrogen and helium.

- Planet in habitable zone. The planet orbit must be in a zone capable of maintaining liquid surface water, which is 0,95 to 1,15 AU for the Sun. The planet must also have a rocky crust, rich in a variety of chemicals, which would have come from a supernova explosion.

- The star must not be too large or too hot. Hot stars would radiate too much ultraviolet light and have lives too short for the evolution of life. Small red dwarf stars would have small radius habitable zones which would cause one face of the planet to constantly face the star (tidal lock). The star should also have a constant energy output. A variable star would repeatedly freeze or boil off the planet's surface water.

- Planet size. A small planet would not be able to retain an atmosphere sufficient to support life or oceans, and would become too cold to maintain plate tectonics.

- Planet axis. The planet should have a tilted axis similar to the Earth on order to provide a climate with seasonal changes. The planet would also need a large moon to stabilize the axis which could otherwise change chaotically. The planet would also need a magnetic field for protection from cosmic radiation.

- Big brother. It has also been suggested that Jupiter has provided the Earth with essential protection from catastrophic bolide impacts.

Charles Darwin (1809–1882) made a huge impact on the natural sciences and in the process stirred up much debate and aroused huge controversy with his ideas. The sciences of geology and palaeontology were well established at the time but it was his ideas on the evolution of living creatures through natural selection that caused all the trouble. Darwin was a man of social standing. His father was a physician, who wished his son to follow in the same profession, but the son had an intense dislike of the sight of blood and the brutality of surgery and chose instead to study the natural sciences. His mother, Susannah Wedgwood, was the daughter of Josiah Wedgwood, the famous manufacturer of exquisite tableware. Charles married his cousin Emma Wedgwood in 1839. The exquisite Portland vase dating from the 1st century is the finest example of Roman cameo glass in existence. It was made from dark blue glass with white figurines. Sometimes called the Barberini vase, it was in the possession of the Barberini family for 150 years. It was acquired by Scottish architect James Byrnes who sold it in the early 1780s to diplomat and archaeologist Sir William Hamilton (1730–1803). Sir William had a large collection of Pompeii antiquities which he sold to the British Museum in 1772. Horatio Nelson was on friendly terms with ambassador Sir William and even friendlier terms with Emma, Lady Hamilton, who bore him his daughter Horatia in 1801. Nelson died in glorious triumph in 1805 at the Battle of Trafalgar when a French sharpshooter in the rigging of the *Redoutable* fired the most famous musket ball in naval history. The musket ball is on display to this day and embedded scraps of gold braid from the admiral's uniform are clearly visible. Nelson's full title at the time of his death was: Vice Admiral of the White, The Right Honourable Horatio Viscount Nelson, Knight of the Most Honourable Order of the Bath. Emma had a practical idea about titles, writing in a letter:

If I was the King of England, I would make you the most noble, puissant Duke Nelson, Marquis Nile, Earl Alexandria, Viscount Pyramid, Baron Crocodile and Prince Victory that posterity might have you in all forms.

The Portland vase was bought by Margaret, Duchess of Portland, and which was inherited by her son the Duke of Portland. After the base of the vase had been broken the duke lent it to the British Museum where it could be appreciated by the public and would presumably be kept in safe custody. The hope was in vain – a drunken visitor to the museum smashed the priceless treasure. The Portland vase had to be painstakingly restored and is now still proudly on display in the museum. In 1790, Josiah Wedgwood started manufacturing highly successful replicas of the vase in black Jasperware. The Wedgwood factory has remained in production to this day manufacturing the most exquisite tableware inspired by the famous Portland vase. More recently the Wedgwood Etruria factory has produced fine china for the more juvenile connoisseur decorated with delightful Beatrix Potter animal illustrations. Josiah Wedgwood was a man of many interests and a supporter of liberal causes. He was a leader in the abolition of slavery movement as well as a supporter of the French Revolution and the American Civil War. Wedgwood manufactured and distributed Jasperware medallions promoting the abolition of slavery. These depicted a shackled slave with the caption: "Am I not a man and a brother?" Wedgwood is well known for his invention of the pyrometer for measuring the high temperatures used in kilns. On the advice of Erasmus Darwin, Wedgwood installed a steam powered engine in his Etruria factory – a very innovative idea at the time. Josiah Wedgwood generously assisted Joseph Priestley with funding for his pioneering chemistry research.

A most significant opportunity arose in 1831 after Darwin had completed his final examinations at Cambridge. He was offered the unpaid position of gentleman's companion to Robert FitzRoy, captain of HMS Beagle, on a two year expedition to chart the coastline of South America. This would give Darwin a valuable opportunity to develop his career as a naturalist. His father disapproved of the voyage regarding it as a waste of time but was persuaded to acquiesce by Josiah Wedgwood II. The expedition stretched into five years and would make substantial contribution to several fields of science. Darwin was not a good sailor finding himself prone to seasickness. Darwin was able to spend two thirds of the time of the expedition exploring on land. He had with him a copy of Sir Charles Lyell's 'Principles of Geology' which he found invaluable. He wrote home that he was seeing landforms 'as though he had the eyes of Lyell'. Darwin studied a rich variety of geological features, fossils and living creatures. He discovered fossils of huge extinct animals and glyptodons (armadillo like animal). He collected an enormous number of specimens which would establish his reputation as a naturalist. His visit to the Galapagos islands provided the clues to the evolution of species due to natural selection. When the Beagle returned to England on October 2, 1836, Darwin was a celebrity in scientific circles. His father arranged investments so that Charles could become a self-funded gentleman scientist.

Darwin joined the influential and advanced literary circle of Harriet Martineau (1802–1876). Harriet, a significant writer of the time, authored the first systematic methodological treatise on sociology. Other prominent members of the circle were: Charles Babbage, George Eliot, Florence Nightingale, Charles Dickens, Thomas Malthus, William Wordsworth and Charlotte Brontë. Lyell introduced Darwin to the Geological Society of London where he read his first paper in January 1837. Darwin presented his mammal and bird specimens to the Zoological Society. The mammalia were taken on by George R. Waterhouse. Lyell introduced Darwin to anatomist Richard Owen who determined that some of Darwin's fossil bones were from gigantic extinct rodents and sloths. Lyell used his presidential address at the February 1837 meeting of the Geographical Society to present Owen's findings on Darwin's fossils. Darwin was elected to the Council of the Geographical Society. Darwin wrote a number of books and articles, many of which concerned shell fish, worms and other invertebrates, but it was his book 'On the Origin of Species' published in 1859 that caused the sensation. He was mindful of the pain and social embarrassment that his book might bring to his wife and withheld his work from publication for many years despite being convinced of the correctness of his views. His wife was devoutly religious and he did not wish her to think that she would have an afterlife in heaven and that he would languish in hell. He need not have worried – Darwin was given a decent interment in Westminster Abbey like few others could hope for. Darwin's hand was forced in 1856 when Alfred Russell Wallace wrote a paper with ideas almost identical to those of Darwin. Charles Lyell read the paper and urged Darwin to publish his theory in order to establish precedence. Charles Darwin and his wife Emma had ten children, three of whom died in infancy. Darwin was distraught with grief when his daughter Annie died in 1851. After this he would go for walks alone on Sundays when his family attended church services.

On Darwin's death, the president of the Royal Society in London, William Spottiswoode, sought permission from the Dean of Westminster Abbey for Darwin to be buried in the Abbey. The Dean, George Granville Bradley, unhesitatingly gave his cheerful assent. Darwin was laid to rest next to John Herschel and close to Sir Isaac Newton with his grave inscribed; 'Charles Robert Darwin, born 12 February

1809, died 19 April 1882.' The Bishop of Carlisle, Harvey Goodwin, in a memorial sermon said:

I think that the interment of the remains of Mr. Darwin in Westminster Abbey is in accordance with the judgment of the wisest of his countrymen ... It would have been unfortunate if anything had occurred to give weight and currency to the foolish notion which some have diligently propagated, but for which Mr. Darwin was not responsible, that there is a necessary conflict between a knowledge of Nature and a belief in God.

In 1888, Darwin's family erected a bronze memorial in the Abbey with a bust in relief simply inscribed 'Darwin'.

A revival of the evolution dispute occurred in Dayton, Tennessee, in July 1925 in the famous 'Scopes Monkey Trial'. This appeared as a Broadway play in 1955 called 'Inherit the Wind' and was made into a movie starring Spencer Tracy and Gene Kelly in 1960. This was an entertainment movie, not a documentary, and bears little resemblance to events both in and out of court and the schoolhouse.

One of the most astonishing scientific discoveries of all time – the ultimate key to all life forms – DNA (Deoxyribonucleic acid) was discovered in the mid 19th century but it was only in 1953 that James Watson, Francis Crick, Maurice Wilkins, Rosalind Franklin, Seymour Benzer et al proposed that it could store genetic information. An early work dealing with the idea that the key to life lay at molecular level was published in 1944 by world renowned physicist Erwin Schrödinger (1887–1961) of wave equation fame, in his book 'What is Life?' This was based on lectures delivered under the auspices of the Dublin Institute for Advanced Studies at Trinity College, Dublin, in February 1943. Schrödinger misbehaved badly with his DNA in Dublin putting two Irish women in serious predicaments. Schrödinger's huge contribution to physical science will be discussed in chapter seven.

In 1962, Francis Harry Compton Crick, James Dewey Watson and Maurice Hugh Frederick Wilkins were jointly awarded the Nobel Prize in Physiology or Medicine for:

Their discoveries concerning the molecular structure of nucleic acids and its significance for information transfer in living material.

The presentation address was given by Professor A. Engström, member of the staff of Professors of the Royal Caroline Institute. Rosalind Franklin would also have been a strong contender for the prize but she sadly died at a young age in 1958, quite possibly as a result of careless use of X-Ray equipment. The Nobel Prize is not awarded posthumously and not to more than three nominees.

Professor Engström concluded his presentation address with:

Dr. Francis Crick, Dr. James Watson, and Dr. Maurice Williams. Your discovery of the molecular structure of the deoxyribonucleic acid, the substance carrying heredity, is of utmost importance for our understanding of one of the most vital biological processes. Practically all the scientific disciplines in the life sciences have felt the great impact of your discovery. The formulation of double helical structure of the deoxyribonucleic acid with the specific pairing of the organic bases, opens the most spectacular possibilities for the unravelling of the details of the control and transfer of genetic information.

It is my humble duty to convey to you the warm congratulations of the Royal Caroline Institute and to ask you to receive this year's Nobel Prize for Physiology or Medicine from the hands of His Majesty the King.

The field of molecular biology is an extremely complex and highly specialised study – only a very elementary outline can be presented here. There is a remarkable similarity between the DNA molecule and a computer program. The foundations of modern computer science were laid down by pioneering scientists Dr John (Janos) von Neumann (1903–1957) and Dr Alan Mathison Turing (1912–1954). Before proceeding further with DNA, let us take a brief look at how data is stored by computers. Computer data can be stored electronically, on magnetic media, or optically on discs. The data is ultimately in the form of binary digits (bits) which is a bi-stable condition of yes–no, on–off but usually given as ones and zeroes. In early computers much use was made of Hollerith punched cards and punched paper tape where the information was carried by hole or no-hole. In order to make sense of ones and zeroes they must be grouped – usually in groups of eight called bytes. An eight bit byte can have values from 0 to 255 giving 256 possible values. The bytes are further grouped into words of usually four bytes each and then larger groups to suit processing, addressing and storage requirements. When the bytes are used for carrying textual information the 256 values provide ample scope to store the alphabet in upper and lower case characters, numbers, punctuation, accents and several other special purpose values. This allocation of numbers to symbols is not fixed and can be defined in many ways to suit foreign alphabets, mathematical and scientific symbols etc. When the storage is used for numeric data there are several formats in which this can be done: direct binary integer values in a group of bytes, plain or compressed numeric characters or a special format (floating point) for values of any size and suited to high speed calculations. Storage used for computer instructions is formatted quite differently. Instructions can occupy a half word, whole word or even a few words. These are divided into a byte for the code of the computer instruction to be performed, some bits to indicate registers to be used and some bytes to carry one or two memory addresses to be accessed.

Due to the sensational publicity given to DNA most readers will be familiar with the twisted ladder of the DNA molecule. Unlike computer memory, the DNA molecule uses four basic units for storing information which are often shown in illustrations as the four aces of a deck of playing cards. Four types of data are essential to the working of the molecule.

The four bases or nucleotides are as follows:

A	adenine	Pairs with thymine in DNA or uracil in RNA
T	thymine	Pairs with adenine
C	cytosine	This molecule has found a place in a new and bizarre branch of computer science – Molecular Computing.
G	guanine	Name derived from guano: sea-fowl excrement forsooth. Guanine is used as an additive to cosmetics and shampoo to give a pearly lustre.

The nucleotides bond together in four ways to make the rungs of the ladder: A+T, T+A, C+G and G+C, so that both sides of the ladder will have all four nucleotides. The backbones of the ladder consist of alternate sugar (deoxyribose) and phosphate molecules. The nucleotides bond to the sugar molecules. It is the sequence of the four types of ladder rungs that make up the program of the molecule in much the same way as ones and zeroes make up a computer program. The DNA molecule is in effect a Turing machine. The basic grouping of the base pairs is in groups of three called codons. The codon corresponds to a byte in computer memory. The three pair codon provides sixty four possible values. Some codons (UAA, UGA and UAG) act as delimiters and serve no other purpose. This corresponds to the NOP (no-operation) instruction of a computer program. When referring to a codon as UGA this implies the base pairs U+A, G+C and A+T but it is sufficient to mention only one side of the ladder.

A most astonishing feature of the molecule is that a trivial chemical reaction or heat can split the molecule right down the middle. The nucleotides are bonded by double or triple hydrogen bonds which can easily be broken. An enzyme called polymerase synthesises new chains of nucleotides. An enzyme called ligase links these fragments into a continuous strand and matches them to the original. The split DNA polymer is thus repaired into two identical molecules. This process can be performed *in vitro* (in glass) so that forensic scientists can increase the quantity of DNA when only an extremely small sample is available. If any mistakes occur in the replication, this will result in a mutation. It should not be supposed that the molecule actually looks like a ladder. If the thread of the molecule could be seen in detail it would appear as a compact conglomeration of atoms. The molecule is of course too thin to be observed in detail by any microscopic means but it can be observed as a thread if heavily stained. The diameter is about two nanometres but the total unrolled length about two metres. The genome occupies only about five centimetres of the molecule – no purpose for the rest of the molecule has been discovered. This is not to say that it is not very useful. This is the part that is used by forensic scientists for DNA fingerprinting. This is most useful for identifying

Chromosomes in various species	
Fruit Fly	8
Dove	16
Earthworm	36
Domestic cat	38
Lab mouse	40
Bread wheat	42
Human	46
Great apes	48
Potato	48
Elephant	56
Horse	64
Dog	78
Goldfish	100–104
Tongue fern	1440

criminals and settling paternity disputes. The DNA molecule would not survive very well as a long thread but is fortunately very compactly coiled up around spool-like proteins in chromosomes known as histones. The spiral structure of DNA was discovered by Rosalind Franklin by means of X-Ray crystallography. The DNA ladder is twisted into a right hand spiral much the same as a right hand screw thread. The DNA polymers are directional; the one end has an exposed hydroxyl group on the deoxyribose and the other end an exposed phosphate group. This directionality is vitally important to the working of the molecule.

It should be noted that DNA is not 'alive' but is only the specification for a life form. Despite its extreme complexity, DNA has a remarkable ability for survival. DNA has actually been extracted from 30 000 year old ground sloth dung and it has even been suggested that DNA can be extracted from million year old samples! Studies of DNA taken from the extinct Mauritian Dodo have showed that it is related to the common pigeon. DNA has also been extracted from a quick-frozen ice age woolly mammoth. A most significant result of a DNA study was announced in 1997 which confirmed that the Neanderthals were a distinct species which had become extinct and did not contribute to the DNA of modern humans. It is estimated that the divergence between humans and chimpanzees occurred 4 to 5 million years ago and the divergence between Neanderthals and modern humans at 550 000 to 690 000 years ago. The Neanderthals inhabited Europe from 300 000 to 30 000 years ago.

In February 2010 a most remarkable study was made of the family of Pharaoh Tutankhamun using DNA and archaeological evidence. It was found possible to extract uncontaminated DNA from within undamaged bone. The study confirmed that Tut's father was the revolutionary Pharaoh Akhenaten and his grandfather Amenhotep III. It was also discovered that Tut's parents were direct brother and sister – something which would not have been noteworthy at the time. The two stillborn infant mummies in the tomb were found to be children of Tut. Tut was no chariot racing young firebrand – a crippling bone disease would have caused him to use a walking stick.

A few pertinent comments on DNA:

Sir Karl Raimund Popper, 1974.

The undreamt of breakthrough of molecular biology has made the problem of the origins of life a greater riddle than it was before: we have acquired new and deeper problems.

Professor Kenneth Nealson, 2002.

Nobody understands the origin of life. If they say they do, they are probably trying to fool you.

Dennis Overbye, 2004.

We emerge inevitably or by luck from the chipping of DNA by cosmic rays, chemical currents in space, the bubbling of volcanic mud.

All plants and animals have chromosomes which form part of the cell nucleus of living tissue. The chromosomes in turn each contain a DNA molecule which defines the functional parts. The DNA molecule in turn contains a vast number of genes along its length which define protein production. The name chromosome (coloured body) is a misnomer – an object of this size is much smaller than light wavelengths and cannot therefore have colour, but they can be microscopically observed if heavily stained. From the table given here, it will be seen that the number of chromosomes bears little relation to the size or complexity of the organism.

The cells of the human body each have twenty three pairs of chromosomes with matching shapes. One of these pairs will be the sex chromosomes which are designated XX in the case of females and XY for males. As the chromosomes come in pairs, the DNA molecules, and consequently genes, from each parent will also be paired. Males and females form sex cells which are contained in ova

The 23 human chromosome pairs		
	Genes	Total bases
1	3 148	247 200 000
2	902	242 750 000
3	1 436	199 450 000
4	453	191 260 000
5	609	180 840 000
6	1 585	170 900 000
7	1 824	158 820 000
8	781	146 270 000
9	1 229	140 440 000
10	1 312	135 370 000
11	405	134 450 000
12	1 330	132 290 000
13	623	114 130 000
14	886	106 360 000
15	676	100 340 000
16	898	88 820 000
17	1 367	78 650 000
18	365	76 120 000
19	1 553	63 810 000
20	816	62 440 000
21	446	46 940 000
22	595	49 530 000
23X 23Y	1 093 125	154 910 000 57 740 000

and sperm in the case of humans. The offspring develops when the chromosomes of the ova and sperm combine. Obviously the number of chromosomes cannot double with each new generation, so that the ova and sperm cells must each have only half the number of chromosomes. When sex cells are formed the process starts with two similar chromosomes which then exchange DNA material, split and then form four individual ovum or sperm cells. During reproduction the crossing over of genes results in offspring having DNA with genes from both parents. The XX and XY arrangement ensures that there will be an equal chance of male or female offspring being produced. In humans, the males and females are physically quite different from head to toe, as well as mentally and emotionally, yet the only genetic difference is in one out of forty six chromosomes. (*Vive la différence!*) Individual variations of a gene are called alleles. The alleles are structurally similar but differ in nucleotide arrangement. The alleles can be either dominant or recessive. For example we can have AA, Aa, aA or aa alleles for a trait. The capital letter denotes dominant and the lower case recessive. If a flower is produced with 'A' representing blue petals and 'a' for white, and both parent chromosomes have Aa alleles, then three out of four flowers are likely to have blue petals. It is interesting to note that a recessive trait can be passed on to offspring without being apparent in the parents. The different alleles in humans determine inherited traits such as hair and eye colour, susceptibility to illness, bodily stature etc. Some traits such as height can also be influenced by environmental factors, such as the need for exercise, nutrition availability etc., other traits such as eye colour are not.

The table given here (next page overleaf) gives some idea of the extreme complexity of the human genome. The values given are estimates based partly on gene predictions. Astonishingly, all this information is stored in molecules too small to be directly seen, even with a microscope. The genome of the single celled Amoeba dubia is estimated to contain more that 200 times the number of base pairs as the human genome.

The early pioneering work on genetics was done by Augustinian monk Gregor Mendel (1822–1884) who used smooth and wrinkly peas to study the way in which dominant and recessive genes were propagated. The peas also expressed other traits in their flowers, leaves and pods. A major discovery achieved by Mendel's work was that traits are not blended from those of the parents, but switched on or off in binary fashion. Some traits in humans do appear to be blended, such as complexion and facial features, but this is due to the traits being carried by a number of genes. Mendel was named Johann but assumed the name Gregor on taking up monastic life. Mendel discontinued his scientific work after being elevated to abbot in 1868. His work went largely unnoticed in his lifetime; the great significance of his discovery was only realized in 1900.

For each successive human gen-eration the offspring will have a new DNA configuration. In addition to the DNA molecules that make up the chromosomes within the cell nuclei, the cells also have many organelles called mitochondria which contain circular DNA loops called mitochondrial DNA. The mitochondrial DNA does not express traits and does not change from one generation to the next unless mutations occur. Mitochondrial DNA is contained in ova but not in sperm so that this DNA can only be transmitted to the next generation by the female parent. Mitochondria therefore provide a useful means of tracing female ancestry back for as many generations as DNA can be found.

The DNA molecule is not directly involved in the expression of genes. The genes are transcribed into a second type of nucleic acid, RNA (Ribonucleic acid) which is typically single stranded and with the sugar ribose instead of deoxyribose. This molecule is much less stable than DNA. Not all parts of a gene are used for encoding products. Regions called introns are removed from the messenger RNA in a process called splicing and regions encoding products are called exons. A significant portion of gene coding is devoted to controlling and switching off protein production. This is somewhat similar to computer data transmission where a significant portion of the data stream exercises controlling, handshaking and data integrity functions. This is curiously referred to as 'line protocol'. The control aspect of genes is obviously necessary – genes responsible for the growth of an ear should not produce fingers, toes or eyes like a Picasso painting.

Some viruses do not have DNA but store their entire genome as RNA. This allows their cellular hosts to directly synthesise their proteins without transcribing DNA. Viruses such as HIV are RNA retroviruses which require reverse transcription of their RNA genome into the DNA of their hosts before their proteins can be synthesised.

The self-repair aspect of DNA is astonishing to the point of miraculous. It has been estimated that the DNA in a single cell can be damaged up to 10 000 times a day by carcinogens and radiation. The DNA can even be damaged by products within the cell. The DNA molecule will take this damage in its stride but occasionally the damage will remain unattended resulting in the start of mutation or cancerous growth.

The discovery of the function of DNA brought with it the urgent requirement to actually read the coding. This has become known as DNA sequencing. Early attempts required the use of highly toxic chemicals and radioactivity. The DNA had to be read in rather short sections and dummy DNA had to be spliced to the ends as the readout of the ends was of poor quality. This was similar to the blank film spliced to the ends of 8mm home movies for threading through the projector. Many improvements have been made to sequencing techniques but the ultimate goal is a method of direct readout. One suggestion is that a split DNA could be passed through a hole and the sequence detected electrically. The difficulties in doing this are mind boggling. The hole would have to be about 1,5 nanometres in diameter which is less than one percent of the wavelength of ultraviolet light. All the operations would have to be done with the aid of enzymes as everything being done would be totally invisible. The electrical measurements would involve the measurement of picoamperes.

The Human Genome Project was launched by the U.S. Department of Energy and the National Institutes of Health in October 1990. The project goals were to:

> Identify all of the more than 20 000 genes in human DNA.
>
> Determine the sequences of the chemical base pairs that make up DNA.
>
> Store the genome information in databases.
>
> Improve tools for data analysis.
>
> Transfer related technologies to the private sector.
>
> Address the ethical, legal and social issues that may arise from the project.

The project was completed ahead of schedule in 2003 making a vast wealth of information available for further research. There is of course no single human genome – each individual has a genome slightly different to that of everyone else.

The human genome was also sequenced privately by Dr. Craig Venter, in competition with the Human Genome Project, using his 'shotgun' method of sequencing. In 2008, Venter was listed by TIME Magazine as one of the 100 most influential people in the world. He received the 2001 Prince of Asturias award for Technical and Scientific Research. In May 2010, the team of Dr. Craig Venter announced the development of the first synthetic cell by reconstructing the genome of a bacterium and producing a synthetic chromosome. This was then transplanted into a cell of a bacterium, effectually reprogramming the software to produce a synthetic bacterium which was able to rapidly reproduce. The Frankenstein bacterium has been named 'Cynthia'. This achievement has met with mixed reactions of jubilation and horror.

July 2012 saw a breakthrough in computational biology. Stanford researchers reported the first complete computer model of an organism – a tiny parasite *Mycoplasma genitalium* – the world's smallest free living bacterium. This organism has only 525 genes. This achievement holds great promise as an aid to understanding cellular function and fundamental biological processes.

A remarkable project was launched in 2005 to study historical human migration patterns by analyzing DNA samples from hundreds of thousands of people from all parts of the world. This is a privately funded collaboration between IBM, the National Geographic Society and the Waitt Family Foundation. This huge anthropology project is expected to run for five years. The samples are taken by means of mouth swabs (buccal swabs). The project also includes the sale of self-testing kits which can be purchased through the post. People purchasing the kits can have the migration of their early ancestors determined either by mitochondrial DNA or chromosome-Y DNA. The mitochondrial DNA trace will give results for female ancestry and the chromosome-Y test will show male ancestry. Women wishing to trace their male ancestry must obtain the swab from a close male family member. Profits from the sale of kits are ploughed into a Legacy Fund which will be spent on cultural preservation projects. This exciting project has already produced fascinating results but sadly it has also met with opposition from people who feel their tribal identity under threat. The Human Genome Diversity Project (HGDP) was originally proposed by geneticist Luigi Cavalli-Sforza. A remarkable discovery was made when the mitochondrial DNA of Cheddar Man was compared with that of living local residents. The DNA was extracted from a molar of a 9 000 year old Briton who was found in Gough's Cave in Cheddar Gorge, Somerset. The tests showed that many present day locals were descended from ancient Britons and not only from later invading foreigners as was widely believed.

The huge advances made in the sequencing and manipulation of DNA have resulted in the science of Genetic Engineering. The use of GM (genetically modified) crops is now widespread. We have here a similar situation to the discovery of nuclear fission. This was hoped to provide wonderful benefits including the availability of cheap and inexhaustible power. These benefits have unfortunately been offset by the horrors of nuclear weapons and large scale radioactive contamination. The manipulation of DNA has the potential of becoming a Pandora's Box of unspeakable horrors. Some may think that using genes from a fish in agricultural crops to improve frost resistance may be a good idea but others will regard this with a frisson of terror. Constructing new life forms at bacterial level has already been attempted

– intervention is clearly required here lest some mini Frankenstein's monster wipe out human life on Earth.

The computer program aspect of the DNA molecule has not been lost on computer scientists. An enormous amount of research is being done on molecular computing using DNA molecules. This is not expected to replace electronic computers so no one need have any fears that their bank balance will depend on a DNA molecule. In the original von Neumann/Turing design the computer instructions would always be performed in sequence, one at a time. The instructions could of course perform decision making and jump to various parts of the program as required and also perform repetitive loops. A way of increasing performance would be to have several processors working together in parallel but only specific applications can be speeded up in this way. By contrast, molecular computing would be massively parallel with thousands or even millions of processes being performed simultaneously. There are very few applications which could benefit from this degree of parallel processing. A likely candidate is encryption cracking. Data encryption has developed to such a degree of sophistication that even with modern computers cracking an encrypted document could take thousands or even millions of years. The molecular computer holds out some hope that this can be performed in a more reasonable time. Another remarkable possibility is the use of molecules for data storage. Information stored in DNA-like molecules would mean that millions of digitised photographs could be stored in a piece of material scarcely large enough to be visible. Finding a workable storage and retrieval method to access this data is a matter still in the realm of science fiction.

In August 2012 it was announced that a team of UCLA scientists had discovered that the four DNA bases, A,T,G and C could influence the shape of metallic nanoparticles. Gold nanoparticles are made by sowing tiny gold seeds in a solution of a gold compound. The seeds are first incubated with short segments of DNA before being added to the gold solution. It was found that strands of repeating A bases produced round gold particles. T bases produced stars, C bases produced flat circular discs and G bases, hexagons. This research holds promise in bio-nanotechnology and applications in catalysis, sensing, imaging and medicine.

THE GEOLOGICAL HISTORY OF THE EARTH

In the early days of geology, there were two schools of thought, the Plutonists, who thought that all rocks were formed through volcanic action, and the Neptunists, who claimed that all rocks were formed through water action. It was Sir Charles Lyell who brought the two schools of thought together when he wrote 'Elements of Geology' in 1838. Lyell's 'Elements of Geology', which deals with stratigraphic and palaeontological geology, was written as an overflow from part of his 'Principles of Geology' which had become too unwieldy. A third and important rock type is the metamorphic. Metamorphic rock can be igneous, sedimentary or even other metamorphic rock which has been transformed under tremendous heat and pressure into a quite different type of rock. This can typically happen deep within the Earth's crust during orogeny (mountain building). Common types of metamorphic rock are: quartz, amethyst, marble, slate, gneiss, graphite, diamond, coal etc. Marble is metamorphosed from sedimentary limestone of marine origin. Marble can be in a variety of colours and also have beautifully textured patterns resulting from various impurities. The wonderful Pentelic and Carrara reserves are of a particularly pure whiteness and almost seem purpose made for the Classical and Renaissance sculptors to give expression to their artistic genius. The metamorphic production of marble from sedimentary marine limestone and its ultimate return to the sea represents a calcium cycle spanning many millions of years. Living organisms can play an important role in the concentration of sparsely distributed minerals.

The best estimate indicates that the Earth originated about 4570 million years ago which is about one third of the age of the universe. Before this time the solar system would have formed from the rich variety of chemicals ejected from a super-nova event. The oldest dated rocks are estimated to be 4100 million years old. Based on the evidence found in the rocks and associated fossils, the Earth's geological history was conveniently divided into five eras by the early geologists, Nicholas Steno, William Smith, Georges Cuvier and Alexandre Broginart, each era sub divided into systems. The eras were given Greek names: Archaeozoic (ancient life), Proterozoic (first life), Palaeozoic (old life), Mesozoic (middle life) and Cainozoic (new life). The early geologists were mostly British hence the predominance of British names of geological periods. 'Cambrian', 'Ordovician' and 'Silurian' were named after ancient Welsh tribes and 'Devonian' was named after the southern county. The 'Permian' was named after the Perm province in Russia and 'Jurassic' after the Jura mountain range of central Europe. The names 'Tertiary' and 'Quartenary' for naming epochs are no longer in use. The geologists were able to identify the eras by studying earth strata but had no means of estimating the time scale involved. The only strongly asserted timescale of the time was contained in a large tome published by Irish Bishop James Ussher (1581–1656) in which, based on Biblical studies, he claimed that the Earth, and presumably the entire universe, was created in 4004 BC. The geologists would have to wait until the twentieth century for accurate rock and fossil dating techniques. The eras determined by the early geologists have subsequently been greatly expanded, subdivided and dated to an astonishing degree of accuracy. The geologists based their work on the assumption that strata would be progressively older as one delved deeper into the earth. Geology is seldom as simple as this. Strata can be heaved up and even inverted, modified by earthquake, volcanic activity and weathering, and tectonic plate activity can heave up mountains and even subduct one continental plate under another.

Plate tectonics is one of the most important and far-ranging geological theories ever proposed. Alfred Lothar Wegener (1880–1930) proposed his theory in 1912 and published his first book on 'Continental Drift' in 1915. His theory was substantiated by matching geological evidence on opposite sides of the Atlantic as well as numerous fossil records indicating moving continents. The theory originally met with ridicule but was supported later by geologist Alexander du Toit (1878–1948) in books published in 1927 and 1937. Overwhelming evidence in support of the theory came later with accurate mapping of the ocean floors showing the mid Atlantic ridge and even video images of magma welling up between the plates. Satellite positioning now provides accurate measurement of the movement of the various plates. The plate movement is astonishingly slow – typically a few centimeters per year. This is not a snail's pace – this is movement similar to the rate of growth of one's fingernails. The study of plate tectonics has developed far further than Wegener could ever have imagined, with complex movements going back many millions of years before the single landmass that Wegener envisaged. It has been established that the Earth's crust is made up of eight major plates and a number of lesser plates, the plates covering the entire planet including the oceans. Diverging plates will cause new crust to be formed – colliding plates can cause mountains to be raised or the one plate can be subducted beneath the other. The boundaries between plates are in many cases areas of volcanic or seismic activity. A very interesting project, the NanTroSEIZE, has been launched to study plate activity beneath the Nankai trough south east of Japan. The drilling ship *Chikyu* can take core samples as deep as six kilometers below the sea bed penetrating the Eurasian Plate and the subducting Phillipine Sea Plate. In September 2012 it was announced that *Chikyu* had drilled 2111 m below the seabed off the Shimokita Peninsula. These plates are moving together at four centimeters per year. The core samples are studied in detail by means of computer aided tomography. A wealth of scientific data is harvested from this project which also may hopefully lead to a means of predicting earthquakes and tsunamis.

Rocks dating from the earliest times of the earth can be found lying about on the surface. Darwin noticed a band of sea shells high in a cliff face in the Andes and sea shells can also be found high in the Himalayas leaving no doubt as to the marine origins of these mountains. Leonardo da Vinci was fascinated by the marine fossils that he saw in the Italian Alps and deduced that these parts were below sea level in prehistoric times. It is not necessary to travel to the ends of the earth to see evidence of the violence and unimaginable time scales in which the earth formed. One can see evidence of boulders moved by glaciers in hot countries where natural ice is unknown or you can see evidence of the bending of strata in the sides of a roadway cutting. There is an awesome display of geological bending on cliff faces along the road to the Cedars of Lebanon. I can remember, when visiting the Graeco-Roman town of Glanum in the south of France, that many of the building blocks had embedded sea shells leaving no doubt as to the sedimentary marine origins of these stones. Old rock in plain view is much in evidence at Cape Town and the Peninsula. The base of Table Mountain, the Malmesbury Group, is late Precambrian greywacke sandstone and slate. The sequence was metamorphosed by heat and pressure and folded so that the layers are now almost vertical. Peninsula granite intruded into the Malmesbury Group 630 million years ago as magma and crystallised deep in the earth. Spheroidal granite boulders are prominent around Llandudno and Simonstown's Boulder Beach which is well frequented by tourists and penguins. The name 'penguin' derives from the Welsh words for 'white head' despite the fact that penguins have black heads. The penguins spend much of their time feeding in the coastal waters and their plumage conveniently keeps them permanently smartly attired for formal dinner. The age of the huge boulders makes the appearance of dinosaurs seem a recent event. The Cape Floral Region of Table Mountain has been declared a Unesco World Heritage site. This area has more plant species than the entire British Isles. Another awesome geological site is the famous landmark in central Australia, 'Ayers Rock'. This mountain sized rock of sedimentary sandstone is actually a syncline of rock folded downwards together to form the vertical strata. The entire rock has acquired a red colour from oxidised iron content.

Not all geological events occur over the passage of millions of years – some major events can occur in days or even hours. Apart from the obvious action of earthquakes, volcanoes and meteorites, an event which can reach catastrophic proportions is the ice dam. When a glacier blocks a valley branching from the side this can cause, over some years, the formation of a dam of immense proportions. Ice does not have the structural strength to form a dam wall but a glacier blocking a side channel will be supported by the opposite side of the glacier channel. Supercooled water under extreme pressure can enter fissures in the ice dam and gradually undermine the entire structure. When failure eventually comes, the result is total and catastrophic. Small ice dams are quite common, and some are visited by tourists, such as those in southern Patagonia and Iceland. Possibly the most sensational ice dam ever, was of the glacial lake Missoula, which covered parts of Montana and Idaho. This lake formed from meltwater from the Cordilleran ice sheet which covered much of Canada. Geological evidence showed that the ice dam and flood had occurred several times, the most recent dam burst occurring 13 000 years ago near the end of the last ice age. The traces of this ice dam burst were first noticed by J. Harlan Bretz in 1920 when he was puzzled by huge sand ripples in the flood channel scablands which can best be seen from the air. The ice dam theory was at first treated with skepticism as it seemed inconceivable that there could have been a large enough flood to cause these ripples. He also noticed several huge glacial erratic rocks which seemed to have come from nowhere. The flood was of immense proportions. It is estimated that the volume of water and ice involved was 2000 cubic kilometres. The flow was 600 metres deep with a speed of 100 kilometres per hour. The Missoula dam burst drained from northern Idaho, across central Washington, down the Columbia gorge and out to the Pacific. Gorges scoured out by massive floods will be of rectangular cross section. Those formed by rivers will be 'V' shaped and glaciers will produce a 'U' shaped gorge. Several cataclysmic floods have occurred in prehistoric times – let us take a look at two others.

One of the busiest waterways on Earth, the English Channel, was also the result of an ice dam but in this case the scouring was not the result of ice collapse.

The freezing of the upper part of the North Sea caused the southern part to rise considerably due to the inflow of major rivers and glacial meltwater. When the water level became high enough to overflow the land connecting England and France the scouring torrent increased until it reached unimaginable proportions. It is estimated that the initial flood took place 400 000 years ago and produced a torrent of up to a million cubic metres of water per second. A second flood is estimated to have taken place 180 000 years ago. This action is well substantiated by geological evidence on the channel seabed as well as the geology of the two landmasses. There are no fault lines between England and France, both landmasses being on the same tectonic plate. This fact has been a major advantage in finding a continuous path in material suited to tunneling for the Chunnel project.

Another ancient flood that can be mentioned here is the repeated flooding and drying of the Mediterranean Sea as shown by geological evidence of core samples. The straits of Gibraltar would be pinched together by tectonic plate activity causing the Mediterranean to dry due to insufficient inflow. The most recent flood is estimated at 8 million years ago. A huge head of sea water would eventually breach the Gibraltar blockage and flood the area with a wave of possibly more than a kilometer in height.

Fossils provide a most valuable tool for studies into the early history of the earth but we are most fortunate that there are any at all. Practically all dead animals are quickly recycled into the food chains – fossilisation is an extremely rare occurrence. If of all the millions of dinosaurs roaming the earth only one became fossilised each year we would have a hundred million dino fossils. In some areas the climate was quite unsuited to fossilisation so that in these parts there are no fossils at all. Fossilisation is so rare that most of the early species of fauna have disappeared without any trace at all. The word fossil means 'dug up' and does not refer only to old bones. Body fossils can contain original material or could be replaced by minerals. The quick frozen ice-age woolly mammoth discovered in Siberian permafrost and found to have freshly eaten food in its belly can also be classified as a fossil. In petrified forests the wood of the trees has been completely replaced by stone. Trace fossils can be footprints left by living creatures or plaster casts such as those of the cavities in ancient Pompeii, where the unfortunate residents met their end. Despite its combustible nature, coal is in geological terms a rock, but it is also fossilised wood. Coal is a valuable commodity, rich in a variety of chemicals, and should never be used as a fuel. This is a matter of the utmost urgency imperiling the future of human civilisation – this matter is discussed further in chapter eight. The burning of coal returns carbon dioxide to the atmosphere which took millions of years to remove.

The study of dinosaurs (terrible lizards) is a quite recent branch of palaeontology. The first nearly complete dinosaur skeleton fossil ever to be discovered was found in Haddonfield, New Jersey, in 1858 by fossil hobbyist William Parker Foulke. He heard that gigantic bones had been seen in a marl pit twenty years before so he organised a dig. What he found would send shock waves around the world; the almost complete skeleton of a huge animal, larger than an elephant, and with lizard-like features. The beast was named *Hadrosaurus* (bulky lizard) *foulkii*. Before this time no one had any idea that the earth had been dominated for a hundred and fifty million years by gigantic lizard-like reptiles. Dinosaur bone fossils have been discovered at various times going back to antiquity, possibly giving rise to legends of dragons and giants but these did not receive methodical scientific study. The first scientifically described dinosaur fossil was named by William Buckland in 1824 and given the species name *Megalosaurus bucklandii*. Another significant early dinosaur discovery was made in 1822 and named in 1825 by physician and geologist Gideon Mantell *iguanodon* as its teeth bore a remarkable resemblance to those of a living iguana. This beast had a mass of over three tons and a length of ten metres.

The term dinosauria was coined in 1842 by Sir Richard Owen from the Greek 'deinos' and 'sauros' meaning terrible or fearful lizard. We will probably never know exactly how many dinosaur species have inhabited the earth. We can get some estimates from Glen Kuban: Based on fossil body parts, the number of genera is given as 775. The estimate of the total is given as 50 000 dinosaur genera and up to half a million species. As much as 90% of these animals, however, lived in non-depositional regions where their fossils would not be preserved, and we are highly unlikely ever to find them. Many of these animals were however small (crow size) and some bird like, which are seldom found due to their small size and fragility. The number of really big dinosaur genera is estimated at one or two hundred.

Dippy, the magnificent 26 metre long diplodocus who has amazed millions of visitors to the London Natural History Museum has recently celebrated his 100th year in the vast central hall of the museum. Dippy was assembled from casts of bones of three dinosaurs found in Wyoming, USA. Dippy's original bones date from the upper Jurassic 150 million years ago. If Dippy were still in the flesh he might not be pleased with the name that he has been given. Diplodocus is from Greek meaning 'double wooden beam'. A more appropriate name would be 'Very-long-neck-and-tail-o-saurus'. Dippy is not alone in the museum – the dinosaur section contains an awesome collection of fossils.

The Brontosaurus (thunder lizard) is the only dinosaur to become extinct in the Holocene epoch – there was never such an animal – the mistake came about by a mismatch of heads on the Apatosaurus.

Dinosaur fossils can be found on every continent on earth and even Antarctica which was at the equator during the Jurassic. A modern dino fossil boom has occurred in China where fossils dating from 130 million years ago are being found in profusion. Feathered dinosaurs have also been found giving weight to the theory that birds evolved from dinosaurs. One of the most remarkable discoveries was several specimens of a Microraptor from 120 million years ago. This creature was from 42 to 83 cm long, completely covered in feathers, and astonishingly, had four wings. The forelimbs were fully developed bird-like wings and the hind limbs were also feathered wings. It appears that this creature was not suited for walking on the ground. The head had a beak with teeth as well as a feathered head crest. It also had a very long bony tail with a feathered fan at the end. Specimens are being found in their thousands including flowers, plants and even mammals. Scientists have unearthed in Yunnan province a 400 million year old bony fish that is linked to all modern vertebrates, including humans.

The Burgess Shale is probably the richest site of Middle Cambrian marine fossils ever discovered. It is situated in the Yoho National Park in the Rocky Mountains of British Columbia, Canada. This World Heritage Site was discovered by Charles Walcott in 1909. At the time of deposition this site was near the equator and at the base of a cliff from where mudflows would cover and preserve the sea life. An astonishing feature of the fossilisation is that much of the shape of the soft tissue of the creatures has also been preserved.

Another most remarkable fossil site is the Messel Pit, near Darmstadt, Germany. This is the richest site on Earth for understanding the living environment of the Eocene epoch, between 57 million and 36 million years ago. In particular, it provides unique information about the early stages of the evolution of mammals and includes exceptionally well preserved mammal fossils, ranging from fully articulated skeletons to the contents of stomachs of animals of this period.

The Messel Pit is a disused quarry near the village of Messel not far from Darmstadt, which was mined for brown coal and bituminous shale since 1895.

Because of its plethora of fossils, it has significant geological and paleontological importance. After almost becoming a landfill, strong local resistance eventually stopped these plans, and the Messel Pit was declared a UNESCO World Heritage site in 1995. Since 1974 amateur collectors were allowed to collect fossils. They developed the 'transfer technique' that enabled them to preserve the fine details of small fossils, the method still employed in preserving the fossils today. Many of the known specimens from the site have come from amateur collectors.

The Messel Pit boasts extensive preservation of structural integrity, even going so far as to preserve the fur, feathers, and 'skin shadows' of some species. The diversity of species is no less astonishing (thanks in part, perhaps, to the hypothesized periodic gas releases). A brief summary of some of the fossils found at the site is as follows:

- Early primate fossil with anthropoid characteristics.

- Over 10,000 fossilized fish of numerous species.

- Thousands of aquatic and terrestrial insects, some with distinct colouration still preserved.

- A plethora of small mammals including pygmy horses, large mice, primates, ground dwellers (hedgehogs, marsupials, pangolins), aardvark relatives, and bats.

- Large numbers of birds, particularly predatory species.

- Crocodiles, frogs, turtles, salamanders, and other reptiles or amphibians.

- Remains of over 30 distinct plant species, including palm leaves, fruits, pollen, wood, walnuts, and grapevines.

Significant scientific discoveries are still being made, and the pit has become an increasing tourism site as well.

One of the most important palaeoanthropological sites is the Sterkfontein cave complex near Johannesburg. This World Heritage Site has been recently developed and is now known as the Cradle of Humankind. In addition to the Sterkfontein cave complex, the Cradle of Humankind also includes the Swartkrans, Gladysvale, Kromdraai and environs in Gauteng as well as the Taung fossil site and the Makapansgat. The Taung site is famous for the juvenile hominid skull discovered by Australian born professor Raymond Dart in 1924. Dart named the genus *Australopithecus Africanus*

(Southern ape of Africa). The great significance of this discovery was for many years bedevilled by confusion arising from the infamous Piltdown hoax.

The Sterkfontein caves were formed within the past few million years in dolomites dating back more than 2 000 million years to the palaeoproterozoic era. The caves were originally used as a limestone quarry and cave infill was removed by blasting. A rich source of fossil remains was discovered in the discarded breccia resulting in the caves being saved from wanton destruction. These caves yielded the first adult ape-man discovered by Scottish medical doctor Robert Broom (1866–1951) in 1936 which was identified as the same genus and species as the Taung skull. In 1946, Dr. Broom and Dr. John Talbot Robinson (1923–2001) discovered a nearly perfect cranium of an Australopithecine which received world interest and was named Mrs. Ples. In 1956 Dr. Robinson published a most significant monograph on the dentition of the *Australopithecinae*. Dr. Robinson began a professorship at the University of Wisconsin in 1963 but continued to make trips to South Africa for further research. A sensational find in 1997 was the complete skeleton of a 3,3 million year old (possibly older) *Australopithecus Africanus* which has yet to be fully released from its concrete hard cave infill. This fossil has been named 'Little foot'. The sequences of deposits in the caves date from 1,5 to 3,5 million years ago. The fossils of the earliest humans date from 100 000 to 250 000 years ago. The caves have produced 500 skull, jaw, teeth and skeletal fossils as well as 9 000 stone tools and thousands of animal fossils including some sabre toothed cats.

A most remarkable discovery was announced on 8 April 2010. Anthropologist professor Lee Berger's nine year old son Matthew found a fossil while on a dig with his father on 15 August 2008. Further excavations revealed an adult female and a juvenile male fossil remains of a new species named *Australopithecus sediba*. This was a most significant find and possibly represents a transitional species between *Australopithecus africanus* and *Homo habilis*. A competition was held to name the juvenile fossil and the winning entry was 'Karabo' which means 'answer' in Setswana. The fossils have been dated from 1,78 to 1,95 million years ago.

The area also contains sites relating to the stone age, iron age and more recently to the Boer War. There are some forty fossil sites of which only thirteen have thus far been excavated. There is also evidence of the first deliberate use of fire dating from 1,3 million years ago. The Cradle of Humankind is being developed to suit the requirements of the sophisticated tourist as well as serious scientific researchers. The Maropeng museum complex offers an underground boat ride on an earth, water, air and fire theme and an excellent fossil collection. If you are in to 2,5 million year old *Australopithecus Africanus* fossils, this is the place to go. The Sterkfontein cave complex is not very far from the Hartbeeshoek satellite tracking station.

A considerable contribution to anthropology was made by three generations of the Leaky family. Louis Seymour Bazett Leaky (1903–1972) had a passion for African pre-history. He had a particular interest in the Olduvai Gorge in Kenya. This is a one hundred metre deep and nearly fifty kilometre long chasm not far from the Ngorongoro Crater originally discovered by German entomologist Wilhelm Kattwinkel in 1911. Leaky's first major discovery was the jaw of a pre-human creature called *Proconsul*. Other famous discoveries made by the Leaky family were: *Austrolopithecus Boisei, Homo Habilis* and the Laetoli footprints.

A most significant find was made by the team of Dr. Meave Leaky at Lake Turkana, Kenya in 1999. She named this *Kenyanthropus platyops* (flat-faced man of Kenya). A 3,5 million year old skull and partial jaw were found which are thought to belong to a new branch of the early human family. Meave is married to renowned British palaeontologist Richard Leaky. Meave is an advisory board member of the IBM-National Geographic Genographic project. Other famous fossil discoveries were:

- Eugene Dubois (1858–1940) discovered the first Homo Erectus.

- Davidson Black (1884–1934) discovered the Peking Man.

- Donald Carl Johanson (1943–) discovered several important hominid fossils in Ethiopia including the partial skeleton 'Lucy'.

It is not necessary to travel to remote inhospitable regions to visit fossil sites. The delightful town of Lyme Regis on the coast of West Dorset was made famous in the early part of the 19th century by palaeontologist Mary Anning who, by 1819, had found an ichthyosaur fossil and later a complete plesiosaur and remains of a flying reptile. Lyme Regis is on the fossil rich coastline which stretches 153 km from Orecombe Point near Exmouth in East Devon to Harry Rocks near Swanage in East Dorset. This coastline has been declared a World Heritage Site. The coast is well known for ammonite fossils which are found in profusion. The fossils come from the crumbling cliffs along the coast which have strata of the Mesozoic era which includes the Triassic, Jurassic and Cretacious periods. It is not necessary, or even safe, to probe the cliff faces; the fossils can be found on the beach after a receding tide. This is a marvellous spot for a family outing.

On 27 October 2009 the Dorset County Council announced the purchase of a fossil skull of a truly enormous pliosaur from a collector who had gathered the fragments over a period of four years. The skull alone has a length of 2,4 m and the entire beast would have been from 10 m to 16 m in length. When fully prepared the fossil will be put on display by the Dorset County Museum. It is thought that the remainder of the pliosaur fossil may still be embedded in the Dorset Jurassic cliffs.

In 1984, palaeontologists David Raup and Jack Sepkoski proposed a hypothesis concerning the major extinctions which have been experienced by the Earth. They identified 12 extinction events with an average periodicity of 26 million years and suggested an extraterrestrial cause. Several astronomers have followed up on this and suggested a red dwarf or brown dwarf star orbiting the Sun in an elongated orbit with an aphelion beyond the Oort cloud. This star has been named Nemesis. On its inward passage towards the Sun the dwarf star could disturb a vast number of Oort cloud objects and draw them towards the inner Solar System. It is quite possible that many of these could impact the Earth and others could become long period comets. It is only a matter of time before the dwarf star hypothesis is confirmed or rejected. Present and future space observatories specially designed for detecting moving objects such as asteroids, comets and trans-Neptunian objects will easily be able to detect the star in visible or infrared wavelengths.

An interesting idea that has gained much undeserved popular support is the theory of earth crust displacement. This theory asserts that the entire Earth's crust is able to move independently on the Earth's core. The prime mover for this to happen is a build-up of ice in Antarctica placing a huge eccentric mass away from the South Pole which can produce enough centrifugal force to cause a movement of the crust. This has been used to explain the cause of ice-ages, floods and other geological cataclysms. A movement of the crust due to a force applied at one place would require that the pressure be transmitted right around the globe as the tension on the opposite side would hardly be likely to make any contribution. If an eccentric mass could really cause any crust movement all that would happen would be the formation of mountains on the upside and the opening of chasms on the downside.

GEOLOGICAL TIME SCALE OF THE EARTH

Start: Millions of years ago	Aeon	Era	Period	Epoch	Major events
4570		Cryptic			
4150	Hadean	Basin groups			Hellfire and brimstone Moon making impact Oldest known rocks
3920		Nectarian			
3850		Lower Imbrian			
3800	Archaean	Eoarchaean			First craton formation Single cell organisms Oxygen producing bacteria First stromatolites
3600		Palaeoarchaean			
3200		Mesoarchaean			
2800		Neoarchaean			
2500	Proterozoic	Palaeo-Proterozoic	Siderian		Banded iron formations
2300			Rhyacian		Bushveld formation
2050			Orosirian		Vredefort and Sudbury meteorites. Atmosphere became oxygenic
1800			Statherian		
1600		Meso-proterozoic	Calymmian		
1400			Ectasian		
1200			Stenian		
1000		Neo-proterozoic	Tonian		
850			Cryogenian		'Snowball Earth' hypothesis
650			Ediacaran		First multi celled animals. Worms and sponges.

542	Phanerozoic	Palaeozoic	Cambrian	Lower	First vertebrates and trilobites. Major diversification of life.

542	Phanerozoic	Palaeozoic	Cambrian	Lower	First vertebrates and trilobites. Major diversification of life.
513				Middle	
501				Furongian	
488			Ordovician	Lower	Diverse invertebrates, corals, brachiopods, bivalves etc. First primitive land plants. Ice Age ends Period.
472				Middle	
461				Upper	
444			Silurian	Llandovery	Vascular land plants. First jawed fish.
428				Wenlock	
423				Ludlow	
419				Pridoli	
416			Devonian	Lower	First seed bearing plants and trees. Early sharks, ray finned and lobe finned fish. Coelacanth
398				Middle	
385				Upper	
360			Carboniferous/ Mississippian	Lower	Large primitive trees, First land vertebrates
345				Middle	
326				Upper	
318			Carboniferous/ Pennsylvanian	Lower	Abundant winged insects and amphibians. First reptiles. Coal forests.
312				Middle	
306				Upper	
299			Permian	Cisuralian	Beetles and flies evolve. Gorgonopsids evolve. Great Permian extinction event destroys 95% of life on Earth.
270				Guadalupian	
260				Lopingian	
251		Mesozoic	Triassic	Lower	First mammals, dinosaurs and crocodiles.
245				Middle	
228				Upper	

200	Phanerozoic	Mesozoic	Jurassic	Lower	Dinosaurs flourish. Conifers, cycads and ferns evolve. First birds and lizards. Breakup of Pangea into Gondwana and Laurasia.
175				Middle	
161				Upper	
145			Cretaceous	Lower	Flowering plants and more insects. Breakup of Gondwana.
100				Upper	
65		Cenozoic	Palaeogene	Palaeocene	Dinosaur Extinction by Chicxulub meteorite. Plants and mammals diversify.
56				Eocene	Modern mammals evolve. Whales diversify and first grasses appear. Start of Ice age.
34				Oligocene	Rapid diversification of mammals.
23			Neogene	Miocene	Mountain building in northern hemisphere. Modern mammal and bird families.
5,3				Pliocene	Intensification of Ice Age.
1,8				Pleistocene	Extinction of many large mammals. Evolution of modern humans.
11 400 years				Holocene	End of most recent ice age and start of modern civilizations.

This table has been constructed as a best attempt from public domain sources to represent the geological time scale.

The names, dates and subdivisions can be expected to change from time to time so that readers requiring accurate information should consult authoritative up to date sources. An internationally accepted naming standard for geological time intervals remains to be agreed upon. Palaeontologists have a system of faunal stages based on fossil records and which are more numerous that the geological stages.

In order to get a better feel for the lengths of the geological eras, let us construct a linear time-line. A useful scale would be to use one millimetre per million years. This will give us a length of 4,57 metres, so if we wish to roll it out on a table we will need two dinner tables placed end to end. The Hadean will occupy the first 95 cm, the Archaean the next 1,3 metres and the Proterozoic the following two metres. We have now used up just over four metres of the time-line representing the Precambrian. The Palaeozoic would occupy the next 29 cm leaving 25 cm for the Mesozoic and Cenozoic. The time span of the earliest human forms to the present would require about two millimetres. The Holocene epoch which would include the Egyptian, Phoenician and Babylonian civilisations and others up to modern times would require less than 12 microns, an eighth of the thickness of a piece of paper, which would be invisible without a microscope. The dividing line between BC and AD would be only two microns from the edge of the chart. On this time-line the dinosaurs would have a space of fifteen centimetres and the trilobites an awesome thirty centimetres. If we wish to have a time-line showing the lifetimes of prominent people, two microns for the AD period would be quite useless. Let us draw another time-line scaled one millimetre per year giving two metres for the AD period. This will give us space to write names and other information and even a few illustrations. The entire time-line would now be 4 570 kilometres long. When describing elapses of time on the geological scale it is convenient to use the measure Ga (Giga anno – one thousand million years). This would then give 4,57 Ga for the age of the Earth and 13,7 Ga for the age of the universe.

The Hadean aeon, resulting from the agglomeration of material from the accretion disc, is the least known of the stages of the earth's development. It was extremely hot. Objects colliding at astronomical speeds lose their kinetic energy which is instantly converted to heat resulting in incandescent temperatures. The moon making impact must also have occurred at this stage. Needless to say, the moon would also have been white hot after the impact. The moon is discussed further in chapter five. The earliest rocks would have formed at this stage but it is not precisely known when the Earth's crust had cooled sufficiently for rocks to form.

To this day the centre of the earth remains hotter than the surface of the sun. Any water arriving in the form of icy comets would remain vapourised in the atmosphere until cooler times. The Hadean is estimated to have lasted 950 million years.

The Archaean (Archaeozoic) aeon has been divided into four eras. This aeon saw the emergence of the first simple single celled organisms and stromatolites – oxygen producing bacteria. By this time the Earth had cooled considerably with an average global temperature of a few degrees below the present average. The first cratons begin to form. These are stable portions of the continental crust reaching down by as much as 200 km into the mantle beneath but are generally twice the thickness of the continental crusts and could possibly be anchored in the aesthenosphere. The Earth's great landmasses each have several cratons. An exciting feature of cratons is that some have deep volcanic kimberlite pipes. It is these pipes which have brought diamonds from deep within the bowels of the Earth to the surface. A famous craton is the Kaapvaal which has the city Kimberley near its centre. The kimberlite pipe here has been excavated to awesome proportions to become the 'Big Hole'. It is interesting to note that the diamonds were crystallised from naturally occurring mineral carbon as trees had not yet appeared on the planet. An astonishing announcement was made by the Harvard-Smithsonian Centre for Astrophysics in February 2004. It was discovered that the star BPM 37093 in the constellation Centaurus, fifty light years away from Earth, was the crystallised remains of a white dwarf star which had become a diamond 4 000 km across. The star has been given the pet name 'Lucy'. The star has a hydrogen atmosphere making it slightly larger than the Sun. This star has no internal energy source so that it is only visible due to remaining heat and reflected starlight. Our own Sun may also become a diamond when it has reached the end of its life cycle. Vast numbers of diamonds larger than the moon out in the cosmos? This is not nonsense – the astronomers and astrophysicists are serious about it.

Another diamond in the sky was reported by Yale University scientists in October 2012. A planet orbiting star 55 Cancri 40 light years away was found to have twice the diameter of the Earth, is very hot and orbits its star in only 18 hours. The likely composition is a surface layer of graphite, a thick layer of diamond, a layer of silicon based minerals and a core of molten iron. The quantity of diamond is estimated at three times the mass of the earth. This is one of five planets orbiting the star.

During the Archaean, the Earth's heat flow was very high due to higher temperatures and the greater concentration of radioactive isotopes. The mantle was much more fluid than at present and there was considerable tectonic plate activity by small protocontinents. The Archaean lasted for 1300 million years.

The Proterozoic aeon saw some remarkable changes. The formation of the Rodinia supercontinent, which extended from the South Pole to the equator. The atmosphere became oxygenic and the average global temperature rose slowly to about five degrees above the present average. Metamorphic belts were created due to mountain building as Rodinia forms and then later breaks up. The palaeoproterozoic saw the most violent transfer of energy in the history of the Earth with the exception of the moon making impact. This was the impact of the Vredefort meteorite – a thousand cubic kilometre asteroid which plunged deep into the Kaapvaal craton and produced a 300 kilometre crater. This meteorite is discussed further in chapter five. There was also a flourishing of life; the first complex single celled life and then the first multi-celled animals. Worm like creatures and sponges also develop. The 'Snowball Earth' hypothesis has been proposed for the neo-proterozoic era. It is claimed that the entire earth froze over due to disruption of warm sea currents and the runaway effect of the reflection of solar heat by the expanding polar ice caps. The Earth eventually broke free from its icy grip due to volcanic activity and re-established ocean currents. The vast amount of heat released by a single volcano in its lava and pyroclastic flows is not sufficient to influence global temperature. Volcanoes can however influence global temperature in two other ways. The incineration of vast forests around the volcano can cause a considerable release of carbon dioxide to the atmosphere contributing to global warming and at the other extreme vast quantities of sulphur released into the upper atmosphere can result in clouds of sulphuric acid droplets which can reflect sunlight causing global dimming. The question of global warming is a matter presently receiving much coverage in the media and possibly represents the greatest ever threat to human civilization, as well as vast numbers of other species, and the effects can be seen daily unfolding in slow motion. For the first time in recorded history we can expect an ice free North Pole, and this can happen within a year. The point of no return threshold is currently set at a further two degrees rise in temperature. This may at first seem a trivial rise in temperature – people living in cold regions may well enjoy going on holiday to a climate which is ten or even twenty degrees higher than at home but this is not what all the fuss is about. The two degree rise will have catastrophic effects world wide and it will be too late to do anything about it. Siberian permafrost will thaw releasing vast quantities of carbon dioxide; rainforest trees will instead of consuming carbon dioxide die off and release the problem gas; glacier and snow coverage will reduce reflecting less heat away from the Earth. The list goes on and on. A sea level rise of half a metre will result in the de-population of many coastal cities worldwide and a rise of a few metres will result in a significant reduction in the land area of the Earth with many low lying countries and islands disappearing beneath the waves. The Earth's atmosphere is made up of 78% nitrogen, 21% oxygen, 0,9% argon and a variable quantity of water vapour. Air contains only 360 parts per million of carbon dioxide – it is vital that this level be maintained but unfortunately the balance is so delicate that it can be increased by human generated emissions. The point of no return is currently estimated at 500 parts per million. The Proterozoic lasted nearly 2 000 million years. The entire age of the Earth up to the end of the Proterozoic, often referred to as Precambrian, lasted just over 4 000 million years – about 88% of the life of the earth and the most exciting creatures to develop were worms.

The remainder of the life of the earth is called the Phanerozoic aeon which is divided into three eras.

The Palaeozoic era saw a major diversification of life. The Cambrian period brought the first vertebrates and trilobites. The trilobites flourished in the seas and at their peak there were 150 families, 5 000 genera and 15 000 species. These are small crab or beetle-like creatures with plenty of legs and are usually less than a centimetre to several centimetres in length. The smallest ever found was less than one millimetre long and the largest a whopping 72 centimetres. The trilobites had compound eyes with from several hundred to several thousand lenses per eye. Compound eyes cannot focus but the trilobites had eyes with lenses of a remarkable multi-focus design. They were cute dear little creatures and their fossils make perfect collectors items. Some collectors have several thousand specimens. The trilobites lived for nearly 300 million years and during this time the Earth could certainly have been called 'the planet of the trilobites'. There is a very slim chance that living trilobites may still lurk in some unknown corner of the oceans. No traces of any creatures leading up to the trilobites have been found but there is nothing exceptional in this – the pre-trilobites, like most other creatures, could have disappeared without trace. About one third of the Rodinia supercontinent broke away and drifted north and west. These included parts of North America (Laurentia) northern Europe (Baltica) Siberia and central Asia. The remains of Rodinia included South America, Africa, Australia, Antarctica, China, India and the Middle East. The continental blocks carried with them broad continental shelves providing shallow seas in which Cambrian marine life could thrive. Other creatures to develop at this time were crabs, lobsters, sponges, starfish, jellyfish, squid and octopus which have survived to the present time. The Cambrian lasted for 54 million years.

The Ordovician period saw much further diversification of life in invertebrates, corals, brachiopods, bivalves, nautiloids etc. The first primitive land plants developed. The Earth experienced a milder, warmer and more humid climate. The period lasted 44 million years and ended with an ice age and the first of the mass extinctions of land and marine life due to glaciation and destruction of marine habitats by falling sea levels. It has recently been proposed that the Ordovician mass extinction was triggered by an immense gamma ray burst from space which caused multiple catastrophes including the ice age. Baltica and Siberia continue to drift toward North America and the central Asian block drifted towards the North Pole. Much of North America lay submerged as a very large coastal shelf. The large remains of Rodinia now called Gondwana stretched from the South Pole to the equator with South America and Africa upside down from their present positions.

The Silurian period saw the development of the first vascular land plants, also the first jawed fish and sea scorpions. Millipedes developed. The trilobites were still going strong. The continents of North America, Baltica and Siberia reassembled into the supercontinent of Laurasia and new mountain ranges were raised. Gondwana, made up of the fused continents of South America, Africa, Antarctica, India and Australia, moved around the globe towards Laurasia from the south and east. The Silurian lasted 28 million years.

The Devonian saw the first seed bearing plants and trees and also the appearance of wingless insects and spiders. Much development in sea life occurred. Early sharks appear as well as ray finned and lobe-finned fish. A most astonishing lobe-finned fish of this period is the Coelacanth (hollow spine – Greek). This fish was thought to have become extinct by the end of the Cretaceous until a living specimen was caught in South African coastal waters in 1938 and called *Latimeria Chalumnae*. This was a fishy story that shocked the world. Subsequently large numbers of living specimens have been observed by divers off the Comores where the local fishermen regarded the Coelacanth as a disgustingly inedible nuisance. The Coelacanth has a remarkable way of swimming with its lobed fins – it almost seems to walk in the water. The large blocks to become China and Tibet broke away from equatorial Gondwana while the rest containing Africa, South America, Antarctica and Australia began to drift over the South Pole. As Gondwana moved north past the pole it pushed the eastern edge of the Pacific plate (Panthalassa Ocean) forming a volcanic arc and earthquake faults along the North American west coast. The Devonian lasted 56 million years and saw the second of the great extinctions resulting in the loss of 70% of species due to severe climatic changes over a period of 3 million years.

The Carboniferous periods saw the development of large primitive trees, land vertebrates, winged insects, amphibians and reptiles. The vast forests laid down the coal beds. The push against North America continued and raised the Rocky Mountains with most of North America remaining as a submerged coastal shelf. Siberia broke away from Laurasia and drifted north. and the central Asian block drifted south from the North Pole. Gondwana collided with Laurasia from the south with areas to become South America and Africa. The Carboniferous lasted 67 million years.

The Permian period saw the development of flies and beetles as well as cockroaches and giant dragonflies with wingspans up to 71 cm. The Permian was named in 1841 by Scottish geologist Sir Roderick Murchison (1792–1871) after the ancient kingdom of Permia in Russia. The Permian ended with the great Permian extinction which destroyed 95% of all life on earth. Marine life was the most severely affected with 96% of species, including the trilobites, being destroyed. Terrestrial vertebrates lost 70% of species. This was also the only mass extinction of insect life. It is currently held that the extinction occurred in three pulses, over a period of a million years, of which the third was the largest. It has recently been estimated that the Earth took ten million years to recover from the extinction.

Several causes of the extinction have been proposed which could include one or more of the following:

- Meteorite (bolide) impacts.
- Vulcanism: it is estimated that flood basalt eruptions in the Siberian Traps covered up to $100\,000\,km^2$ with lava and produced vast volcanic degassing.
- Release of methane hydrates by gasification from sea floor.
- Sea level change.
- Anoxia (oxygen starvation) in seas.
- A gamma ray burst from a supernova event within the Milky Way galaxy. (A highly speculative hypothesis).
- Destruction of coastlines by consolidation of landmasses.

The newly assembled supercontinent of Pangaea lay across the equator stretching almost from pole to pole. The former Laurasia to the north and the African continental block to the south formed the Tethys Ocean, which opened to the east. Most of the Earth was occupied by a single sea, the Panthalassa Ocean. At this time, the area of the landmass was greater than that of the ocean. The Permian lasted 48 million years.

The Triassic was an exciting time which saw the appearance of the dinosaurs, crocodiles and mammals. This period also saw the splitting up of Pangaea into separate continents. This was also a very watery period characterised by extensive sedimentation, sandstones and shales.   Towards the end of the Triassic the fourth of the great extinctions occurred, although the smallest of them all. Almost a quarter of all families became extinct. It is thought that this extinction was caused by a drastic increase in rainfall. The northward movement of Gondwana pushed Siberia to the North Pole. Further plate activity caused more of North America and Alaska to rise from the sea. The Tethys ocean expanded and Cimmeria (Turkey, Iran and Afghanistan) continued to drift north towards Laurasia. Antarctica was at the equator.   Chinese researchers have traced exceptionally well preserved fossils found in the Guizhou Province to the Middle Triassic (235-242 Million years ago). A remarkable find was the first flying fish, 153 mm long, *Potanichthys xingyiensis* which has a pair of primary wings and a smaller pair of auxiliary wings. The Triassic lasted for 51 million years.

The Jurassic period is probably the best known due to the efforts of the movie makers. This was a great time for dinosaurs and also for birds, mammals, coniferous trees, cycads and also spiral-like sea creatures called ammonites. The dinosaurs were the most spectacular of all forms of animal life and dominated the animal world for a hundred and fifty million years. A fascinating creature of this period is the *Archaeopteryx lithographica* (ancient wing from the printing stone) found in Jurassic Solnhofen limestone in southern Germany. Seven specimens have so far have been found showing that feathered dinos appeared long before flying reptiles and birds. The early flightless feathered dinos had soft fluffy down feathers for warmth – flight feathers appeared later. A University of Alberta team found that the early cretaceous fossilized *Sinocalliopteryx* (Chinese beautiful feather), a feathered but flightless raptor-like predator, had three undigested birds in its belly. The *Sinocalliopteryx*, found in the Yixian Formation, was about two metres long. Two specimens of another early cretaceous feathered dino found in the Yixian Formation were named *Caudipteryx* (Tail feather). This curious mixture of reptile and bird features was about one metre from beak to tail tip. This creature, thought to be omnivorous, had a toothed beak, substantial muscular legs for fast running and a short bony tail tipped with a fan of feathers. The *Archaeopteryx* looks like a bird at first sight; it has wings and feathers, but it also has teeth, three claws on each wing, a bony tail and most of the skeletal features are those of a lizard, particularly the neck, which attaches at the back of the skull instead of from below. Although the *Archaeopteryx* fossils were found in Germany, Europe did not exist at that time but was an area of islands and shallow seas. Fossil remains of several other feathered dinos have been found. Geologically the Jurassic period also saw some violence, widespread igneous activity, the crust of the Earth literally burst open, allowing massive lava flows over hundreds, or thousands of kilometres all over the earth. A major rift from ancestral Mexico to the Tethys ocean formed the ancestral Gulf of Mexico and the North Atlantic Ocean. Complex patterns of movement, subsidence and uplift occurred in Asia and Europe. Much of central Europe, northern Africa and the Middle East remained submerged as shallow seas. India drifted away from Africa, and Antarctica centred on the South Pole. Australia separated from Antarctica and drifted northwards. The Jurassic lasted for 90 million years.

The Cretaceous period, as the name implies, was characterised by calcified sediments, sandstones, limestones, marine clays and conglomerates. Flowering plants and more insects appeared. New types of dinosaur (T-Rex etc.) appear as well as modern crocodiles and sharks. Primitive birds gradually replace pterosaurs and marsupials appear. Break up of Gondwana starts. India and Australia continue to move northwards and opened the Indian Ocean. The Cretaceous lasted for about 80 million years. A most remarkable fossil from this period was discovered in Montana on the last day of a Summer expedition in 2000. The 77 million year old fossil is of a juvenile seven metre long duck billed brachylophosaurus. The most remarkable aspect of this fossil is that it is the fossilized remains of a complete mummified dinosaur. It has been determined that the dino had a soft skin with five sided scales and a small sail frill on his back. The actual soft tissue has been replaced by minerals but the detail remains astonishing: muscle, nails, foot pads and a beak. The stomach even contains the dino's last meal; a salad of ferns, conifers and magnolias and even the pollen of more than 40 different plants. This dino has been named Leonardo from graffiti found at the site.

The Palaeogene period (Cenozoic era) started with a catastrophic event: the

Chicxulub meteorite struck the Earth off the coast of Yucatan causing a lengthy world wide blockout of sunlight, which in turn caused the destruction of food chains and sadly the end of the dinosaurs together with half of all genera. Meteorites are discussed further in chapter five. The Palaeogene saw the rapid diversification of modern animals, mammals, whales, plants and grasses. Reglaciation of Antarctica and ice age starts. Continental reassembly continues and the modern patterns of the Earth's landmasses finally appear. The Palaeogene lasted for about 42 million years.

The Neogene period, the last of the Cenozoic era, starts with the Miocene epoch of mountain building in the northern hemisphere. Modern mammal and bird families appear. The Pliocene sees an intensification of the ice age. The Pleistocene epoch sees the extinction of many large mammals and ominously the evolution of modern humans who have left us examples of their artistic efforts in caves worldwide. The catastrophic Toba volcanic eruption occurred some 75 000 years ago. This is the most violent volcano known to science which released a vast quantity of sulphur dioxide resulting in a global coverage by sulphuric acid droplets in the upper atmosphere. This resulted in global cooling and the increased ice and snow coverage caused runaway cooling by reflection of more heat from the sun. The end result was the start of an ice age and a drop of five degrees in ocean temperature. The last geological epoch, the Holocene, takes us back only 11 400 years. This brought the end to the most recent ice age and the start of the modern civilisations. The latest human biped, the *Homo Sapiens*, is the only biped to inhabit the planet who has the intelligence, power and irresponsibility to artificially cause a mass extinction of life. Not all the specimens are bad: some have produced some good literature, music, art, philosophy, mathematics and scientific discovery.

Let us close this chapter with Hamlet mulling over the nature of the most complex of bipeds:

> ... *What a piece of work is a man! How noble in reason! How infinite in faculty! In form, in moving, how express and admirable! In action how like an angel! In apprehension how like a god! The beauty of the world! The paragon of animals! And yet, to me, what is this quintessence of dust? Man delights not me ...*

THE SCALE OF THINGS

When speaking in cosmological terms it is all too easy to mention galaxies and quarks in the same sentence without regard for the unimaginable range of scale that this implies. On looking at the table given here it may seem a hopeless task attempting to visualise the vast range of dimensions but all is not lost. Let us attempt the impossible in two stages:

1. Going up

We start with a fairly large galaxy, such as our own Milky Way, of 100 thousand light years diameter. Let us first make a model of the galaxy. A convenient material for the stars would be 1 mm grains of birdseed. How much do we need? About 400 cubic metres, so if you have a 10 cubic metre truck we will need about 40 loads. We can now spread the birdseed out into a large disc and for added realism we can add some spiral streamers at the edge. How large must the disc be? Here is the catch – our galaxy model will have to have nearly the same diameter as the Sun. Let us first look at the Solar System. Our own planet Earth would be an invisibly small speck about 12 cm from the grain of birdseed representing the Sun and the furthest planet Pluto a much smaller speck about four metres away. In this model our nearest neighbour star would be about 27 km away. With the galaxy so sparsely populated with stars it is obvious that few stars can be seen as having any size at all. It is only their intense brightness which makes them visible as light point sources. One might think that it is quite impossible to detect planets in orbit around other stars but astronomers have actually succeeded in this by observing the slight wobble to the star's position caused by large orbiting planets! Observing the stars with a powerful telescope has the advantage of making more stars visible but point source stars which are already visible will actually appear smaller due to the sharper focus. Using huge modern telescopes and interferometric techniques it is actually possible to measure the sizes of some stars. One might think that if galaxies are so sparsely spread out, two galaxies could pass through each other without a single star colliding with another.

This would be true if the galaxies could move quickly through each other but as the galaxies must necessarily move slower than the speed of light it would take millions of years for them to pass through each other and gravitational attraction would make a chaotic jumble of the stars. The average density of matter in a galaxy is very much less than the density of a puff of smoke! We now scale this galaxy down to a 1 cm shirt button size smudge of light containing thousands of millions of stars. The universe is not 'round' but it is difficult to imagine it as anything other than a sphere. The extent of the universe is currently considered to be about 13,7 thousand million light years which is the length of an imaginary straight line which would meet itself end to end. With the same scale as the smudge the universe would have a size of 1,37 km filled with thousands of millions of other smudges. This universe is expanding at the speed of light which means that our scaled down model would be expanding at the rate of 1 mm per ten thousand years. On the galactic scale the speed of light is unimaginably slow!

Let us also take a look at the total number of stars in the visible universe – 3 x 10^{22}. (thirty billion trillion). This number is humanly incomprehensible but let us try to get a grasp of it. We again take 1 mm grains of birdseed so that we can pack 1000 grains into a 1 cm cube. A cubic metre will then contain 10^9 grains and a cubic kilometre 10^{18} grains. To represent all the stars in the universe we will need 30 000 cubic kilometres of birdseed. Let us now spread this birdseed out on a sphere the size of the Earth – and we will get a layer sixty grains deep. There really are a lot of stars out there and quite possibly a similar number of planets. There is another super-large number that we can consider – Avogadro's constant: 6.022×10^{23}. This means that there are 20 times as many atoms in 12 grams of carbon-12 as there are stars in the universe.

2. Going down

Present estimates give the size of a superstring as 10^{-31} metre. Let us take the superstring as a 1 mm grain. The next stage is the quark which would then have a scaled up size of 10^{10} metre. Going further the proton would be about 10^{13} metre and an atom 10^{18} metre which is more than a hundred light years! One's first knee jerk reaction to figures like these is to think that they are absurd – but even if the superstring is estimated too small by a factor of a million the scaled up proton would still be about as large as the Earth. Let no one be under any illusion about how much 'solid' matter there is in an atom! With the universe extending over such a vast range of scale let us take a look at how much of it we are able to observe. Animals have evolved with five senses which equip them superbly for the basic functions of food and mate acquisition and escape from predation. Human nature being what it is we have an insatiable urge to know what lies beyond and how things work. Do we really need to meddle with things that animals don't know about? Yes we do!

On the macro side we can do quite well with visual observation. On a clear moonless night we can see quite a lot of our own milky way galaxy and even a few other galaxies as faint clouds of light. After this, optical equipment makes a huge difference, and the vast number of stars in our galaxy becomes apparent. We have made tremendous advances since Galileo's home made instrument. The photographic results from the Hubble space telescope are mind boggling. There are also the huge new groups of terrestrial optical telescopes and radio telescopes to bring us vast amounts of data from the early life of the universe. Descriptions of several well known telescopes are given in chapter six.

On the downward side our visual powers of observation are severely limited. We can easily read printing in 4 point type but anything much smaller than that

Universe		1,296	x 10^{26}
Yottametre	Ym	1	x 10^{24}
Megaparsec		3,08	x 10^{22}
Zettametre	Zm	1	x 10^{21}
Galaxy		1	x 10^{21}
Light year		9,46	x 10^{15}
Astronomical unit		1,5	x 10^{11}
Sun diameter		1,39	x 10^{9}
Earth diameter		1,27	x 10^{7}
Human		1,8	x 10^{0}
Micron	μm	1	x 10^{-6}
Light wavelength		5,5	x 10^{-7}
Angstrom unit Å		1	x 10^{-10}
Atom		1	x 10^{-10}
Femtometre	fm	1	x 10^{-15}
Proton		1	x 10^{-15}
Quark		1	x 10^{-18}
Zeptometre	zm	1	x 10^{-21}
Yoctometre	ym	1	x 10^{-24}
Superstring		1	x 10^{-31}
Planck length		1.616	x 10^{-35}

Some sizes of familiar objects and measures in metres

will require magnification. Optical equipment can help tremendously with seeing things as small as microbes but is limited by the wavelength of visible light – there is no way for light to resolve images smaller than the wavelength. The next step is to use an electron microscope – electrons have an effective 'wavelength' much shorter than that of visible light but this still doesn't permit observation of images as small as atoms. There are other specialised techniques for seeing even smaller but at best they can only detect atoms as humps on a surface – hardly picture gallery stuff and we are not very far down the scale on the small side of the universe. Research into particle physics does not rely on image observation – such research is done by accelerating particles to tremendous speed, letting them collide head on and then analysing the trajectories of the collision debris. The LHC (Large Hadron Collider) is described in chapter six. For smaller than this we have to rely on esoteric theories and mind boggling mathematics.

A full list of SI prefixes is given in the appendix. (yotta, yocto, zetta, zepto etc.)

To get some grasp of the scale of the Planck length let us consider the following: We start drawing a line and for every second since the Big Bang we make the line longer by a million Planck lengths – how long will the line become? $4{,}73 \times 10^{-12}$ metres, which is smaller than the size of an atom! And how large is the atom? Let us run through again. For every second since the Big Bang we put an atom into a teaspoon – how much do we get? If the atoms are actually sugar molecules there would not be enough to sweeten a cup of coffee. Does all this suggest that the universe is not very old? Let us run through once more but this time we make a stack of printer/copier paper with one sheet for every year of the age of the universe. The first million years will give us a stack of 100 metres. We will end up with a stack of paper of about 1400 kilometres.

The small size of the superstring, which is estimated by leading academics to be equivalent to one Planck length, is so awesome that it is worth looking at another practical example. We will however use a larger size of superstring, 10^{-31} metres, in order to conform to theory presented elsewhere in this book. We arrange the superstrings in rows and layers like eggs in an egg crate. This is of course an impossible state for matter to exist in but it is a useful arrangement for us to get an idea of their size. We now take the strings from all the matter in the Sun and arrange them in our egg crate. What size of crate do we get? It will be about the size of an atomic nucleus. How about all the material from a galaxy? This would give a huge increase to our crate but it would still be not much larger than an atom. If we take the material from all the galaxies in the universe this would make an even greater difference but it may be just large enough to see without a microscope! The size of superstrings may be pushing the limits of credulity and absurdity but what is even more astonishing is that the strings are hollow resembling empty egg shells more than whole eggs! After looking at the shockingly small size of the Planck length it seems natural to assume that there can surely be no smaller number to describe any part of the universe. This is not the case; there is a yet smaller number that we need to mention – much smaller! The Planck time. This is the time required for light to travel the distance of one Planck length: $5{,}39072 \times 10^{-44}$ second. This is the ultimate unit of time representing the step from one time instant to the next and the time for existence to annihilation of matter energy in a time slice. This unit of time is so small that any attempt to illustrate its size by means of a scaled up practical example would be futile. The Planck units of measure are further described in the appendix under the heading 'The Beautiful Natural Planck Units of the Universe'. We should perhaps mention here that an individual superstring occupies several million time slices as it burrows its way along the time vector. There is yet another super-small number: the ratio between gravitation and electrostatic force between electrons – an awesome 10^{-43}. This number is humanly incomprehensible. Let us take a more practical look at this ratio: we take a balance beam and apply the electrostatic force between two electrons one centimeter from the fulcrum. We now take two electrons with the same spacing and apply the gravitational attraction between them at the other end of the beam. What length of beam will we need to balance these forces? The beam will need to be many times longer than the size of the entire universe! Does all this mean that it is impossible to know exactly what nucleons, quarks and superstrings look like? Yes? No! Read on and hold tight for the rest of a mental roller coaster ride!

THE HEAVENLY BODIES OF THE UNIVERSE

On our quest of perception of the universe we now take a look at the physical objects which constitute the cosmos. The variety of objects is huge. We give here an alphabetical list with brief descriptions of the more popularly known objects as well as mentioning some of the people and events involved in their discovery.

Asteroid

Asteroids are small mostly rocky remains of the original accretion disc of the Solar System. There is much confusion with a term describing these bodies. William Herschel proposed the term 'asteroid' (star like object) but Giuseppe Piazzi, who discovered Ceres, preferred the term 'planetoid'. They have also been commonly known as 'minor planets'. In 2006, Ceres was upgraded to 'dwarf planet' and Pluto was downgraded to 'dwarf planet' making the term 'minor planet' seem inappropriate to asteroids. These bodies are to be found in the asteroid belt between the orbits of Mars and Jupiter. Some have been known to move inside the Earth's orbit and even as far out as the orbit of Saturn. Asteroid 'belt' is something of a misnomer – the asteroids are widely spaced and their combined mass is a small fraction of the mass of Earth's moon. There are, as can be expected, noticeable concentrations of asteroids at the Lagrangian points of the orbit of Jupiter as these are reasonably stable positions where an object can share the orbit of a planet. So far 155 368 asteroids have been catalogued but the total could possibly number millions. The asteroids generally move at about 20 km/sec in their orbits around the Sun. The giant asteroid Ceres is about 1000 km across but the rest are much smaller ranging down in size to pebbles. Ceres, the first asteroid to be discovered, was discovered by Giuseppe Piazzi in 1801. A remarkable photograph of Ceres, taken by the Hubble Space Telescope, shows that it is not at all like the usual idea of an asteroid but is a smooth sphere like a moon or mini-planet and may actually contain water ice. Sixteen asteroids are more than 240 km across, most of which were discovered in the nineteenth century. If one of these were to plunge into the Earth this would certainly cause a mass extinction. The large asteroids are of sufficient mass to have a measurable influence on the orbits of the planets. It is quite commonplace for small asteroids to be disturbed in their orbits by Jupiter or comets and strike the Earth. Asteroids which are on target for Earth are called meteoroids, those that burn up in the atmosphere are called meteors and those that make it all the way to Earth are called meteorites.

The Galileo Jupiter spacecraft, launched in 1989, was the first to visit asteroids when it passed by Gaspra in 1991 and Ida in 1993 where it discovered for the first time an asteroid orbited by a moon, or rather, a small rock.

The first space mission designed to collect samples from an asteroid was launched by JAXA (Japan Aerospace Exploration Agency) on 9 May 2003. This is a mission of great scientific significance as it will give information of the original accretion disk material of the solar system. Hayabusa (peregrine Falcon), designed by ISAS (Institute of Space and Astronomical Research) made its rendezvous with asteroid 25143 Itokawa in September 2005 and landed in November 2005 to collect samples. The samples were to be collected by firing pellets at the asteroid and then collecting the scattered material. Collecting samples by means of drilling could not be employed as there is no convenient means of securing the vehicle to the asteroid. Problems were encountered with the sampling but it is expected that some material did find its way into the collection chamber. The spacecraft is remarkable in that it used electric propulsion for much of the travel to the asteroid and also for its autonomous navigation technique required because of the long delay of radio signals between Earth and the spacecraft. The ion engine first ionises the propellant gas xenon and then emits ions giving an efficient, albeit weak, propulsion system. The craft began its return to Earth on 24 April 2007 and made a safe landing in the Australian Outback on 15 June 2010.

The robotic Lincoln Near-Earth Asteroid Research project (LINEAR) has of June 2006 discovered 67 820 asteroids. This observatory is described further in chapter six.

There is a most remarkable mission concept being planned by NASA – the ANTS mission (Autonomous Nano Technology Swarm). This is specifically intended for data gathering where a single spacecraft would be ineffectual. A sub-mission would be the Prospecting Asteroid Mission (PAM). A transport ship from Earth (ant habitat) would stop at a Lagrangian point and then release up to 1000 spacecraft which would use thrust from solar sails for mobility. A large number of these craft can be expected to be lost due to a variety of mishaps. Up to 80% of the craft would be workers each carrying a single specialised instrument (magnetometers, spectrometers, cameras etc.). The coordinator ants will decide which asteroids are worth investigating and organise the workers accordingly. The key feature of this mission is its autonomy – it would be impossible to control the ants from Earth due to time delay. The ANTS mission has an envisioned time frame of 2020-2025.

The DAWN mission, part of NASA's Discovery Mission Program, was launched on 27 September 2007. DAWN is destined to rendezvous and orbit two of the largest asteroids, Ceres and Vesta, in 2011 and 2015 respectively. Vesta in July 2011 and Ceres in February 2015. The craft took a gravity assist flyby of Mars on 4 February 2009 to reach the asteroids. Photos returned by DAWN show a spherical planet-like Vesta with two large impact craters as well as minerals distributed across the surface. Dawn has a high quality camera, along with a back-up; a visible and near-infrared mapping spectrometer to identify minerals on the surface; and a gamma ray and neutron spectrometer to reveal the abundance of elements. Dawn has obtained more than 20 000 images of Vesta and millions of spectra.

This mission is unusual in that the craft will go into orbit about the asteroids instead of the usual flyby and will take a spiral path to the targets. A remarkable feature will be its xenon ion propulsion system which produces 90 mN of thrust. (A feeble thrust by terrestrial standards but adequate for the purpose) The spacecraft was loaded with 425 kg of xenon propellant. Dawn detected a band of discolouration around the Vesta equator caused by the evaporation of volatile matter, possibly water. A rich harvest of previously unobtainable data is expected from the mission.

ESA is considering a space mission to study the effects of a spacecraft impacting an asteroid. This is called the Don Quijote (Don Quixote) space probe proposed for launch in 2013 or 2015. The mission consists of an impact module Hidalgo and an orbiting observer spacecraft Sancho Panza. After the impact the orbiter will release the Autonomous Surface Package which will enter the impact crater and investigate the surface.

Black Hole

Black holes are the ultimate horrors of the universe. Huge neutron stars which have become so massive that they draw in everything in the vicinity, stars by the thousand, and become so massive that not even light can escape from their all consuming gravity. A black hole has been detected at the centre of our own Milky

Way galaxy. A black hole formed from a large star would have a size of less than 20 km. A supermassive black hole containing the material of an entire galaxy could be as large as the solar system. If the 'material' at the core of a black hole can become compressed to quark level, as has been speculated, then the size of the black hole will be reduced by a factor of tens of thousands. It has even been suggested that black holes could be as small as atoms! We need not look into much detail here as black holes are a matter of advanced theoretical physics in the extreme. There is a little ray of hope – it is thought that there is in fact a way for black holes to release energy by means of trapping 'virtual particles' and in effect 'evaporating'.

Scientists of The University of Cambridge have used infrared surveys to discover a new population of enormous rapidly growing supermassive black holes in the early universe – about 11 billion years ago. The most extreme object in the study is the direction of constellation Virgo. This monster has 10 billion times the mass of the Sun and 10 000 times the mass of the supermassive black hole at the centre of the Milky Way, making it one of the most massive black holes ever detected. Research indicates that there could be as many as 400 of such giant black holes in the visible part of the universe.

Brown Dwarf

These are small stars, much larger than gas giant planets such as Jupiter but too small to support a continuous nuclear fusion reaction at their cores. Brown dwarfs are the runts of the litter which will gradually grow dimmer until they are no longer visible. These stars are not wholly without interest. Due to their low luminosity they are convenient targets in the search for extra-solar planets.

Centaur

The astronomical body known as a centaur has nothing to do with the mythical man-horse combo that abducts fair damsels and then gallops away. The centaur is an object which has strayed from the Kuiper belt and taken up an erratic orbit between Saturn and Neptune. These are potential short period comets which may eventually take an elliptical orbit around the Sun if they do not collide with a planet or plunge into the Sun. Some well known centaurs and their discovery dates are:

Chiron 1997, Pholus 1992, Chariklo 1997 Asbolus 1995, Nessus 1993, Hydonome 1995, Therus 2001, Elastus 1999, Pelion 1998, Okyrhoe 1998, Cyllarus 1998 and Bienor 2000.

Depending on their behaviour, the centaurs will eventually be reclassified as comets or asteroids if they do not destroy themselves or are expelled from the Solar System. There is also a constellation named Centaur. This is indeed named after the mythical man-horse and contains the well known stars Proxima Centauri and Alpha Centauri which are our closest neighbour stars in our galaxy.

Cepheid

Cepheids are large stars with a mass of from 5 to 20 times the mass of the Sun. These are stars which have used up their supply of hydrogen and started fusing helium. Their most significant property is their instability which causes them to brighten and dim with precise regularity. The periodicity of these variable stars is determined by their mass and can be from 1 to 100 days. The pioneering work on cepheids was done by Henrietta Leavitt between 1893 and 1906. She produced a catalogue of 1777 variable stars in the Magellanic clouds. Since then the study of cepheids has been long and complex. The original cepheid – Delta Cephei was close enough to take a parallax measurement of its distance. This paved the way for the measurement of the distances to remote galaxies by observing the brightness and periodicity of cepheids. The most useful light for observing cepheids is in the infrared but this is inconvenient due to atmospheric filtering. Space astronomy has a huge advantage here. The Hubble Space Telescope has measured the distance to a cepheid in galaxy M100 – an awesome distance of 56 million light years!

Comet

Comets are the most spectacular naked eye astronomical objects visible. In the middle ages they were viewed with terror – sure omens of terrible things to come. The most obvious feature of a comet is its tail. This forms when the comet approaches the Sun and volatile matter is sublimated off and dust particles are also released. The tail of gas is susceptible to pressure from the solar wind causing it to point away from the Sun even when the comet itself is moving in the same direction. The dust portion of the tail can separate from the gas tail giving the comet two tails.

The actual nucleus of the comet is difficult to see as it is surrounded by a coma of tail material. Long period comets (more than 200 years) originate in the Oort cloud about 50 000 astronomical units from the Sun. These have extremely elongated elliptical orbits which do not necessarily lie close to the ecliptic or conform to the general rotation of the Solar System. Short period comets (under 200 years) come from the Kuiper Belt beyond the orbit of Neptune about 30 to 50 astronomical units from the Sun. There are at least 70 000 objects there with a diameter of more than 100 km. The composition of comet nuclei is about half frozen water and the rest dust. Comets lose significant quantities of material with each pass around the Sun making them less spectacular with each appearance. The most spectacular comet event ever recorded occurred in 1994 when more than 20 large fragments of the Shoemaker-Levy comet plunged into Jupiter. The comet had been captured into orbit around Jupiter, a most unusual event, and in a close approach in 1992 was fragmented by the Jovian gravity. The final plunge of the fragments, some more than 2 km across, took place from the 16th to the 22nd of July. This event was recorded by nearly every telescope on Earth and in space and many thousands of sensational photos were taken. Halley's comet is probably better known than any other. Its appearances have been noticed for more than two thousand years but it was only after Edmond Halley suspected that the 1456, 1531, 1607 and 1682 appearances were of the same comet that its periodicity was established. Halley unfortunately died in 1742 and was therefore unable to see his calculation confirmed when the comet returned as predicted on Christmas eve of 1758. It again appeared in 1835, 1910 and 1986. A famous appearance occurred in 1066 just before the battle of Hastings bringing good luck to some and appallingly bad luck to others. Halley's comet is prominently featured at the battle on the Bayeux Tapestry, an astonishing historical document of some 70 metres in length. It is not actually a tapestry but an embroidery – a copy has been made by nimble fingered ladies and is kept on display at Reading. The military fortunes were reversed at the battle of Agincourt in 1415 but in this case there was no comet involved – the calamity was the result of a very much down to earth matter – sticky mud and a lack of crowd control. An entire generation of French nobility was wiped out by a much smaller army of poorly paid ruffians who suffered negligible loss.

Shakespeare let Henry V put it this way:

O God! thy arm was here;
And not to us , but to thy arm alone,
Ascribe we all.
When, without stratagem,
But in plain shock and even play of battle
Was ever known so great and little loss
On one part and on the other?

The military fortunes were later again to be reversed but this time by the efforts and visions of a teenage peasant girl *Jehanne la Pucelle*.

Close up images of the Halley's comet nucleus were taken by the Giotto spacecraft in March 1986 and its size determined as 16 km long and 8 km wide. If an object of this size were to strike the Earth this would be bad news on an astronomical scale. There is little chance however that a comet would be able to strike the Earth. Comets are very fragile objects consisting largely of water ice and would probably fragment and evaporate in the atmosphere – only a very large comet would have sufficient material for a large portion to survive the plunge through the atmosphere. There is a popular misconception, originating in crime fiction, that a bullet made from ice could be used with effect when fired from a pistol. This is clearly not possible – ice becomes liquid water under pressure so that an ice bullet discharged from a pistol would become vapour on leaving the muzzle.

A most remarkable robotic space probe, ROSETTA was launched by ESA on 2 March 2004 which will explore and go into orbit around comet 67P/Churyumov-Gerasimenko and make observations for two years during the comet's approach to the Sun. The probe will also release a small lander, named Philae, packed with scientific instruments which will be the first landing of a space craft on a comet. The landing is expected to take place in November 2014. Two harpoons will be fired into the comet to secure the lander from bouncing off. The probe was actually intended for another comet but missed its launch window and had to be re-scheduled. The probe has a highly complex mission planned: a Mars flyby in February 2007, an Earth flyby in November 2007, a flyby of asteroid 2867 Steins in September 2008, an Earth flyby in November 2009, a flyby of asteroid 21 Lutetia in July 2010, a hibernation from May 2011 to January 2014, comet approach January–May 2014, comet mapping August 2014, comet landing November 2014, comet escorting around the Sun from November 2014 to December 2015.

Another exciting mission to a comet is STARDUST which was launched on 7 February 1999. It made a close approach to comet WILD 2 on 2 January 2004 and collected samples of particles in a special aerogel collector. Aerogel is the lightest solid substance ever made – similar to glass but 99,8 percent air. The mission returned as planned with samples to Earth in January 2006. The comet surface observed by the craft was quite different to the usual 'dirty snowball' description showing a cohesive surface with spires, pits and craters. On 4 July 2005 the Deep Impact mission achieved a spectacularly successful climax when the impact probe smashed into comet Tempel 1 at a velocity of 37000 km/h. The probe took photographs of the comet until seconds before impact. The flyby module was equipped with wide and narrow angle cameras with which it took sensational photos before and after impact. The event was also photographed by the Hubble, Chandra, Spitzer and SWAS telescopes as well as the Rosetta spacecraft on its way to comet Churyumov Gerasimenko. The James Clerk Maxwell telescope also observed the event, in submillimetre heat wavelengths, using its bolometer array.

Constellation

A constellation is more of an idea than an object. Constellations are easily recognisable patterns of stars which have been given names for easy reference. Mention Orion or Scorpio and this will give an easy way of referring to a part of the sky as well as the time of year for convenient observation. The constellation patterns are of course as seen from the Earth – seen from any other angle they would be quite unrecognisable. There are a great many constellations but those of the zodiac are probably the best known. The zodiac, from the Greek meaning 'circle of animals', comes to us from ancient Babylon from where it has been taken to China, Egypt and Greece. The Chinese animals are completely different to the western zodiac and in ancient Egypt the animals were replaced by Egyptian deities. There is of course not the slightest reason to suppose that the star patterns actually look like anything in particular – this is a case of facts being pressed into a particular purpose. The zodiac is a band of sky of about eight degrees on either side of the ecliptic which includes the paths of the Sun and the planets. The signs of the zodiac as known in the West are all of Greek mythological origin and which have been given Latin names. These names are one of the few remnants of ancient astrology which have found a place in modern astronomy.

The ancient Egyptians had a well developed astronomy as can be seen at the pyramids of Giza. The layout and sizes of the three large pyramids match the three central stars of Orion in position and magnitude. Orion was named Osiris by the Egyptians. The narrow shafts projecting through the great pyramid are also of astronomical significance pointing to star positions with great precision. The task of cutting these shafts during the construction of the pyramid must have been enormous. The purpose of the shafts remains a mystery as they do not penetrate to the outside of the pyramid. The shape of the pyramids gives them the same height and base perimeter as hemispheres showing the builders sensitivity to the fitness of things. One of the shafts of the Queen's chamber of the great pyramid points, with the constellations correctly positioned, to the star Sirius (Dog Star) in the constellation Canis Major. Sirius was endowed with particular significance by the ancient Egyptians. This is the brightest star in the night sky with a mass twice that of the Sun and twenty times the brightness. This star is 6,7 light years away from the Earth.

In 1834, F. W. Bessel noticed that the star was being perturbed by an unseen star orbiting it giving it an apparent spiral motion. This companion star was detected in 1862 and photographed for the first time in 1970. Sirius B is particularly difficult to photograph due to the extreme differences in brightness between the two stars and is best detected using an underexposed photograph. There is a spectacular Hubble photograph of the Sirius binary pair. It has since been discovered that the companion star Sirius B is a white dwarf star, extremely hot and with a density of 200 000 times that of the Earth and a diameter of only 11 700 km and orbiting Sirius A in 50,1 years. The orbit is highly elliptic varying from 8,1 to 31,5 AUs. The star also has a rapid spin of 23 r.p.m. Sirius also has an Oort cloud of icy comet material similar to that of the Solar System. A slight perturbation of Sirius B is indicative of the existence of a third star Sirius C orbiting with a period of 6 years. For a planet with liquid water to exist in orbit around Sirius A, this would have to orbit 4,6 AUs from Sirius A with a period of 6,8 years. It is very unlikely that any major non aquatic life forms could exist here. The perturbations caused by the elliptical orbit of Sirius B would cause extreme variations in climate precluding the evolution of above-water life forms.

We now need to consider the Dogon tribe of Mali in West Africa. An unlikely venue for astronomical matters? No, not at all! The Dogon claim to have ancestry in ancient Egypt and secret astronomical traditions going back 5000 years! Of the Solar System, they knew of the four large moons of Jupiter and the rings of Saturn millennia before Galileo, but even more startling are their claims about Sirius. This they claim has two orbiting companion stars. The one a small star of extreme heat and density and with a rapid spin and also a highly elliptical orbit with a period of 50 years. The other companion is a small dim star, presumably a 'brown dwarf'. There are also planets orbiting Sirius and it is from here that visitors came to Earth in ancient times. These people were described as hideous beings of ichthyoid appearance (yeech). A great deal has been written about the Dogon mystery so let us leave the matter here.

The Ancient Greeks also had a well developed astronomy. Eratosthenes was successfully able to determine the diameter and circumference of the Earth. Anaxagoras of Clazomenae (500–428 BC.) and Aristarchus of Samos (c. 310–280 B.C.) had theories of a heliocentric universe similar to that of Copernicus. The most astonishing astronomical invention of ancient Greece was surely the orrery type of device recovered from a wreck off the island Antikythera. The wreck was discovered in 43 metres of water by a sponge diver who came up gasping that he had found a heap of naked dead women! It was a treasure vessel laden with bronze and marble statues, jewellery, wine and significantly the Antikythera mechanism. The instrument was housed in a wooden box the size of a shoebox and the surviving part contains some thirty finely crafted bronze gears incorporated into a complex mechanism of epicyclic and differential gearing. The instrument was made to show the positions of the Sun, Moon and five inner planets with respect to the zodiac for any chosen date. This instrument was manufactured in 87 B.C. and lost at sea in 76 B.C. Another remarkable scientific construction in ancient Greece was the Tower of the Winds supervised by Macedonian astronomer Andronikos of Kyrrhos. The octagonal horologion was constructed in 50 BC of Pentelic marble with a height of 12 m and a diameter of 8 m. The structure was topped with an octagonal pyramid of marble slabs and surmounted with a weather vane bronze figure of sea god Triton. The eight sides were decorated with sculpted panels depicting the wind deities of the eight winds. The Triton would point with a wand to the panels indicating the direction of the wind source. The horologion contained a mechanised 24 hour clepsydra driven by water from the Acropolis. The interior of the horologion was built in the Doric style and the exterior was of the Corinthian order.

Galaxy

Galaxies are the main building blocks of the universe – awesome clouds of thousands of millions of stars arranged in slowly rotating discs often with spiral streamers at the edge. Until the early part of the 20th century galaxies were thought to be nebulae of gas illuminated by starlight. A typical galaxy such as our own Milky Way has a diameter of about 100 000 light years and contains about 400 thousand million stars. In 1783, William Herschel discovered that the Solar System was moving at 20 km/sec within the galaxy towards a point called the Solar Apex. The Milky Way galaxy takes about 220 million years to turn through one revolution. The stars in a galaxy are extremely sparsely spread out – several light years between stars. The average density of a galaxy is only a few atoms per cubic centimetre – less than the best vacuum that can be achieved on Earth. It is currently estimated that there are 350 thousand million galaxies in the universe. It is thought that most galaxies started forming about 13 billion years ago. The formation of stars, galaxies and clusters of galaxies is a complex matter which will not be described here. The formation of galaxies is greatly influenced by the distribution of dark matter – much work is being done to map the distribution of this mysterious substance. The future development of the universe is expected to be greatly influenced by another mystery – dark energy.

Globular Cluster

These are groups of up to a million old stars within a space of up to 100 light years across. These are not galaxies but only groups contained within a galaxy. Some 158 clusters have been found in our own Milky Way galaxy but other larger galaxies have many more. The clusters do not have the flat disc shape of galaxies.

Magnetar

Magnetars or soft gamma repeaters are neutron stars with exceptionally strong magnetic fields and which can emit huge repetitive bursts of soft gamma rays. Soft here means lower frequency and hard means higher frequency. (Refer to the description of a neutron star). In a single burst of gamma rays, the magnetar can release as much energy as the Sun in 1000 years! The magnetic field is so immense that it slows the star and in the process emits the gamma bursts. It is estimated that there are a few hundred million magnetars in the Milky Way. The magnetic field strength is unbelievable – 80 gigatesla. Thousands of millions of times stronger than can be made on Earth. The strongest man made sustained magnetic field ever achieved was about 45 tesla. The magnetar's magnetic field is so strong that atoms at the surface are deformed into needle shapes. The magnetic flux intensity of the Earth at the poles is about 60 microtesla. The magnetic flux density used in electric motors and transformers is typically from 1 to 2 tesla. The renowned Russian nuclear physicist Andrei Sakharov is well known for developing the most powerful nuclear device ever detonated. For his work on the Tokamac nuclear fusion reactor he developed exploding pumped flux compression generators. These achieved pulsed magnetic fields of 2500 tesla and currents of 100 million amps. The early pioneering work on magnetism was done by Johan Carl Friedrich Gauss (1777–1855), Wilhelm Eduard Weber (1804–1891) and Alexander von Humboldt (1769–1859). In March 1979, a most interesting event occurred. A massive gamma ray burst swept through the Solar System. This was not detected on Earth due to the protection of the Earth's atmosphere but it was recorded by nine spacecraft which were in orbit at the time. Gamma detectors are small and no problem to include in the spacecraft instrumentation. The burst was so strong that it went right out of range of the instruments but the precise times of the measurements were recorded. From these times and the positions of the spacecraft it was possible to calculate the direction of the radiation and pinpoint its source. The burst was found to be from the magnetar of a supernova event in the Great Magellanic Cloud 180 000 light years away. The Great Magellanic Cloud is a dwarf galaxy which orbits our own Milky Way.

Meteor

Meteors, or shooting stars, are a very common feature of the night sky but are most spectacular when seen as a meteor shower. Meteors are usually small extraterrestrial fragments of rock and even submillimetre grains which are burned up as they plunge through the Earth's atmosphere. A spectacular shower occurred in November 1833 when countless meteors fell like flakes in a snowstorm. Meteor showers occur when the Earth passes through the path of a comet which has left vast quantities of rock and dust in its trail which have been released by the heat of the Sun. Meteor showers fall in parallel streams to Earth as do sunbeams, and like sunbeams they appear to come from a point due the perspective of observation. The point of the shower source is called the radiant. Showers are named after the constellations from where they appear to come. Orionid meteors appear to come from constellation Orion. These occur in mid October and originate from debris left by Halley's Comet. During the shower there may be as many as 30 meteors per hour. Another well known shower is the Leonid appearing to come from constellation Leo. These occur in mid November and are debris left by comet Temple-Tuttle. In 1966 a shower estimated at 150 000 meteors per hour was observed. The most spectacular meteor event of recent times was the explosion over the Tunguska river in the skies of Siberia in the early morning of the 30th June 1908. It is estimated that this was a rocky asteroid of about fifty metres in diameter which exploded six kilometres above the ground with the violence of a hydrogen bomb. The explosion felled 80 million trees and devastated an area of 2150 square kilometres. Reindeer herders 30 km away were blown into the air and people 60 km away were thrown to the ground. Trees directly below the explosion were stripped bare and those further away fell pointing away from the blast. Eyewitness reports described a huge fireball streaking obliquely across the sky leaving a long white trail. Such an object is known as a detonating bolide, which could be either an asteroid or a comet. The huge explosion was followed by the sounds of many impacts as rocks fell to the earth over a wide area. There was no central impact crater as the asteroid had fragmented in the atmosphere. The long oblique path of the asteroid contributed to its very high temperature and fragmentation from thermal stress which in turn caused the explosion by instantly heating a very large volume of air. The explosion was similar to a weapon known as a fuel-air bomb where fuel is sprayed into a large volume of air and then ignited causing a vast explosion much the same as in an internal combustion engine.

It has also been suggested that the Tunguska event was caused by a comet but this is most unlikely as a comet would melt and the water and dust blown away before it could reach a high temperature. It is impossible to heat an object consisting largely of ice to incandescent temperature. There are many other theories about the cause of the Tunguska explosion, in varying degrees of absurdity. The craziest suggestion of all is that the explosion was caused by Nikola Tesla irresponsibly fiddling with electrical apparatus thousands of kilometres away in America.

Meteorite

Meteorites are asteroids or comet debris that have been disturbed from their orbits, plunged through the Earth's atmosphere and made it all the way to Earth. Most meteorites are of rocky material but about five percent are of iron and nickel. Many metallic meteorites have been found but those of rocky material fragment, often becoming indistinguishable from the material of the impact site. Metallic meteorites can be seen in many geological museums around the world. Meteorites that cause any serious damage are rare, most of them plunging into the sea. There are however several reports of people, cars and buildings being struck. There have of course been some major meteorite strikes which have caused cataclysmic damage. It is now practically certain that the dinosaur extinction was the result of an impact by the Chicxulub meteorite 65 million years ago. There is also world wide evidence of iridium deposits resulting from this meteorite confirming a global blocking of sunlight for a lengthy period. This must have been a most unusual meteorite – iridium is very rare in the earth's crust and also unusual to be found in high concentration in a meteorite. The meteorite also contained an unusual concentration of platinum and osmium (Platinum group metals). Most meteorite impact sites disappear in time due to weathering but those on the Moon remain permanently for all to see. The entire lunar surface is pock marked by meteorite strikes; the Earth, being larger, must have had many more. The largest crater on the Moon, which is also the largest crater in the Solar System, is the South Pole Aitken basin on the far side of the Moon. This is 2240 km in diameter and 13 km deep. There are a few hundred identifiable meteorite craters on the Earth – we give here a list of some of the largest.

The Vredefort and Sudbury meteorites were by far the largest and oldest known meteorites to strike the earth but this was in the Palaeo-proterozoic era before the evolution of single-celled life and could not therefore have caused a mass extinction. Although the Vredefort meteorite is the largest known meteorite to have struck the earth there may well have been even larger impacts of which the traces have become obliterated over time. It is estimated that the Vredefort meteorite had a diameter of more than ten kilometres (about 1000 cubic kilometres of rock) which slammed deep into the basement granite of the Kaapvaal craton. This resulted in an astrobleme (star wound) of 300 kilometres in diameter and excavated about 70 000 cubic kilometres of material and probably increased the oxygen content of the earth's atmosphere. Ejecta from the impact was scattered over an area 700 km across and included individual blocks of rock up to 4 km across. The extreme temperature generated would have evaporated as much as 70 cubic kilometres of rock. It is estimated that the impact was a magnitude 14 seismic event. The Richter scale is logarithmic which means that each increase of a unit on the scale represents a tenfold increase in amplitude measurement. A magnitude of 14 would be one hundred million times as violent as a major earthquake of magnitude

Meteorite craters larger than 50 km		
Meteorite	**Crater size in km**	**Age – millions of years**
Vredefort, South Africa	300	2023
Sudbury, Ontario	250	1850
Chicxulub, Yucatan	170	65
Popigai, Russia	100	37,5
Anicouagan, Quebec	100	214
Acraman, South Australia	90	590
Chesapeake Bay, Virginia	90	35,5
Puchezh- Katunki, Russia	80	167
Morokweng, South Africa	70	145
Kara, Russia	65	70,3
Beaverhead, Montana	60	600
Tookoonooka, Queensland	55	128,5
Charlevoix, Quebec	54	342
Siljan, Sweden	52	361
Kara-Kul, Tajikistan	52	<5

six as reported a few times a year in the media. The corresponding energy increase in a unit increase on the scale would be 31. Seismic events greater than magnitude 10 can only be caused by impacts from space. The periphery of the astrobleme is no longer visible on the surface due to erosion but the central uplift dome is still very apparent and is a site of great geological interest. This can be seen as a circle of mountains 180 km in diameter which was declared a Unesco World Heritage site in July 2005. One of the most remarkable features of the impact is that it lifted and also preserved one of the richest series of goldfields on earth. Some 50 000 tons of gold have been produced by mining operations in the area. An unusual aspect of the gold mining is that uranium is produced as a by-product. At the time of impact, the Kaapvaal craton was on the opposite side of the globe to where it is now.

An interesting theory has been proposed concerning the seismic effects of meteorite impacts – Antipodal Focusing. It has been suggested that seismic waves originating at an impact site and travelling through the Earth's mantle can focus at the opposite side of the globe resulting in earthquakes and eruptions. The sharpness of focus is still a matter of debate. It has been proposed that the vast basaltic lava flows of the Siberian Traps during the Permian Period were the result of antipodal focusing from an exceptionally large bolide impact on the opposite side of the globe.

A most remarkable meteorite fell to Earth on 28 September 1969 near the town of Murchison in Australia. This has become known as the Murchison meteorite. It was seen, shortly before noon, by many observers as a fireball streaking across the sky, which fragmented and was dispersed over a wide area. More than 100 kg of material was recovered. The meteorite contained an exceptionally large number of organic compounds, including more that 100 amino acids, and also components of the genetic code suggesting an extraterrestrial source for the key components of life.

Moon

The delightful, traditional German folk song 'Dear Moon' will be well known to youngsters taking elementary music lessons:

Guter Mond du gehst so stille in den Abendwolken hin,
bist so ruhig und ich fühle, daß ich ohne Ruhe bin!
Traurig folgen meine Blikke deiner stillen heitern Bahn:
O wie hart ist mein Geschikke, daß ich dir nicht folgen kann.

There is very little difference between a planet and a moon. Planets orbit stars and moons orbit planets. Moons are very plentiful in the solar system – seven of the planets are known to have moons and Jupiter, at the latest count, has a whopping 63. Even an asteroid has been found to have a tiny moon. If the four large moons of Jupiter were to orbit the Sun independently they would quite likely be regarded as dwarf planets. There are nearly a hundred moons orbiting the solar system planets and most have been given names, with one notable exception – the moon of planet Earth. This is rather like having a pet cat and calling him 'the cat'. The word 'Moon' is ultimately derived from the Proto-Indo-European root *me-* which gives us time interval words such as Monday, measure, month and menstrual. The terms *Moonlight, Clair de Lune* and *Mondschein* all mean the same thing and all three have romantic connotations in music and literature. Moonshine also means the same thing but has connotations poles apart – with illicit liquor. The sight of the Moon usually engenders a feeling of tranquility but the 'raging moon' of Welsh poet Dylan Thomas is an exception:

In my craft or sullen art
Exercised in the still night
When only the Moon rages
And the lovers lie abed.

The Moon also has many associations with bad things such as lunacy and werewolves. Beware the loup-garou!

From ghoulies and ghosties
And long-leggedy beasties
And things that go bump in the night,
Good Lord, deliver us!
Scottish prayer.

At this point one cannot help but think of the incident in the boyhood of Nikola Tesla when the faithful family horse saved his father from being devoured by a pack of ravenous wolves. In the southern part of wildebeest country there are no werewolves – instead they have a rapacious little devil called a *tokoloshe!* As a traditional protection against this nasty little fellow a significant section of the population raise the levels of their beds by placing discarded paint cans or bricks under the bed legs. Curiously, no businessman has yet seen the manufacture of long-leggedy beds as a business opportunity. The nursery rhyme *Jack and Jill went up the hill* has its origins with the waxing and waning of the Moon in Norse mythology. This connection was discovered by Sabine Baring-Gould (1834–1924) who did wonderful work collecting and writing down English folk songs thereby saving a valuable cultural heritage for posterity. He had, unfortunately, to moderate many of the folk songs which contained words unprintable at the time. Baring-Gould is well known for his strident and rousing hymn 'Onward! Christian soldiers' which comes with a matching tune by Arthur Sullivan. Baring-Gould also has the distinction of composing the hymn tune 'Now the day is over...' which has been found to be the key to the enigma of Edward Elgar's 'Enigma Variations'.

The Earth's moon has a diameter of 3 474 km and orbits the Earth at an average distance of 384 403 km (perigee 363 104 km, apogee 405 696 km). The Moon always presents the same face to the Earth due to tidal braking of any other rotation that it once had, but it has a slight libration, which is an oscillation of the face presented to the earth which allows 59% of the surface to be seen from Earth. The Moon orbits the Earth every 27,3 days (sidereal period) and the Sun, Earth, Moon configuration repeats every 29,5 days (synodic period). The mean orbital velocity is 1,03 km/sec, the maximum surface temperature is 123°C and the minimum temperature -233°C. The Moon is too small to retain an atmosphere. The pull of gravity at the surface is only 17% of that on Earth. Swiss mathematician Leonhard Euler made a considerable contribution in his 775 page book 'Theoria Motus Lunaris' published in 1753. It was from this work that the first lunar tables were constructed for use in navigation.

The origin of the Moon has been debated and disputed for ages but it is generally now accepted that a massive body about the size of Mars accumulated at one of the Lagrangian points of the Earth's orbit, then escaped due to impacts, and slowly drifted closer to Earth dealing it a glancing blow which resulted in the Moon. This is supported by the fact that moon rocks have the same oxygen isotopes as those on Earth and the lack of a large iron core makes it unlikely that the Moon formed independently. This moon making planet has been given the name Orpheus (also called Theia). The corner cube reflector installed on the Moon's surface has permitted precise laser beam measurements to be made of the Moon's orbiting distance. It has been found that the Moon's distance is increasing by nearly four centimetres per year. In the early days of the Solar System, the Moon was very much closer to the Earth than it is now. This has been confirmed by geological evidence indicating that sea tides in the past were very much greater and more frequent than they are now. For the Moon to be moving further from the Earth it must necessarily be increasing in energy. There is a very simple explanation for this. Due to the rotation of the Earth, the tidal bulge will always be slightly ahead of the Moon position, causing a very slight but continuous increase in the Moon's orbital energy, while at the same time decelerating the rotational speed of the Earth, but this effect is almost imperceptible (about 2 milliseconds per day per century). It is interesting to note that by increasing the orbital energy of the Moon this will result in a higher orbit but will also paradoxically cause a reduction in the orbital speed and angular velocity.

Do we really need the Moon? And how we do! The Moon is not there solely for the benefit of honeymooners. The Moon plays a vital role in the lives of innumerable land and sea creatures. Many sea creatures depend on spring tides for spawning and farmers are very fussy about lunar cycles when it comes to seed planting. Vineyard farmers are also careful to harvest their grapes according to the lunar cycle. Trees to be used in the manufacture of pipe organs are felled according to the phase of the Moon. There is a practical reason for this as it ensures an absence of woodworm from the timber. The Moon is actually essential for human life on Earth. Without the stabilising effect of the Moon, the Earth's spin could drift erratically, causing the polar ice caps to drift about and also drastically reduce the land area available. It is doubtful whether human life could have evolved under such conditions.

Romeo had more confidence in the beneficence of the moon than Juliet:

> *Rom.* Lady, by yonder blessed moon I swear
> That tips with silver all these fruit-tree tops, -
> *Jul.* O! swear not by the moon, the inconstant moon,
> That monthly changes in her circled orb,
> Lest that thy love prove likewise variable.

The cynic may see here a suggestion that Romeo's passion and silvery tip were regulated by Juliet's ovulatory cycle. The silvery tipped fruit tree idea is also taken by the bawdy nursery rhyme:

> *I had a little nut tree and nothing would it bear,*
> *But a silver nutmeg and a golden pear.*
> *The King of Spain's daughter came to visit me,*
> *And all for the sake of my little nut tree.*

> *The King of Spain's daughter had trinkets in her hair,*

Like my silver nutmeg and my golden pear.
When she went back to Spain she was merry as can be,
For she had a little seed from my little nut tree.

The trinkets refer to good luck charms which are quite common in Hispanic countries. These are in the shape of a small fist with the thumb grasped between the next two fingers. This nursery rhyme probably refers to the visit by Juana of Castile to Henry VII in 1506. Nothing of any dynastic significance resulted from this visit. The king, no doubt, greatly treasured his gold and silver family jewels. At this time it was fashionable for kings to wear codpieces of more than ample proportions. Shakespeare's tragic teenage love drama revolving about the feuding Montecchi and Capuleti families of Verona was written many years later. I can remember seeing a Russian ballet performance of 'Romeo and Juliet' where effective use was made of the grasped thumb gesture to portray the confrontational dialogue between Abraham and Sampson:

Abr.   Do you bite your thumb at us, sir?
Sam.   No, sir, I do not bite my thumb at you, sir; but I bite my thumb, sir.

The grasped thumb gesture has largely given way to a two word insult of four and three characters respectively.

The well-loved English nursery rhymes, which are taught to children from a tender age, must seem utterly incomprehensible to foreigners – many of the rhymes refer to plague, violent death, murder, rape, prostitution, brothel keeping, erectile dysfunction and political scheming. To the romanticist 'leaves glistening sliver in the moonlight' remain a beautiful sight.

The moon, being our closest neighbour in space, has been the object of more space missions than any other. More than sixty lunar space missions have been launched. Most of these were for scientific data capturing and photographic mapping of the lunar surface. The spacecraft terminated their missions by impacting the moon or going into lunar or solar orbit. The manned moon missions were possibly the most daring and adventurous missions of discovery of the twentieth century.

On 13 September 1959, The Luna 2 spacecraft launched by the USSR became the first man made object to reach another body in the universe when its Lunar Lander impacted the surface of the Moon.

On 7 October 1959, the Luna 3 returned the first photographs of the far side of the Moon and on 3 February 1966, the Luna 9 performed the first soft landing on the Moon and returned the first photographs of the surface.

On 21 December 1968, the Apollo 8 mission was launched, manned by Borman, Lovell and Anders. The spacecraft made ten orbits around the moon making the crew the first astronauts to see the far side of the moon.

On 3 March 1969, Apollo 9, the first manned test of the lunar module was launched without actually landing on the Moon. McDivitt and Schweickart manned the lunar module and Scott the command module.

On 19 May 1969, Apollo 10, a full rehearsal of a lunar module landing was launched with the Lander coming within 16 km of the lunar surface. Photos were taken of the Apollo 11 landing site. This mission was manned by Stafford, Young and Cernan.

On 16 July 1969, Apollo 11, the first manned mission to the lunar surface was made by Armstrong, Collins and Aldrin. Armstrong and Aldrin remained on the surface for 20 hours and took a 2 hour moonwalk. This mission was one of the greatest engineering achievements of the twentieth century and captivated public imagination worldwide. Neil Armstrong, the first man to set foot on the Moon, died on 25 August 2012 aged 82.

On 14 November 1969, Apollo 12, a second landing mission on the moon was launched. Conrad and Bean collected 31 kg of lunar samples and also retrieved parts of the unmanned Surveyor spacecraft. The command module was manned by Gordon.

On 11 April 1970, Apollo 13 was launched but the mission nearly ended in total disaster. The craft was severely damaged by an oxygen cylinder explosion. The stricken craft passed around the Moon for a gravity assist to return to Earth and made a safe splashdown after exceptional efforts by the crew and control centre staff.

On 31 January 1971, the Apollo 14 was launched and became the third successful moon landing mission. It was manned by Shepard, Roosa and Mitchell. Rock and soil samples were collected using a tool cart.

On 26 July 1971, the Apollo 15 mission manned by Scott, Warden and Irwin was launched. This was the fourth successful lunar landing. Scott and Irwin were the first astronauts to use the Lunar Roving Vehicle.

On 16 April 1972, The Apollo 16 mission manned by Young, Mattingly and Duke was launched. Young and Duke used the Lunar Roving Vehicle to explore the lunar highlands.

On 7 December 1972, the Apollo 17 was the last manned lunar mission to be launched. This was manned by Cernan, Evans and Schmitt. Cernan and Schmitt completed three moon walks and collected rock and soil samples. They used the Lunar Rover to cover more than 30 km of territory. It is now more than 35 years since anyone has set foot upon the Moon.

On 27 January 1973, the USSR launched the Venera 8 which made a soft landing on the Moon and deployed the second automated rover which travelled 37 km exploring the lunar surface. The Luna 24 returned soil samples to Earth in August 1976.

In January 1994, the Deep Space Program Science Experiment (DSPSE) was launched from the Vandenberg Air Force Base. This is also known as the Clementine Mission. This was designed to test lightweight miniature sensors and advanced spacecraft components and also to perform extensive mapping of the moon. The mission returned 1,8 million digital images – a spectacularly successful result. The mission and the people associated with the program were honoured with seven prestigious awards. Radar data acquired by the spacecraft indicated the presence of water ice at the base of a South Pole crater of the moon. A remarkable aspect of the Clementine mission is that it was developed in only 22 months at a cost of only $80 million.

SMART-1 (Small Missions for Advanced Research and Technology) is the first mission of the European Space Agency to the Moon. Nine European countries and the US contributed to the mission. The mission, launched by Ariane-5 rocket on 27 March 2003 entered lunar orbit in November 2004. The mission was designed to test solar electric propulsion and other deep space technologies. The prime contractor was the Swedish Space Corporation, Solna, Sweden. The elliptical orbit ranged from 300 km to 10 000 km above the lunar surface. The Moon was mapped and studied in great detail and also the darker regions of the South Pole were studied where craters may actually contain ice. The mission ended on 3 September 2006 when the spacecraft impacted the lunar surface in the Lacus Excellentiae region.

JAXA's first large lunar explorer Kaguya was launched on 14 September 2007. The mission's objectives are to understand the Moon's origin and evolution. The fourteen instruments on board will study global data on lunar chemical distribution, mineral distribution, topography and gravity field. Kaguya has made use of the legacy of the SMART-I mission.

China's first Moon mission Chang'e was launched on 24 October 2007. Three days later it returned its first image of the lunar surface. Chang'e has spectrometers to map the chemical composition of the Moon's surface, a laser altimeter to map the topography and a camera for taking images of the surface.

The Indian Space Research Organisation (ISRO) launched the Chandrayaan-1 mission on 22 October 2008. Chandrayaan is Sanskrit for 'moon craft'. The mission is intended to survey a complete lunar map of chemical characteristics and three dimensional topography. The high resolution sensing equipment will operate in the visible, near infrared, and soft and hard X-ray frequencies. The 90 kg scientific payload contains six Indian instruments and also five of foreign origin. The six Indian instruments are:

- The Terrain Mapping camera (TMC) – High resolution mapping
- The Hyper Spectral Imager (HySI) – Mineralogical mapping
- The Lunar Laser Ranging Instrument (LLRI) – surface topography
- An X-ray fluorescence spectrometer – a collaboration between Rutherford Appleton laboratory, UK, ESA and ISRO.
- A High Energy X-ray/gamma ray spectrometer (HEX)
- Moon Impact Probe (MIP) a small satellite of instruments to impact the moon.

The foreign scientific payload will include:

- The Sub keV Atom Reflecting Analyser (SARA) from ESA.
- The Moon Mineralogy Mapper (M3) from Brown University and JPL.
- A near infrared spectrometer (SIR-2) from ESA.
- S-band miniSAR from APL at the John Hopkins University to map polar ice.
- RAdiation DOse Monitor (RADOM-7) from Bulgaria.

On 14 November 2008 the Moon Impact Probe separated from the Chandrayaan-1 Orbiter and struck the south pole of the Moon ejecting underground soil which could be analysed for the presence of lunar water ice.

ISRO is planning the Chandrayaan-2 mission for 2016 which will include a motorized rover for collecting samples and then transmitting results of chemical analysis to the mother spacecraft.

Chandrayaan-2 is a joint lunar exploration mission proposed by the Indian Research Organisation (ISRO) and the Russian Federal Space Agency (RKA). The mission will include a lunar orbiter and rover made in India and a lander provided by Russia. Data of the analysed soil samples will be sent to Earth by the Orbiter. The mission has a planned lifetime of one year. The launch date might slip due to the lack of test data which would have been provided by the lost Phobos-Grunt mission. The orbiter will carry five payloads and the rover two. Due to weight restrictions the mission will not be carrying any foreign payloads.

NASA's Lunar Precursor Robotic Program (LPRP) was launched on 19 June 2009 and will prepare for future lunar spaceflights. The two modules are: the Lunar Reconnaissance Orbiter (LRO) and the Lunar CRater Observation and Sensing Satellite (LCROSS). LRO will orbit for a year and gather high resolution images of the lunar surface. LCROSS will explore a permanently shaded area of a lunar pole by impacting a 2000 kg vehicle into the dark crater and then the ejecta will be observed from a 'shepherding' spacecraft and hopefully confirm the presence of water ice. The LCROSS impact took place on 9 October 2009. The impact plume rose 16 km above the lunar surface. On 21 Oct 2010 it was announced that LCROSS had detected that the lunar surface was rich in a variety of chemicals including water in the form of pure ice crystals.

Two proposed UK Moon missions, Moonlite and Moonraker, are scheduled for a possible launch by 2013. The Moonlite mission will fire 'darts' containing instrument packages two metres deep into the lunar surface. A successful test of the darts was performed on 6 June 2008. The Moonraker will be a lander mission.

Nebula

There are several types of nebulae but all are essentially clouds of gas or dust or the remnants of a supernova explosion. Galaxies and star clusters have erroneously also been referred to as nebulae in the past.

- **Emission nebulae** are clouds of high temperature gas of which the atoms have been energised by ultraviolet light from a nearby star.

- **Reflection nebulae** are clouds of dust illuminated by starlight and are also potential sites for star formation.

- **Dark nebulae** are simply clouds of dust blocking the view of whatever lies behind.

Planetary nebulae have nothing to do with planets. They are expanding shells of gas emitted from a star near the end of its life and usually have a spherical appearance. They are typically less than one light year across. Until quite recently there has been no possibility of measuring the distance of planetary nebulae in our galaxy but this can now be done by utilizing the very high resolution of the Hubble Space Telescope to measure the angular rate of expansion of the nebula – this typically requires measurements of milliarcseconds per year! Spectroscopic observation can measure the actual rate of expansion using the Doppler effect. The distance of the nebula can then be calculated from the angular and actual expansion of the nebula. Planetary nebulae are amongst the most spectacular astronomical targets ever photographed.

Neutron Star

Here we are dealing with ultimate extremes. The most significant feature of a neutron star is that its gravity is so immense that the protons and electrons of its atoms are compressed into neutrons! This core material is effectively a single continuous atomic nucleus called neutronium. This results in some startling figures. The final mass of a neutron star is about 1,4 times that of the Sun and with a diameter of only 20 km. This means that it has an immense density of 10^{17} kg.m^{-3} equivalent to a cubic kilometre of iron compressed down to a 4 centimetre cube. This object can have a rotational speed of from one revolution per 30 seconds up to several hundred revolutions per second. The Sun has a rotational speed of one revolution per month. In some papers it is claimed that a neutron star can have a rotational speed of over 40 000 r.p.m. This is faster than a dental turbine drill or ten times as fast as a car engine at cruising speed – can you believe this? – I can't. A quote from Thomas Hardy is appropriate:

Though a good deal is too strange to be believed,
nothing is too strange to have happened

The star's magnetic field strength is equally immense – 100 megatesla which is many millions of times stronger than any that can be produced on Earth. The surface gravity is also extreme – thousands of millions of times stronger than the gravity on Earth. Neutron stars start off as stars with a mass of 15 to 20 times the mass of the Sun. Smaller stars will become white dwarfs and larger ones black holes. The iron core of the star will grow until it reaches a mass of about 1,4 times that of the Sun – the Chandrasekhar mass which is required for the collapse of atoms into neutrons. It is thought that with increased size the neutron star can collapse even further to become a 'quark star' or a 'strange star' before it becomes a black hole, but this remains a theoretical possibility. The density of the neutronium at the core of a neutron star is unbelievable – the density of the core of a quark star, if it exists, would be unbelievably unbelievable. The finer details of a neutron star are a matter of advanced theoretical physics.

Planet

Planets can be almost any objects orbiting stars which are not large enough to become stars themselves. There is at present no clearly specified definition of what constitutes a planet but it is generally accepted that a planet is any star-orbiting body large enough to form itself into a sphere under its own gravity. Many planets orbiting other stars have been found and the list will certainly grow longer. On 19 October 2009 it was announced at the ESO/CAUP conference that 32 new extrasolar planets had been discovered by the HARPS project bringing the total number of known extrasolar planets to some 400. This number has since risen to 777 and will certainly rise further. It may well be quite commonplace for stars to have orbiting planets. The planets orbiting the Sun are a very mixed assortment. The Earth is unique in its exceptionally rich mixture of materials in its crust, abundant water, and oxygen rich atmosphere. No other object in the Solar System can offer anywhere near the same hospitality. The order of the planets starting from the Sun is:

Mercury, Venus, Earth, Mars, Jupiter, Saturn, Uranus, Neptune and Pluto.

Pluto does not actually conform to the accepted idea of a planet and was reclassified on 24 August 2006 to 'Dwarf Planet'. It is quite easy to remember the order of the planets by means of a mnemonic. The first letters of the words correspond to the first letters of the planet names.

My vivacious energetic mama just served us nine pizzas.

This will not always be correct as Pluto has a highly elliptic orbit with its perihelion closer to the Sun than Neptune. Dutch astronomer Gerrit Peter Kuiper (1905–1973) made a major contribution to Solar System astronomy. Among his achievements were: determined that Venus and Mars had atmospheres of carbon dioxide; discovered an atmosphere on Saturn's moon Titan; discovered that Saturn's rings were made of water ice; discovered Uranus's fifth moon Miranda; discovered Neptune's second moon Nereid and also suggested the existence of the Kuiper belt.

There is a very strange rule about the distances of the planets from the Sun. The Titus-Bode Law. In 1755, Johan Daniel Titus found that the distances of the planets from the Sun followed a simple arithmetic series. This was published by Johan Elert Bode in 1772. The law is very similar to a Pascal triangle. Curiously the law requires that there be a planet in the position of the asteroid belt. The asteroid Ceres qualifies but is a bit small to be regarded as a planet. If all the asteroids could be combined in the orbit of Ceres this would do nicely. Astonishingly, the law is quite accurate despite there being no convincing explanation as to why it works. The distance of a planet from the Sun is determined by the perihelion or aphelion velocity – what else? This is an infuriating problem – when complete information about other solar systems becomes available it will be most interesting to see how the Titus-Bode shapes up with these. There are of course forces controlling the orbits of planets forming from an accretion disc. The innermost orbit is determined by the Roche limit. Satellites with little tensile strength will be torn apart by tidal forces inside the Roche limit so there is no possibility for planets to form from the accretion disk in this zone. It is possible for solid rocky objects or small particles to orbit within the Roche limit. Planets forming further out will of necessity have widely spaced orbits. Closely spaced planets would greatly disturb each other's orbits causing either a collision or one of the planets to plunge into the Sun. There is never more than one planet on a particular orbit. It is possible for more than one planet to form on the Lagrangian points of an orbit but these will eventually be disturbed and collide to form a single planet or possible drastic change of orbit. The law skips Neptune and fits Pluto as the planet after Uranus. Quite remarkably, the law can also be applied to the moons of Uranus.

Bode's law starts with the simple series:

0, 3, 6, 12, 24, 48, 96, 192, ...   Doubling up as you go. Add 4 to each term to give:

4, 7, 10, 16, 28, 52, 100, 196, ... Now divide the terms by ten to give the distances in astronomical units.

	Measured average distance	Titus Bode distance	Discovered
Mercury	0,387	0,4	
Venus	0,723	0,7	
Earth	1,000	1,0	
Mars	1,524	1,6	
Ceres (asteroid)	2,8	2,8	1801
Jupiter	5,203	5,2	
Saturn	9,539	10,0	
Uranus	19,18	19,6	1781
Neptune	30,06		1846
Kuiper Belt	30 to 50		
Pluto	39,44	38,8	1930

The fit is astonishing to say the least. Who knows how good the fit was in the early days of the Solar System? The planets are continually influenced by each other and lose energy due to tidal effects. Impacts with large meteorites and comets would also cause orbit disturbances. It is quite easy to make an hypothesis explaining the absence of Neptune from the Titus-Bode law. Pluto and its moon Charon are correctly placed for the 9th orbit but these are an odd pair for a planet and its moon. Charon is too large to be a moon of Pluto which is in turn too small to be a planet. Pluto and Charon could well be former moons of Neptune. If Neptune had actually been in orbit where Pluto is now and accumulated a considerable amount of mass from the Kuiper belt, this could have slowed it sufficiently to take a smaller orbit where it is now, leaving Pluto and Charon behind as binary moons. The highly elliptic orbit of Pluto, which has its perihelion within the orbit of Neptune, is also indicative of some major disturbance. If this is indeed what happened then there is all the more need to put the Titus-Bode law on a sound scientific footing. Two other unexplained puzzles of the Solar System are the slow retrograde rotation of Venus and the horizontal axis of rotation of Uranus.

The analytic calculation of the orbits of the planets remains a theoretically impossible problem. In 1609, Kepler published his 'Astronomia Nova' proposing that the planets move in elliptical orbits around the Sun which was a great improvement on the circular orbits proposed by Copernicus. Kepler is also famous for his discovery that for all planets, a line from the planet to the centre of the Sun will sweep equal areas in the same time. Fortunately the Sun contains nearly all the mass of the Solar System and the planets have very widely spaced orbits so that the simple elliptical and independent orbit approach will give quite good results. In actual fact the planets do influence each other and this must be taken into account when precise calculations are required. The famous three body orbiting problem has been tackled by some of the most eminent mathematicians in history and even with only three bodies an analytical solution is only possible after simplifying the problem to movement in a single plane and assuming one of the bodies to be of negligible mass which is hardly a solution as this reduces it to a two body problem. In 1772 mathematician Leonhard Euler published his 'Theorea Motuum Lunae' in which he tackled the three body problem. This was also taken further by Laplace, Lagrange and Poincaré . Pierre-Simon Laplace wrote his most important work 'Traite di Mecanique Celeste' dealing with the planetary orbits. This was in five volumes, the first two being published in 1799. The Laplace vector is a line from the centre of the Sun to the centre of the ellipse of an orbiting planet which is a measure of the eccentricity. For a circular orbit this vector will be zero. In the case of the Earth's orbit this vector can change over time due to the gravitational influence of Jupiter. It has been proposed that variations in the Laplace vector could have caused the periodic ice ages experienced by planet Earth. The Solar System contains a large number of bodies with orbits at various inclinations to the ecliptic as well as a large number of moons and other material orbiting the planets, and which is further complicated by tidal effects in non-rigid bodies. The movement of all this is essentially chaotic. Sir Isaac Newton discovered that a very large and a small body acting upon each other by gravity would produce circular, elliptic, parabolic or hyperbolic orbits (conic sections) but of the three body orbiting problem he despaired:

This exceeds, if I am not mistaken, the force of any human mind

Precise calculations of the planetary positions are nevertheless required and this is achieved by the perturbation approach. The orbits are calculated using a simplified elliptical approach and adjustments are then made to account for perturbations caused by other planets. When comets make their periodic approach to the sun they can be greatly affected by planets which happen to be in the vicinity of their paths. Edmond Halley calculated the perturbations of the orbit of his famous comet when calculating the date of its return. Friedrich Wilhelm Bessel (1784–1846) made huge contributions to astronomy, celestial mechanics and mathematics. He was one of the first astronomers to consider all the possible observational, meteorological and instrument errors that could influence final astronomical observation results. His refinement of historical astronomical observation data was sensational. Bessel used parallax to determine the distance to 61 Cygni announcing his result in 1838. John Herschel, when he learnt of Bessel's achievement, described it as:–

... the greatest and most glorious triumph which practical astronomy has ever witnessed

Bessel was a brilliant mathematician who is well known for a class of functions known as Bessel functions. This was probably based on previous work by the Bernoullis, Euler and Lagrange. These functions were more fully developed in a study of planetary perturbations. Bessel functions are now widely used in applied mathematics, physics and engineering. The number crunching power of computers provides a means of obtaining phenomenally accurate planet positions without orbit calculations. Using an accurately measured starting position and accurate values of the planetary masses and velocities, a mathematical simulation program can compute the movements of the bodies. A simulation program can, of course, also make short work of the infuriating three body problem. The positions of the planets with respect to the zodiac are of course bread and butter to astrologers but let us not get embroiled in this matter here. There is, however, a non-astrological matter concerning planetary conjunctions which needs to be mentioned. A conjunction of planets including Jupiter can cause slightly increased tidal movements in the Earth's crust resulting in seismic activity where tectonic plate forces have reached a point of instability.

Let us take a brief look at the planets of our own Solar System.

Mercury

This is the closest to the Sun and the smallest except for Pluto. Mercury has a mass of only 5,5% of the Earth, a diameter of 4878 km and has a highly elliptical orbit varying from 46 to 70 million kilometres from the Sun. The planet orbits the Sun in 88 days and has a slow rotational period of 59 days. The temperatures are extreme, varying from 425°C by day to −180°C by night. There are craters at the poles which are never penetrated by sunlight so it is quite possible that these could contain permanently frozen water ice. The planet has a large iron core and a very modest magnetic field. The planet is too hot and too small to retain an atmosphere. It has been calculated that the perihelion of Mercury should advance by 500 seconds per century due to gravitational effects from the other planets but the measured advance was found to be 543 seconds. It was initially thought that there was a small unseen planet affecting the orbit of Mercury. The discrepancy was precisely accounted for by Einstein's general theory of relativity, providing additional proof of Einstein's theory. The relativistic effects on the other planets are negligible. Mercury remains the least explored of the Solar System planets. The first space probe to observe Mercury was the NASA's Mariner 10 which performed its mission in 1974–1975 using a gravity assist by doing a flyby of Venus. The spacecraft took 10 000 images of the planet covering 57% of the surface. The craft is now in solar orbit. The current Mercury mission is MESSENGER (MErcury Surface, Space ENvironment, GEochemistry and Ranging) which was launched on 2 August 2004. The craft made an Earth flyby in January 2008, a Venus flyby in October 2008, another Venus flyby in September 2009 and successfully entered a Mercury orbit on 18 March 2011. The mission objective was to study Mercury's chemical composition, geology and magnetic field. MESSENGER ended its primary mission on 17 March 2011 after having collected some 100 000 images. The extended mission is scheduled to last until march 2013.

The most ambitious and exciting mission to Mercury will be ESA's 'Cornerstone' BepiColombo mission which is being planned in co-operation with Japan's ISAS/JAXA. The launch is planned for August 2015 and will go into Mercury orbit in August 2019 and remain in orbit for one year until 2022. The mission will utilise solar electric propulsion to reach Mercury and also utilise five gravity assists, one from Earth, two from Venus and two from Mercury. This path from Earth to Mercury is a matter of extreme complexity. The Sun's enormous gravity presents a huge challenge in placing the satellite in a stable orbit around Mercury. The objectives of the mission include: Origin and evolution of a planet close to its parent star,

Mercury's magnetosphere and vestigial atmosphere, planet's form, interior, structure, geology, composition and craters and also a test of Einstein's General Relativity. The mission will consist of two separate spacecraft orbiting the planet. ESA is building the MPO (Mercury Planetary Orbiter) and ISAS/JAXA will construct the MMO (Mercury Magnetospheric Orbiter). Both of these craft will carry awesome payloads of highly specialised scientific instruments. The craft will also have to endure temperatures as high as 350°C. Mercury has no stable atmosphere; the gaseous exosphere is so rarefied that the neutral atoms of its constituents never collide. Atoms in the exosphere include oxygen, hydrogen, neon, sodium and potassium which were detected by Mariner 10. The mission is named in honour of Giuseppe (Bepi) Colombo (1920–1984) of Padua University. Bepi explained, as an unsuspected resonance, Mercury's peculiar habit of rotating three times in every two revolutions of the Sun.

> I am very pleased that we have given the name BepiColombo to our Mercury Cornerstone. Bepi was a great scientist, a great European and a great friend; we could do no better than name one of our most challenging and imaginative missions after him.
>
> Roger Bonnet, Director of ESA Science Program (1982–2001)

Venus

Venus is the Roman goddess of love and the astrological symbol for Venus is also used as the symbol for 'female' but there is nothing at all romantic or feminine about this planet. Venus is seen as an evening or morning star as it is closer to the Sun than the Earth and will normally be the brightest object in the sky. It is a beautiful sight but best admired from a safe distance. It has a diameter slightly less than that of the Earth but here the similarity ends. Venus is hot, very hot – a temperature range of from 480°C down to –33°C. The atmospheric pressure is tremendous – about 90 times that of the Earth and the atmosphere is equally unpleasant – about 95% carbon dioxide and the rest mostly nitrogen. There is a thick fast moving cloud layer about 50 km above the surface consisting mostly of sulphuric acid droplets. Venus has a year length of 224,7 Earth days but the Venusian day is very long, 243 hours and it turns the wrong way. Venus orbits the Sun at an average distance of 108,1 million km. There is at present no scientifically conclusive explanation for the retrograde rotation (with apologies to Velikovsky). Venus and Mercury are the only Solar System planets which do not have any moons. An interesting aspect of the rotation of Venus is that it will always present the same face to the Earth at closest approach due to an orbital resonance.

On 5-6 June 2012 a most remarkable event occurred – a transit of Venus across the face of the Sun. This event occurs in pairs separated by 8 years and the pairs are separated by 105 or 121 years. The previous occurrence was in 2004 and the next will be in 2117. As Venus ingresses or egresses the disc of the Sun, the dense atmospheric layers diffract the sunlight, giving the 'Arc of Venus' which appears as a thin ring of fire encircling the planet. The event was observed by the Hinode and Solar Dynamics spacecraft as well as several terrestrial coronagraphs.

Several Venus flyby missions have been launched since February 1961. Since 1971, twenty seven space missions to Venus have been launched – 19 by the USSR and 7 by the USA. These were designed for gathering scientific data and mapping the surface. Several of the craft which approached the surface were crushed by the tremendous atmospheric pressure. The latest Venus mission, the Venus Express, was launched by ESA on 9 November 2005 using a Soyuz-Fregat rocket fired from the Baikonur Cosmodrome in Kazakhstan. After a 105 day journey it was placed in a nine day Venus orbit and later into a 24 hour polar orbit. The craft was controlled and monitored from the ESA Control centre in Darmstadt, Germany. The mission is equipped with an impressive assortment of specialised scientific instruments. The design of the craft has been largely influenced by features of the Mars Express and Rosetta spacecraft. The mission is planned to remain active until 31 December 2014.

Earth

This is the most pleasant and comfortable place to be in the entire Solar System, and possibly even in the universe, as there is no known or envisioned way of visiting any other solar system. Earth is the only planet in the Solar System with a name not derived from Greek/Roman mythology. (With the exception of the trans-Neptunian objects which are not regarded as planets). It is only since the time of Copernicus that the Earth has been regarded as a solar system planet. The Earth has an equatorial diameter of 12 756 km and orbits the Sun at an average distance of 149,6 million km. This is also the original definition of an Astronomical Unit. The temperature range is from 58°C down to -88,3°C but very few people would ever have experienced these extremes. The temperature of the Earth's atmosphere is very dependent on the small proportion of carbon dioxide. Without any carbon dioxide in the atmosphere the surface temperature would be 35°C lower causing the oceans to freeze. An increase of carbon dioxide in the atmosphere would also be disastrous causing melting of the ice caps, raising of the ocean level and reduction of land area, not to mention catastrophic climate change. The Earth is the only object in the Solar System to have abundant surface water. It is also the only planet in the Solar System to have a crust of tectonic plates which can move about on the hot molten mantle beneath. The plates are in constant motion, albeit by only a few centimetres per year. There are eight major tectonic plates and some twenty lesser plates. The only other solar system body known to have tectonic plate activity is Jupiter's moon Ganymede. The Earth has large inner and outer cores of iron. Electric currents in the outer core produce the earth's magnetic field. The magnetic field provides us with essential protection from cosmic rays. The magnetic field of the Earth is a highly complex matter. There have been many reversals of the poles and the poles can also migrate over great distances. There are also many places on the Earth where the magnetic field is erratic making the use of a compass for direction finding quite useless. The temperature at the centre of the inner core is 7500 K which is hotter than the surface of the Sun. It has been speculated that there might be a significant core of uranium at the centre of the earth sustaining a continuous source of nuclear energy. The Earth's atmosphere provides us with essential protection from dangerous radiation and small meteorites. Another unique feature of the Earth is its oxygen rich atmosphere which is kept in balance by the abundant plant life on the planet.

Since the launch of the Sputnik satellite on 4 October 1957, many thousands of Earth orbiting satellites have been launched. The original Sputnik satellite was placed into orbit by a R-7 launch vehicle designed for carrying a nuclear warhead. The Earth satellites have been designed for a wide variety of purposes such as TV and voice communications, Internet, weather mapping, GPS, geological surveys, topographical mapping, vegetation surveys, forest fire surveys, espionage etc. The list is almost endless. Satellites designed for astronomical purposes are described in chapter six. Satellite orbits can be equatorial, polar or geostationary. Equatorial orbits have the advantage of a velocity assist by utilising the rotation of the Earth but cannot scan all parts of the Earth's surface. Polar orbits are required for satellites which need to observe all parts of the Earth. Geostationary satellites are placed in high orbit so that they appear motionless in relation to the Earth as is required for communication satellites. It is not possible to give a full account of earth orbiting satellites here – let it suffice to mention a few. The first is a proposed mission intended, after several delays, for launch in 2014. This is the KEO satellite. The name KEO is a grouping of the letters K, E, and O which are the three most frequently used sounds in common to the most widely spoken languages in use today. The project is supported by UNESCO, Hutchison Whampoa (a Chinese Fortune-500 company) and the European Space Agency. The satellite is of the nature of a time capsule which will remain in Earth orbit for 50 000 years and then return to Earth. Finding funding for a project with a payback after 50 000 years must present some unique problems. The capsule will carry a diamond which encases a drop of human blood and samples of air, sea water and earth. There will also be an engraving of human DNA. The satellite will also carry an astronomical clock that shows the rotation rates of several pulsars, photographs of people of all cultures and a contemporary 'Library of Alexandria'– an encyclopaedic compendium of current human knowledge. The loss of the original Library of Alexandria is possibly the worst case of vandalistic knowledge destruction in recorded history. A most remarkable feature of the satellite will be its capacity to carry written messages from every human inhabitant of the Earth. The messages and library will be carried on glass-made radiation-resistant DVDs. Instructions will be included to show the finders of the satellite how to build a DVD reader. The outer casings of the satellite are designed to produce a spectacular fireball on re-entry into the atmosphere indicating to the creatures inhabiting the Earth at that time that it is time to do something about it.

Many time capsules of a down to Earth nature have been made and hopefully preserved in safe places. There has been much speculation of a wishful and wondrous capsule concealed in ancient times, either within the Giza Sphinx or buried deep beneath its paws. This idea has started to wear rather thin as all predictions of a revelation date have come to nothing.

A famous capsule called the 'Crypt of Civilisation' was prepared in 1936. This was 6177 years after the establishment of the Egyptian calendar so the opening date was set at 6177 years later in 8113. The Crypt of Civilisation is a sealed, airtight chamber at Oglethorpe University, Atlanta, Georgia. If the chamber were opened today, people below retirement age might consider it a relic from the Dark Ages. It dates from a time before plain paper copiers, Kodacolour film, transistors, microelectronics, FM radio, TV, computers, magnetic tape recording, DVDs, lasers, GPS, cellphones and thousands of other everyday items. Sound recordings were on fragile 78 r.p.m. records read by means of a stylus running in a groove. It is also interesting to note that at that time literature and music had already reached as high a level as they are ever likely to go. The stainless steel door of the crypt was finally welded shut at a ceremony in 1940. An appeal was made to the people of this planet that the crypt remain inviolate.

The International Space Station (ISS) is the successor to several previous space stations: The Russian MIR 2, the US Space Station Freedom, the European Columbus and the Japanese Experiment Module Kibo. Assembly of the space station commenced in 1988 and has been continuously staffed since November 2000. By July 2008 the Space Station was about 75% complete. The complete Space Station will have fourteen pressurized modules with a combined volume of some 1000 cubic metres. The ISS is powered by solar panels with an area of 375 square metres. The ISS is in a low orbit between 278 and 425 kilometres above the Earth. The slight atmospheric drag requires the craft to be boosted in its orbit several times a year. Unlike other space missions with clearly defined objectives, the ISS is a general purpose facility with a capability for a wide range of scientific experiments. The primary research facility is NASA's Destiny Laboratory Module. The Columbus module research facility was designed by ESA. The ISS is a project of collaboration between the US, Russia, Japan, Canada and eleven ESA European countries. Brazil, Italy and China also have interests in the project. On 9 May 2009 the ISS crew of three was increased to six when a SOYUZ TMA-15 capsule docked. This was the first time that all five partner agencies were represented by crew members on the orbiting outpost. On 16 May 2011 the Alpha Magnetic Spectrometer was carried aloft on the last flight of space shuttle Endeavour and installed on the ISS on 19 May 2011. This instrument is designed to search for various types of unusual matter by measuring cosmic rays. By July 2012 it was reported that the AMS-02 had recorded over 18 billion cosmic ray events. This is the most sophisticated particle detector ever sent into space. The total mass of the instrument is 6 717 kg of which 1 200 kg is due to the huge $Nd_2Fe_{14}B$ permanent magnet required to deflect the charged cosmic rays. Only rays passing from top to bottom are recorded – about 1000 per second.

Scientific goals include searching for: Antimatter, Dark matter, Strangelets and Space radiation environment. The AMS-02 has a planned mission life of 10 years or more. The total cost has ballooned to $ 1,5 billion.

The Earth's magnetosphere is well known for its ability to deflect charged 'solar wind' particles to the magnetic poles causing aurora flares visible in the night sky. The THEMIS mission (Time History of Events and Macroscale Interactions during Substorms) is designed to study energy releases from the Earth's magnetosphere known as substorms, magnetic phenomena that intensify auroras near the Earth's poles. THEMIS launched on 17 February 2007 originally comprised a fleet of five spacecraft of which two were subsequently placed in orbits at Lagrangian points of the Moon. The satellites each have a mass of 77 kg, 49 kg of fuel and a power consumption of 37 W.

The satellites each carry identical instrumentation: a fluxgate magnetometer, an electrostatic analyser, a solid state telescope, a search-coil magnetometer and an electric field instrument. From 15 February to 15 September 2007 the five craft coasted in a string-of-pearls configuration after which they were repositioned for data collection in the magnetotail. In 2007 a most remarkable result was achieved when evidence was found of "magnetic ropes" (Birkeland Currents) connecting Earth's upper atmosphere directly to the Sun proposed by Kristian Olaf Bernhard Birkeland in 1908. Initial proof of Birkeland's currents came in 1965 with the launch of a US Navy probe which carried a magnetometer above the ionosphere. Birkeland currents are of a class of plasma phenomenon called z-pinch because of azimuthal magnetic fields produced by the current which pinch the current into a filamentary cable.

Birkeland was nominated seven times for the Nobel physics prize. It was estimated that the ropes pump 650 000 amperes into the Arctic! This enormous current must of course be immediately discharged elsewhere as the Earth's capacitance could not contain the charge even for a microsecond. On 26 February 2008, THEMIS probes were able to determine, for the first time, the triggering event for the onset of magnetospheric substorms. This was found to be a magnetic reconnection event 96 seconds prior to Auroral intensification.

In 2001 the ISS received the prestigious Prince of Asturias Award for International Co-operation. The 2008 award in Communications and Humanities went to Google. Tiangong-1 is China's first space station. Launched on 29 September 2011 it was the first operational component of the Tiangong program which aims to place a modular station in orbit by 2020. Tiangong-1 will be deorbited in 2013 and later replaced by the larger Tiangong-2 and Tiangong-3 modules. The unmanned Shenzhou 8 successfully docked in November 2011 and the manned Shenzhou 9 mission docked in June 2012 with three astronauts aboard, one of whom was China's first female astronaut in space.

Tiangong-1 has a pressurized volume of 15 cubic metres divided into two primary sections: a resource module which mounts its solar panels and propulsion systems, and a larger habitable experimental module.

The number of artificial objects orbiting the Earth has been steadily growing over the years and has reached over 8000 at the latest count and most of it potentially hazardous junk. There are at present over 2500 satellites (operational or otherwise) orbiting our planet.

Mars

This is probably, apart from the Earth, the best known of the planets. A tremendous amount has been written about the inhabitants, civilisation and industry on this planet, and all of it nonsense. Mars is the Roman god of war and the astrological symbol is also used for 'male'. The planet is easily identified by its red colour. It has a diameter of 6 785 km and a temperature range from 270° C down to -123° C. The atmosphere is very unpleasant consisting of 95,3% carbon dioxide and 2,7% nitrogen. There are polar ice caps consisting of frozen water and carbon dioxide. The day length is very similar to that of the Earth at 24 hours and 37 minutes but the year is much longer at 687 days. The orbiting distance from the Sun is from 205 to 249 million km. There is much speculation about the history of Mars – there are many geological features indicating weathering by torrents of water. Mars has two moons called Phobos and Deimos.

Several Mars probes have been launched since 1960, however initial attempts met with little success.

Mariner 4 was launched by NASA on 5 November 1964 and arrived on 14 July 1965. This was the first successful Mars flyby mission and returned 21 close up photos of the Martian surface. A remarkable aspect of the mission was that the mission plan was modified after launch to perform the radio-occultation experiment.

Mariner flyby missions 6 and 7 (identical spacecraft) launched in 1969 returned 75 and 126 images, respectively.

The failed USSR Mars 2 Orbiter/Lander launched in 1971 was the first man made object to impact the Martian surface.

The USSR Mars 3 Orbiter/Lander was launched in 1971. The orbiter transmitted 8 months of data. The Lander was only able to send 20 seconds of data.

NASA's Mariner 9 Orbiter launched in 1971 returned 7329 images. These covered about 80% of the Martian surface and included images of the moons Phobos and Deimos. Atmospheric pressure was measured and found to vary between 2,8 and 10,3 mbar. The Mariner 9 mission exceeded its expectations in every way and paved the way for the Viking program.

The USSR Mars 5 mission launched in 1973 survived 9 days returning 60 images.

NASA's Viking 1 Orbiter/Lander launched in 1975 located a landing site for the Lander and achieved the first successful landing on Mars.

NASA's Viking 2 Orbiter/Lander, also launched in 1975 returned 16 000 images as well as extensive atmospheric data and results of soil experiments.

The Phobos-1 and Phobos-2 spacecraft were launched by the USSR on the 7th and 21st July 1988, respectively. The mission objectives were to: conduct studies of the interplanetary environment, perform observations of the Sun, characterize the Martian plasma environment, conduct surface and atmospheric studies of Mars and study the surface composition of the moon Phobos. Phobos-1 operated normally until communications were lost on 2 September 1988. Phobos-2 operated normally until communications were lost before the launch of two Phobos landers. A strange twist to the mission was that a large elongated unidentified object was photographed close to Phobos. A photograph of the Martian surface showed an unexplained elongated shadow.

Mission Mars Global Surveyor launched by NASA on 7 November 1996 returned more images than all previous Mars missions.

The Mars Pathfinder mission was launched by NASA on 4 December 1996 and reached Mars on 4 July 1997. The Pathfinder entered the Martian atmosphere using an entry capsule, a parachute, and was slowed by solid rockets and cushioned on impact by large air bags. This was the first mission to send a rover to the Martian surface – A competition to name the rover was held resulting in the name Sojourner. The Lander returned 16 500 images and the rover 550. The Sojourner was a six wheeled vehicle 65 cm long, 46 cm wide, 30 cm tall and had a speed of 1 cm per second. Both lander and rover were equipped with several instruments for analyzing the atmosphere, climate, geology, rock composition and soil. The lander returned 8,5 million measurements of atmospheric pressure, temperature and wind speed. The mission far outlived its planned survival time on Mars. Remarkably, this mission cost less than $150 million.

The 2001 Mars Odyssey mission was launched by NASA on 7 April 2001 and went into Mars orbit on 24 October 2001. This is a robotic mission using spectrometers and imagers to search for past or present water or volcanic activity. The spacecraft also functions as a relay station for communications between Martian rovers and Earth. The primary instruments of Odyssey are:

- THermal Emission Imaging System (THEMIS).

- Gamma Ray Spectrometer (GRS) – including High Energy Neutron Detector (HEND) from Russia.

- Mars Radiation Environment Experiment (MARIE).

The Mars Express mission is the first planetary mission by ESA. It was launched on 2 June 2003 from Baikonur Cosmodrome in Kazakstan using a Soyuz-Fregat rocket. The mission utilized the fact that Mars was at its closest approach to Earth in 60 000 years. The Mars Express Orbiter carried the Beagle-2 Lander but this was unfortunately lost on arrival. The Beagle-2 was named after the ship of Charles Darwin's voyage of discovery. The prime contractor was Astrium, Toulouse, France, which led a consortium of 24 companies from 15 European countries and the US. Some of the objectives of the mission are: the high resolution imaging of the entire surface, a map of the mineral composition of the surface and structures of the sub-surface to a depth of a few kilometers.

The Orbiter instrumentation includes the following:

- High Resolution Stereo Camera (HRSC)
- Energetic Neutral Atoms Analyser (ASPERA)
- Planetary Fourier Spectrometer (PFS)
- Visible and Infrared Mineralogical Mapping Spectrometer (OMEGA)
- Sub-surface Sounding Radar Altimeter (MARSIS)
- Mars Radio Science Experiment (MaRS)
- Ultraviolet and Infrared Atmospheric Spectrometer (SPICAM)

On 2 October 2007 the Mars Express performed a close flyby of Phobos coming within 140 km of the moon.

Twin Mars exploration rovers were launched by NASA in 2003. SPIRIT was launched on 10 June 2003 and OPPORTUNITY three weeks later. SPIRIT Landed on 4

January 2004 in Gusev crater. This landing site has been named 'Columbia Memorial Station' in memory and honour of the seven astronauts who perished in the Columbia Shuttle disaster. OPPORTUNITY landed on 25 January 2004 in Meridiani Planum on the opposite side of the planet. The descents were slowed by parachutes, rockets and finally airbags to cushion the landings. The rovers performed large numbers of successful scientific investigations of rocks, soil samples, craters and hills and took a large number of spectacular images. The magnificent full circle panoramic images are well worth seeking out on the Internet. The SPIRIT rover's rock abrasion tool successfully ground a hole into a rock named Adirondack exposing the interior to examination by the rover's imager and spectrometers mounted on the robotic arm. SPIRIT took images of the Martian moons and also a transit of the Sun by moon Deimos as well as an eclipse of Phobos by Mars.

The rovers were provided with hazard cameras, two front and two rear as well as the following instruments on the robotic arms:

- Mössbauer Spectrometer – for close up mineralogy investigation.
- Alpha Particle X-Ray Spectrometer – for close up analysis of elements in rocks and soil.
- Magnets – for collecting magnetic dust samples.
- Microscopic Imager – for close up high resolution images of rocks and soil.
- Rock Abrasion Tool – for exposing fresh interior of rocks for examination.

On 24 January 2007, both rovers received computer software upgrades greatly increasing their autonomy. The rovers remain operational having exceeded their planned mission lifetimes many times over.

NASA's Mars Reconnaissance orbiter (MRO) was launched on 12 August 2005 and was successfully inserted into Mars orbit on 10 March 2006. At this time there were five other active spacecraft either in orbit or on the surface of Mars – Mars Global Surveyor, Mars Express, Mars Odyssey and two Mars Exploration Rovers. This is a multipurpose mission to map the Martian landscape with its high resolution cameras and to choose landing sites for future missions. The MRO will use its onboard instruments to study the Martian climate, weather, atmosphere and geology and search for signs of liquid water at the poles and underground. After its main science operations are completed the extended mission is to relay communications to lander and rover probes active on Mars. It is expected that the

MRO will relay more data back to Earth than all previous missions combined.

The MRO has an impressive array of specialized instruments. Three cameras, two spectrometers, a radar and science facility instruments. The High Resolution Imaging Science Experiment camera (HiRISE) is a 50 cm reflecting telescope, the largest yet carried on a deep space mission. The HiRISE can produce stereo pairs of images from which topography can be calculated to within 25 cm.

The Context Camera (CTX) is a Maksutov Cassegrain telescope with a 35 cm focal length. By February 2010 it had mapped 50% of the Martian surface.

The Mars Colour Imager (MARCI) is a wide angle, low resolution camera that views the Martian surface in five visible and two ultraviolet bands. It has a 180 degree fisheye lens.

The MRO has provided many astonishing views of Mars including some dust devils and a twister. On 23 March 2008 HiRISE took spectacular stereo images of moon Phobos.

The MRO's radar measurements of the north polar cap revealed that the cap contains 821 000 cubic kilometers of water ice. It was also found that recently excavated craters on the Martian surface have revealed water ice. Once exposed, this ice gradually sublimates away.

On 4 August 2011 the MRO detected what appears to be flowing salty water on the surface or subsurface of mars. NASA announced that in September 2012 the MRO had detected a snowfall of frozen carbon dioxide at a polar cap.

The MRO was built by Lockheed Martin Space Systems at a cost of $720 million. It was launched on an Atlas V-401 rocket at Cape Canaveral Air Force Station. The launch mass was 2 180 kg which included 1 159 kg of hydrazine monopropellant. The solar panels provide the craft with 2 kW of power. The craft has twenty rocket engines for orbit insertion, trajectory corrections, and attitude control as well as four reaction wheels for precise control.

The robotic Phoenix spacecraft was launched by NASA on 4 August 2007 and made a successful landing on Mars on 25 May 2008 near the North polar region. The lander returned 48 images back to Earth within two hours of landing. The Mars Reconnaissance Orbiter was able to photograph the Phoenix descending by parachute using its HiRISE camera. This is a mission under the Mars Scout Program. This multi-agency program is headed by the Lunar and Planetary Laboratory at the

University of Arizona under the direction of NASA. The program is a partnership of universities from the US, Canada, Switzerland, Denmark, Germany and the UK, and also NASA, the Canadian Space Agency, and the aerospace industry.

The mission is designed for a duration of 90 Martian days. The landing was similar to that of the Viking program using landing rockets to slow the descent. This has caused some controversy due to the possible contamination of the landing site. The chosen landing site is in Green Valley in Vastitas Borealis which contains the largest concentration of water ice outside of the poles. Three orbiting Mars satellites were placed on 25 May to track the Phoenix on entering the atmosphere until after landing. In addition to a sophisticated array of instruments, the lander is also equipped with a robotic arm, which can extend 2,35 m from the craft to dig half a metre below the surface. Soil and water ice samples excavated will then be analysed by instruments aboard the lander.

NASA's Mars Science Laboratory (MSL) was launched on 26 November 2011 on an Atlas V 541 rocket from Cape Canaveral and the Rover named CURIOSITY made a spectacular landing on 6 August 2012. This was the most challenging space mission landing ever attempted. The overall objectives include: Investigating Mars' habitability, studying its climate and geology, and collecting data for a manned mission to Mars. The mission is planned to operate for at least a year. The complete spacecraft had a mass of 3 893 kg consisting of several components: an Earth-Mars fuelled cruise stage, the entry descent landing system of 2,401 kg plus 390 kg of propellant and the 899 kg mobile rover with integrated instrument package. The rover is much larger than that of any previous mission and also has more advanced instruments than any previous mission to Mars. The descent stage included a backshell containing a 16 m supersonic parachute which was deployed when the vehicle slowed to Mach 1,7 and the 4.5 m heatshield fell away. At about 1,8 km altitude the descent stage and rover dropped out of the aeroshell and was further slowed to a halt by rockets. The sky crane then lowered the rover about 7,7 m to a soft landing on Mars. The rover deployed its wheels on the way down and severed the sky crane's cables on landing. The sky crane then flew off to a crash landing 650 m away. The landing sequence required six vehicle configurations, 76 pyrotechnic devices, the largest supersonic parachute ever built and more than half a million lines of code. The Mars Reconnaissance Orbiter was able to acquire an image of the MSL descending under parachute. The chosen landing site was Gale Crater which contains a mountain of layered rock rising 5.5 km and much material of geological interest. On landing the rover replaced its flight software with its full surface operations software. The rover has wheel treads which leave tracks which can be used to determine the distance travelled. The pattern represents the Morse code for "JPL". The rover will analyse dozens of scooped up soil samples and powders drilled from rocks. Soon after landing Curiosity discovered a dry river bed.

The CURIOSITY landing site has been named "Bradbury Landing" in honour of sci-fi author Ray Bradbury. The estimated total cost of the MSL project is $2,5 billion.

The instruments aboard the rover include:

MastCam –	A high resolution stereoscopic multi-spectra colour camera with a video capability.

Mars Hand Lens Imager (MAHLI) – A robotic arm mounted camera for microscopic images of rock and soil.

MSL Mars Descent Imager (MARDI) will take about 500 high resolution colour Images of the landing site and surrounding terrain.

ChemCam –	A remote Laser-induced breakdown spectroscopy (LIBS) system. This can target a rock from up to 13 metres away, vapourising a small amount of underlying material and then collect a spectrum of light emitted by the vapourised rock using a micro-imaging camera with an angular resolution of 80 microradians.

The laser-induced breakdown spectroscopy (LIBS) instrumentation was developed by the Los Alamos National Laboratory and the French CESR Laboratory.

Alpha-particle X-ray spectrometer (APXS) will irradiate samples with alpha particles and map spectra of emitted X-rays to determine elemental composition.

CheMin –	Chemistry and Mineralogy X-ray Diffraction/X-Ray Fluorescence Instrument will quantify minerals and mineral structure of samples.

Sample Analysis at Mars (SAM) will analyse organics and gases from atmosphere and solid samples. The Tunable Laser Spectrometer is capable of measuring the isotope ratios of carbon and oxygen in CO_2.

Radiation Assessment Detector (RAD) will characterize the broad spectrum of radiation found near the surface of Mars for determining the viability and shielding needs for human explorers.

Dynamic Albedo of Neutrons (DAN) is a pulsed neutron source and detector for measuring hydrogen or ice and water at or near the Martian surface.

Rover Environmental Monitoring Station (REMS) is a meteorological and ultraviolet sensor provided by the Spanish Ministry of Education and Science with the Finnish Meteorological Institute as a partner.

MSL Entry Descent and Landing Instrumentation (MEDLI) installed in the heat shield of the MSL entry vehicle. The acquired data will also support future Mars missions.

Hazard Avoidance Cameras (Hazcams) will be mounted in pairs fore and aft of the rover and used for autonomous hazard avoidance.

Navigation Cameras (Navcams) mounted in a pair on the mast to support ground navigation.

Phobos Grunt (Фобос Грунт) was a planned sampling mission to the Martian moon Phobos launched by Russia on 9 November 2011. The name 'Grunt' means 'soil' of which a sample was to be collected and taken to Earth. The mission would also perform numerous studies of the Martian atmosphere, dust storms, radiation environment and plasma. This promising mission sadly failed soon after launch and plunged back to Earth on 16 January 2012.

The ExoMars mission will consist of two modules: the Orbiter with Descent and Landing Demonstrator to be launched in 2016 and a mission featuring two Rovers to be launched in 2018. The missions will be carried out in cooperation with NASA. After 9 to 10 months the spacecraft will enter a Mars orbit suited for releasing the Descent Module and then take an orbit best suited for operation as a data-relay satellite. The primary mission objectives are:

- To study the biological environment of the Martian surface, and to search for possible Martian life, past or present.
- To characterise the Mars geochemistry and water distribution.
- To identify possible surface hazards to future human missions.
- To improve the knowledge of the Mars environment and geophysics.

The Rover and Lander will carry a large selection of specialised scientific instruments as well as a drill for extraction of sample material.

Some future Mars mission proposals are:

   – Further missions of the NASA Scout Program.
   – The NASA Astrobiology Field Laboratory.
   – The NASA Mars Sample Return Mission.

Jupiter

By Jove – this is a big one! Jupiter has about 2,5 times the mass of all the other Solar System planets combined and a diameter of 142 796 km. The length of year is 11,86 Earth years and the length of day a bit less than 10 hours at the equator – the orbital speed is 13,07 km/s. Jupiter is not a solid body so that all parts do not rotate at the same speed. Jupiter is a gas giant, more like a star than a planet, containing 82% hydrogen and 17% helium by mass. Jupiter would have to have about 100 times more mass to become a star. The outer part is composed mainly of liquid molecular hydrogen and the interior liquid metallic hydrogen. Metallic hydrogen is under such tremendous pressure that it becomes a conductor of electricity and heat – it is not possible to produce hydrogen in this state on Earth. The central temperature has been estimated to be between 13 000 K and 35 000 K and the central pressure to be 100 million Earth atmospheres. Jupiter has a strong magnetic field – about 10 times the strength of the Earth field and a magnetosphere extending 7 million kilometres which interacts with the solar wind. Jupiter has an internal heat source radiating more heat than it receives. The heat source drives the violent storms at the surface. The huge red spot storm is about three times the size of planet Earth and has been a permanent feature of the planet ever since it could be observed. Jupiter has a faint ring but nothing as spectacular as the rings of Saturn. Jupiter has a large assortment of moons but the best known are the four Galilean moons first observed by Galileo. These moons can be observed even when using fairly modest optical equipment. The moons are the largest in the Solar System ranging in size from 3 000 to over 5 000 km.

The moon Io is the most volcanically active object in the Solar System powered by huge tides produced by Jupiter. The tides can produce a surface rise and fall of as much as 100 metres.

The moon Europa is the smoothest solid body in the Solar System being covered in heavily cracked water ice. This moon is thought to harbour a saltwater ocean sandwiched between a 20 km layer of surface ice and a rocky core below.

The moon Ganymede is the largest moon in the Solar system with a diameter of 5 270 km. This moon is covered with impact craters. It was thought that Ganymede was the only Solar System object other than the Earth to have evidence of plate tectonics. It has now been proposed by UCLA scientist An Yin that the 4000 km Mars canyon *Valles Marineris* is a boundary between tectonic plates.

The moon Callisto, like Europa, is covered by water ice and is pock marked by impact craters.

NASA's Pioneer 10 Mission launched in 1972 was the first spacecraft to pass through the Asteroid Belt and on to take the first close up images of Jupiter. After the flyby it continued in the direction of constellation Taurus. It will take the spacecraft two million years to reach Aldebaran in Taurus. Pioneer 11 was launched in 1973. It returned pictures of Jupiter's Red Spot as it took a gravity assist to reach Saturn where it took close up pictures of the planet and its rings. Pioneer missions 10 and 11 each carry gold anodized aluminium plaques with graphic messages to any extraterrestrials who may be able to intercept the spacecraft. The plaques included nude human male and female figures with a slight concession being made to modesty in the case of the female. It seems astonishing that anyone could think that the extraterrestrials might be offended by a lack of modesty in the figures from Earth. If ET does manage to capture one of the craft he will be left wondering how Earth people manage to reproduce.

On 18 July 2012 it was announced that the unexpected slowing of Pioneer 10 and 11 was due to the slight effect of heat pressure on the craft. The heat is due to electrical currents in the instruments and the thermonuclear power supply.

Voyager 2 was launched by NASA on 20 August 1977 followed by Voyager 1 on 5 September 1977. These craft made use of the rare geometrical arrangement of Jupiter, Saturn, Uranus and Neptune, which made it possible to visit them over a 12 year span, instead of 30, both arriving at Jupiter in 1979. The spacecraft returned photographs and much information of Jupiter and its many moons.

NASA's Galileo spacecraft which was launched on 19 October 1989 consisted of an orbiter and a descent probe. The primary mission was to explore Jupiter, its moons and rings in more detail than was done by the Voyager missions. The spacecraft arrived at Jupiter in December 1995. The Descent Probe was deployed which gave much information on the atmospheric pressure, density and composition. It also measured the planet's radiation belts. Low oxygen content and little lightning indicated a surprisingly low concentration of water. The probe came to a spectacular end as it heated to incandescent temperature. The Orbiter was expected to survive for up to two years but continued to function for eight years. The orbiter returned spectacular photos of the Galilean moons as well as a wealth of other information on Jupiter, the moons and rings.

The orbiter detected sub-surface layers of liquid salt water on Europa, Ganymede and Callisto and also that Io, Europa and Ganymede have metallic cores. Ganymede was also found to have a magnetic field. The Orbiter was able to make a direct observation of a fragment of the Shoemaker-Levy comet plunging into the Jovian atmosphere.

Saturn

This is the sixth planet from the Sun and the second largest after Jupiter. Its rings make it the most easily recognised astronomical object. In any logo or graphic depiction of astronomical matters there will invariably be a drawing of Saturn and its rings. It has an average distance from the Sun of nearly ten astronomical units; $1,513 \times 10^9$ km aphelion and $1,353 \times 10^9$ km perihelion. The orbital period is 29,657 earth years with an orbital speed of 9,69 km/s. The planet has an equatorial radius of 60 268 km and a polar radius of 54 364 km. The planet is less dense than water and the surface gravity is slightly less than that on Earth. The planet consists of 96% hydrogen and 3% helium. It is thought that the planet has a small rocky core surrounded by a layer of liquid metallic hydrogen and an outer layer of liquid molecular hydrogen and helium. The temperature at the centre is thought to be 12 000 K. Saturn has an internal heat source as it radiates 2,5 times as much heat as it receives from the Sun. The rings, which extend 6 630 km to 120 700 km above Saturn's equator were objects of much confusion to Galileo. Dutch physicist, astronomer and mathematician Christiaan Huygens (1629–1695), using a superior telescope to that of Galileo, was the first astronomer to observe that Saturn was surrounded by thin flat rings. The rings are thin indeed compared to their immense size; only about 20 metres and consisting mostly of water ice fragments. Huygens discovered Titan in 1655. It is currently thought that Saturn has more than 60 moons; of these some 52 have been given names. The largest moon, Titan, is an object of much scientific interest. It is larger than planet Mercury and is the only solar system moon to have an atmosphere. Giovanni Domenico Cassini (1625–1712) is well known for his discovery of the Cassini division in Saturn's rings. He was also the first to discover four of Saturn's moons and the Great Red Spot of Jupiter. With the aid of his colleague Jean Richter who he sent to French Guiana he was able to use parallax to determine the distance to Mars and hence determined the dimensions of the solar system. He was the first to make successful measurements of longitude by the method suggested by Galileo, using eclipses of the moons of Jupiter as a clock. Cassini was a professor of astronomy at the University of Bologna and later director of the Paris Observatory. He was also engaged in major engineering projects.

The Pioneer 11 mission launched by NASA on 6 April 1973 arrived at Saturn in September 1979 after a flyby and gravity assist from Jupiter. The Jupiter gravity assist accelerated the craft to 175000 km/hour. The craft transmitted astonishing images of Saturn and returned much information of Saturn, its rings and moons. The craft came within 13000 km of Saturn and is now moving in the direction of constellation Aquila – it will take four million years for the craft to reach a star in the constellation.

The Voyager 1 mission launched by NASA on 5 September 1977 also took a gravity assist from Jupiter and did a flyby of Saturn in November 1980. Nine hundred images of Saturn's moons were transmitted and unexpected discoveries were made of the complexity of the rings and the nature of Titan's atmosphere.

NASA's Voyager 2 mission was launched on 20 August 1977 and after taking a gravity assist from Jupiter, arrived in August 1981 to do a flyby of Saturn. This craft returned 1150 images of Saturn's moons. The Voyager missions have been operating continuously for 35 years and are about to pass through the Heliosheath and enter interstellar space.

The Cassini-Huygens Mission, a Joint flagship-class venture by NASA, ESA and ASI was launched on 15 October 1997. This is a combined orbiter and probe mission, the Cassini Orbiter being supplied by NASA and the Huygens Probe by ESA. Due to the vast distance of Saturn from the Sun, it was not possible to use solar panels to power the satellite. Power was provided by means of three RTGs (Radioisotope thermoelectric generators) fuelled by 32,8 kg of plutonium. This gave rise to controversy due to the risk of disastrous radioactive contamination in the event of the launch rockets exploding in a failed launch. The launch mass was a massive 5574 kg, of which 3132 kg was due to the monomethyl hydrazine and nitrogen tetroxide fuel for the bi-propellant thrusters.

The mission carried a payload of twelve specialized scientific instruments.

The complex route of *Cassini* to Saturn involved two gravity assist flybys of Venus on 26 April 1998 and 24 June 1999. It then took another gravity assist flyby from the Earth on 18 August 1999 also making a close approach to the Moon. It next performed a flyby of asteroid 2685 Masursky on 23 June 2000, also taking images which indicated a diameter of 15 to 20 km. It reached Jupiter in December 2000 taking 26 000 detailed high quality colour images during the flyby and also made several discoveries concerning storms in the atmosphere.

On 10 October 2003, the Cassini science team announced the excellent results of the occultation experiment using radio waves from the craft, which passed close by the Sun, to prove Einstein's General Theory of Relativity.

Cassini arrived at Saturn's moon Phoebe in June 2004 making close up studies during the flyby. It eventually, after a seven year voyage, arrived at Saturn in July 2004 passing through the gap between the F- and G-rings and going into orbit around the planet. Only a day after arrival, *Cassini* performed the first of 24 planned flybys of the moon Titan taking radar and optical images as well as topographical data. The Huygens Probe was released on 25 December 2004. It descended by parachute through the atmosphere of Titan, landed on solid ground and returned 350 images. The probe mission was controlled autonomously by the probe support equipment on the orbiter as it could not be controlled from Earth due to the long delay in communications. It typically takes about 80 minutes for signals to travel between Earth and Saturn depending on the orbital positions of the planets. During flybys of moon Enceladus, *Cassini* observed water-ice geysers suggesting that Enceladus is supplying ice particles to Saturn's E-ring.

On 16 April 2008, the Cassini mission was extended by two years as the Cassini Equinox Mission which will allow a considerable number of additional orbits of Saturn and flybys of moons Titan, Enceladus, Dione, Rhea and Helene. In February 2010 this was further extended as the Cassini Solstice Mission. This allowed another 155 orbits of Saturn, 54 flybys of Titan and 11 flybys of Enceladus. The gallery of close up colour photos of several of the moons is phenomenal. Cassini has so far sent back some 444 gigabytes of scientific data including more than 300 000 images.

The current end of mission plan is a fiery plunge into Saturn in 2017. The magnificent and hugely successful Cassini Mission is a masterpiece of state of the art space technology.

Uranus

This is the seventh planet from the Sun and is the first to be recognised as a planet by means of telescopic observation. Sir William Herschel announced its discovery in March 1781. The planet had been noticed many years before but was barely visible without a telescope and did not attract much attention. Flamsteed had noticed it several times but catalogued it as a star. Herschel at first thought that it might be a comet but it had no coma or tail; its nearly circular solar orbit confirmed that it was a planet. Herschel named the planet *Georgium Sidus* in honour of King George III. In recognition of his achievement the king gave Herschel an annual stipend of £200 on condition that he move to Windsor so that the Royal Family could look through his telescopes. This name was not popular outside Britain and soon the name Uranus proposed by astronomer Bode became widely accepted.

Uranus has an equatorial diameter of 51 118 km, an aphelion of 20,093 AU and a perihelion of 18,375 AU. The orbital period is 84,323 years, the synodic period 369.66 days and the orbital velocity 6,81 km/s. The most remarkable feature of the planet is its axis of rotation which is at only 6,48 degrees to the Sun's equator so that with a pole pointing toward the Sun its rings and planets are seen to move in circles when viewed from Earth. At equinox the rings and moons will appear to move in a vertical line. The planet is quite unlike the gas giants; it has a rocky inner core, an icy mantle in the middle and an outer envelope of hydrogen and methane. The cyan colour of the planet is due to the presence of methane in the atmosphere. This is the coldest planetary atmosphere in the Solar System with a minimum temperature of only 49 K. There are thirteen faint but distinct rings, the brightest being known as the epsilon ring. Uranus has 27 known moons, many of which have been named after characters in the works of William Shakespeare and Alexander Pope. The largest moon has a diameter of 1 578 km and has been named Titania. Other significant moons are Miranda, Ariel, Umbriel and Oberon.

The NASA Voyager 2 launched in 1977 performed a close flyby of Uranus in January 1986 making several discoveries and measurements before continuing on to Neptune.

Neptune

This was the first planet to be discovered mathematically. Perturbations of the orbit of Saturn led to the discovery of this planet to within one degree of the predicted position.

Neptune was noticed by Galileo on two occasions but he mistook it for a star. An early prediction of the planet was made by John Couch Adams of Cambridge in 1843. An accurate prediction of the planet's position was made in 1846 by Urbain Le Verrier of Paris who urged Johann Gottfried Galle of Berlin Observatory to search for the planet. Galle found the planet within one day of receiving Le Verrier's letter.

Neptune is not visible without a telescope. It is the eighth, and with the demotion of Pluto, the outermost planet of the Solar System. Neptune has an equatorial diameter of 49 528 km, an aphelion of 30,44 AU and a perihelion of 29,766 AU giving it a nearly circular orbit. The orbital period is 164,79 years, the synodic period 367,49 days and the orbital velocity 5,43 km/s. Neptune has the distinction of having the strongest winds of any object in the Solar System, which can be as high as 2 100 km/h. Neptune is similar to Uranus in that it is composed mainly of rock and ices. The atmosphere contains 80% hydrogen and 19% helium and traces of methane give the planet a blue colour. The planet has a faint ring system consisting of ice particles and with other material to give them a reddish hue. The rings are named Adams, Leverrier, Galle, Lassell and Arago. Neptune has 13 moons, the largest by far being Triton, discovered by Lassell only 17 days after the discovery of Neptune. Triton is one of the coldest objects in the Solar System with a temperature of only 38 K. Triton is expected to eventually spiral in towards the planet until it is torn apart as it approaches the Roche limit. Nereid was the second moon to be discovered and was found to have a highly elongated orbit. Neptune has a profound influence on objects in the Kuiper Belt setting up many orbital resonances. Neptune also possesses a number of Trojan objects occupying the L4 and L5 Lagrangian points of its orbit.

The only spacecraft to visit Neptune was Voyager 2 which performed a flyby in 1989. This mission gathered a wealth of information on Neptune, its moons and rings and provided 9000 images. The spacecraft relied largely on pre-loaded commands as signals from Earth required 246 minutes to traverse the distance.

Pluto

This is the dwarf planet of the Solar System discovered in 1930 by Clyde W. Tombaugh at Lowell Observatory. Pluto's existence was predicted by erroneous calculations based on the motions of Uranus and Neptune but fortunately these led to this dwarf planet's discovery anyway. It has a diameter of only 2 306 km and a highly elliptical orbit with a perihelion of 4436 and aphelion of 7375 million km. with an average of about 40 astronomical units. It has a very slow rotation with a day length of 153,28 hours. Pluto has a very thin atmosphere of nitrogen, methane and carbon monoxide. Pluto is in fact a TNO (Trans Neptunian Object) but has for many years been known as the ninth planet of the Solar System.

In February 1999 Pluto moved further from the Sun than Neptune and will remain so until April 2231. Two additional small moons orbiting Pluto have been detected by the Hubble in May 2005. These objects are estimated to be between 45 and 160 kilometres in diameter and orbiting 44000 kilometers from Pluto, about twice the distance of Charon and have been named Nix and Hydra. It was announced on 11 July 2012 that the Hubble had photographed two further moons of Pluto designated P4 and P5. It is remarkable that these objects could be detected at all due to their small size and great distance from the Sun. Pluto and its four satellites remain an astronomical puzzle as it seems improbable that an object as small as Pluto could have any moons. A theory proposed many years ago is that Pluto is in fact a discarded moon of Neptune.

On 19 January 2006, NASA launched its New Horizons mission on a Pluto and Kuiper Belt flyby. The spacecraft reached Jupiter in February 2007 for a gravity assist, taking images of the planet during the flyby, and is expected to arrive at Pluto in July 2015 where it will transmit images and scientific data back to Earth. It is expected to come within 10000 km of Pluto and 27 km of Charon. The craft will then continue to the Kuiper Belt where it will return data of one to three objects of more than 35 km in diameter. This phase is expected to last from 5 to 10 years.

<u>Trans Neptunian Objects (TNOs)</u>

These are a large variety of objects lying beyond the orbit of Neptune. These include the Kuiper Belt and Scattered Disk objects up to a distance of 70 AU from the Sun and the Oort Cloud extending as much as 50 000 AU. The first TNO to be discovered was the dwarf planet Pluto in 1930. Since 1992 some 1075 objects have been discovered of which only 132 have well defined orbits allowing easy observatory recovery. Some of largest and brightest TNOs are as follows:

Name	Size km	Average Orbit AU
Eris	2400	67,7

Pluto	2306	39,4
Charon	1205	39,4
2003 EL61	2000	43,31
Sedna	1800	486,0
2005 FY9	1600	45,66
Quaoar	1290	43,58
Orcus	1100	39,34
Ixion	980	47,30
Varuna	780	42,90

Sedna is a very recently discovered addition to the Solar System which is too small to be regarded as a planet. Sedna is smaller than Pluto and may well be the first member of the inner Oort Cloud to be discovered. The orbit of Sedna is extremely elongated and has a period of 12 059 years. This is by far the darkest and coldest part of the Solar System, Sedna has a temperature which never rises above 33 K – from here the Sun will appear as little more than a bright star in the sky. Sedna's highly elliptical orbit has an aphelion of 975,6 AU and a perihelion of 76,16 AU.

The outer Oort cloud is a hypothesized source of long period comet material which has never actually been observed. This is a spherical cloud surrounding the Solar System containing as many as 10^{12} comet nuclei greater than 1,3 km. These objects may be separated by tens of millions of kilometers and have a total mass of 40 times the Earth mass. This vast cloud may extend as much as a quarter of the distance to our nearest neighbour star Proxima Centauri.

Pulsar

Pulsars are neutron stars with a most remarkable property. They have jets of particles streaming out from their magnetic poles at nearly the speed of light. The jets produce very powerful beams of light or in some cases also X-rays. A pulsar emitting only gamma rays has recently been discovered by the Fermi spacecraft. As on Earth, the magnetic poles and rotational poles do not necessarily coincide, so that the pulsar spins about like a lighthouse. If the beam sweeps the Earth the light is seen as a regular pulse. Pulsars were first discovered in 1967 and some 700 have so far been found.

Quasar

This is an acronym for Quasi Stellar Radio Source. Quasars are found at the centres of very distant galaxies indicating that they were formed at the early stages of the universe. The quasar appears to be radio emitting activity around a supermassive black hole with thousands of millions of times the mass of the Sun. The extreme distance of the Quasars makes them convenient reference beacons of fixed points in space.

On 28 July 2012 it was announced that astronomers had observed the heart of a quasar with unprecedented sharpness – two million times sharper than human vision. Astronomers linked the APEX in Chile to the Submillimetre Array in Hawaii and the Submillimetre Telescope in Arizona by means of Very Long Baseline Interferometry and observing in the 1,3 mm wavelength. The baseline lengths were 9447 km from Chile to Hawaii, 7174 km from Chile to Arizona and 4627 km from Arizona to Hawaii. The angular resolution obtained was phenomenal 28 microarcseconds. The observations represent a new milestone towards imaging supermassive black holes by the envisioned Event Horizon Telescope array.

Star

The one star of which we are all intimately aware and totally dependant on is the Sun. Astronomically this is classified as a yellow dwarf though there is nothing small about this monster; there are in fact far many stars smaller than the Sun than larger. It has a diameter of 1,392 million km and a mass of $1,99 \times 10^{30}$ kg consisting of 70% hydrogen and 28% helium. The temperature at the core is 15 million degrees and 6000 at the surface. The source of a star's energy was unknown until Sir Arthur Eddington (1882–1944) proposed that the energy was due to matter annihilation. The nuclear fusion reaction at the core of the Sun is tremendous. In every second 512×10^9 kg of hydrogen is fused into 508×10^9 kg of helium and the remaining 4×10^9 kg is released as energy. This energy is emitted as gamma rays which are repeatedly absorbed and re-emitted until they emerge mostly as visible light and heat at the surface. The pressure at the core is a staggering 250×10^9 Earth atmospheres. Not only does the Sun continuously provide us with heat and sunlight, it is responsible for practically all our energy requirements. Even fuels such as wood, coal and oil provide indirect solar energy. Another vital benefit from the Sun is powering the Earth's weather systems. The Sun contains 99,8% of the mass of the solar system.

There appear to be limits to the sizes in which stars are able to form. A star can have as little as a tenth or as much as 150 times the mass of the Sun. Stars begin their lives in giant molecular clouds (GMC). Several spectacular photographs of GMCs have been taken, which are also known as star nurseries. A star nursery can be as small as 50 LY and as large as 300 LY across and can contain enough material to form up to ten million stars with a mass equal to that of the Sun. As with the Big Bang, there must be some unevenness in the cloud for clumping to take place but there are several ways in which this can happen, so it would be almost impossible for the cloud to be uniformly smooth. As the cloud begins to clump and fragment, large rotating superhot spheres known as protostars form. The protostars are not readily visible as they will usually be obscured by material of the GMC but are sometimes seen in silhouette. With stars, size means everything. At the bottom end of the range we have brown dwarfs which are too small and cool to start a nuclear fusion reaction and lead a life of obscurity. Larger protostars approaching the mass of the Sun will reach a core temperature of about 10 MK which is sufficient to start a hydrogen fusion reaction. The star will reach a stable state when the radiation pressure of the nuclear reaction balances the weight of the outer star material. New stars can have a mass as low as 8,5% to 20 times that of the Sun which will determine its temperature and colour which could be from cool red to hot blue. The Hertzsprung-Russel diagram provides a convenient way of classifying stars at the various stages of their evolution. Stable stars fusing hydrogen will occupy a position on the main sequence zone of the diagram where they will spend the major part of their life cycle. Our own Sun is a main sequence star at about the midpoint of its life cycle. When a star has consumed most of the hydrogen at its core and started fusing helium it is no longer considered to be a main sequence star. Depending on the mass of the star, the reduced hydrogen fusion will cause the star to compress by gravity until opposed by electron degeneracy pressure, or the temperature rises to 100 MK which will start a helium fusion reaction. Electron degeneracy pressure is the resistance against compressing protons into neutrons due to the Pauli exclusion principle. The fate of a small star which has stopped fusing hydrogen has not been observed as this can take many times longer than the present age of the universe. A star fusing helium will cause an accelerated fusion in the hydrogen containing layer around the core. The lower gravitational pull on this layer will cause it to expand and cool resulting in a red giant star such as Aldebaran in Taurus or Arcturus in Bootes. The life cycles of large stars are highly complex and are not described here.

The following forms of star are described under their own headings: Brown Dwarf, Cepheid, Magnetar, Neutron Star, Pulsar, Quasar, Supernova and Wolf Rayet.

Supernova

Supernovas form a crucial part of our existence. It is here that elements heavier than helium are formed which provide the materials from which planets and all forms of life are made. At the beginning of a star's life cycle the nuclear fusion reaction at its core will fuse hydrogen into helium releasing vast amounts of energy. Stars with a mass ten or more times that of the Sun will carry this process further, fusing successively heavier elements until iron is produced. Iron is of particular significance here. Elements lighter than iron can be fused under pressure releasing energy in the process. Elements heavier than iron can also be fused under pressure but will require an external input of energy. The fusing of the heavy elements is a considerably more complex process. Heavy radioactive elements can spontaneously undergo fission resulting in lighter elements and releasing vast amounts of energy. As the fusion fuel becomes exhausted the star will initially expand becoming a Red Giant and then rapidly collapse producing the heavy elements. It will then explode with unbelievable violence ejecting vast amounts of matter including the very rich assortment of chemical elements – this is the supernova. The supernova remnant material can expand into a shell many light years across. If the heavy iron core of the supernova has a mass of about 1,4 times the mass of the Sun it will have reached its Chandrasekhar limit and will contract under gravity to form a neutron star of immense density.

A star with forty or more times the mass of the Sun has an even more spectacular prospect – the hypernova! Unlike the supernova, the outer layers of the hypernova are not blasted away - the star continues to rapidly collapse until the energy of the star is released as a gamma ray burst. This is a spectacular event of unimaginable proportions. The burst is so violent that it will release many times the power of an entire galaxy. The total energy released is many times the energy radiated by a smaller star over its entire life cycle. The super-dense core of the hypernova will collapse itself into a black hole. Can a hypernova occur in the Milky Way? Yes! Hypernovas are detected every day in other galaxies as gamma ray bursts but if one occurred close to home we would never know about it. It is impossible to see it coming as the gamma rays arrive at the speed of light and the blast would cause an instant mass extinction of all life on one side of the planet. Life on the other half would quickly follow suit.

In 1942, J.J.L. Duyvendak, Nicholas Mayall and Jan Oort deduced that a 'guest star' documented by Chinese astronomers in 1054 was a supernova of which the remnant is now visible as the well known and beautiful Crab Nebula. Some 75 'guest stars' have been recorded by Chinese astronomers between 532 BC and 1054 AD, At its brightest, the July 1054 supernova appeared four times as bright as Venus and was visible in daylight for 23 days. It gradually faded but remained visible at night for more than a year.

The year 1054 was also significant for an event on the other side of the world. Malcolm Canmore with the help of Earl Siward led an army against King MacBeth of Scotland and defeated his forces at the battle of Dunsinane Hill. MacBeth escaped but three years later was again defeated and this time killed on the 15th of August 1057 at the battle of Lumphanan. King MacBeth and his Queen Grosch ruled wisely for 17 years and lived in the fortified castle at Dunsinane (Dunsinnan) North of Perth. King MacBeth's rule was so secure that he was even able to undertake a pilgrimage to Rome. Shakespeare's powerful dramatic masterpiece 'Macbeth' bears little resemblance to historical events of the time. MacBeth and his Queen were certainly not the fiendish tyrants that the Bard would have us believe.

On 14 November 1572, Danish nobleman Tycho Brahe (1546–1601) noticed a new star which appeared and remained visible for 18 months. Brahe made meticulous measurements of this star which became brighter and was visible in daylight for two weeks. Brahe made a considerable contribution to astronomical measurement. An unfortunate event occurred when Brahe challenged a fellow student to a duel in a dispute over who was the better mathematician. Brahe's nose was partly cut off and it is said that he wore a gold and silver replacement. The supernova is a type 1a which is a supernova triggered by material donated by a companion star. In this case the companion is known as Tycho B. The Brahe supernova is 10 000 light years distant, has a gas shell of 10 light years diameter and is expanding at 9 000 km/sec. The Brahe supernova remnant shell has been studied in detail with the aid of space and terrestrial telescopes.

On 9 October 1604, German astronomer Johannes Kepler observed and documented his famous supernova. Recent observations from space observatories have provided a wealth of information about the remnants of this object. Studies have been made by the Chandra (X-Ray), Hubble (Visible) and the Spitzer (Infrared) space observatories. The combined colour enhanced image is spectacular. The remnant cloud is about 14 light years wide, expanding at 6 million kilometres per hour and is about 13 000 light years from Earth. The wide spectrum observation has given much information about the mechanism and chemical materials of this spectacular event.

The following are some well known supernovae:

Year	Observation
185 AD	China
386	China
393	China
1006	China, Japan, Korea, Europe, Arab lands
1054	China, Japan, America
1181	China, Japan
1572	China, Japan, Europe, (Tycho Brahe supernova)
1604	China, Japan, Europe, Korea (Kepler supernova)

From this list it may appear that the Chinese astronomers were the only ones consistently on the lookout for supernovas over the past two millennia. A more likely explanation is that the Chinese astronomers were the only ones to adequately preserve their records. The Chinese even have a recipe for making 100 year old preserved eggs. The record of the 1054 observation in America survived as it was inscribed on stone and is still quite legible.

The Kepler supernova is the most recent to be seen in the Milky Way however many hundreds of supernovas have been observed in other galaxies in recent years. A potential supernova that is easily visible is Betelguese in constellation Orion. This star has become a red giant which is in the process of fusing heavy elements and has a diameter of about 300 times the diameter of the Sun. When the supernova explosion takes place if will be easily visible in daylight and will cast shadows at night.

Wolf Rayet Star

Wolf Rayet stars are very rare; only about one in ten million is a WR. These stars are massive with more than twenty times the mass of the Sun. They are also very hot with a temperature from 25 000 to 50 000 degrees. These stars live fast and die young – they have very strong stellar winds which carry off substantial quantities of mass. They carry on fusing after hydrogen and helium, to carbon, nitrogen, oxygen all the way up to iron until they end up as supernova explosions.

EYES ON THE UNIVERSE: THE TELESCOPES

On our quest of perception through the universe we now consider the most obvious means of observation – sight, but we will explore the entire electromagnetic spectrum of which visible light forms a very small part. The most obvious instrument to augment our powers of observation, the telescope, can not only surpass the light gathering area of our eyes by a factor of millions but can also gather light energy over extended periods of time and also hugely magnify the target area of observation. In addition to the visible spectrum, the modern terrestrial and space instruments can also operate in all wavelengths from radio waves to gamma rays. We present here an overview of astronomical instruments followed by a catalogue of well known telescopes and the lives and times of several of the prominent personalities involved. The electromagnetic spectrum is described in some detail in the appendix.

There has never been a better time than the present for astronomy. Since the instruments of the sixteenth century, telescopes have advanced in power and quality beyond recognition. The early instruments were all of the refractive lens type even though concave reflecting mirrors had been known since antiquity. Mirrors remained impractical due to a materials problem. They had to be made from metal (speculum metal) as there was no means of providing a satisfactory reflective coating on glass and the metal would require frequent re-polishing which would damage the optical surface.

Refractive telescopes have serious disadvantages for astronomy. There is a practical size limit to lenses as the lens can only be supported at the edge and this would result in distortion of a heavy lens due to its own weight as it is turned to different angles. The glass quality is extremely important and correction of chromatic aberration adds extra complexity to the optics. The perfection of reflective glass coating opened a whole new era for astronomy making the huge modern telescopes possible. Things were also good for amateur astronomers. It became quite an easy albeit tedious task to grind and polish one's own telescope mirror by using a glass blank to grind the glass objective. Mirror coatings are usually of aluminium, silver or gold. Aluminium is the coating of choice for telescopes operating in the visible wavelengths but does not have good reflectivity in the infrared. Gold is well suited for infrared wavelengths but falls off in reflectivity in blue and green wavelengths. Silver is useful for infrared telescopes as it can substantially reduce infrared radiation from the mirror. There are two types of mountings available for telescopes. In the past, astronomical telescopes were provided with equatorial mountings which allowed a simple means of keeping the stars stationary by a motorised compensation for the Earth's rotation. This type of mount remains in use for smaller telescopes. The other type of mounting is the altazimuth providing separate horizontal and vertical movements. Computer control has made this type of mount the obvious choice. Survey theodolites have altazimuth mounts in order to be able to precisely read the angles of the azimuth and altitude circles. Camera tripods also have a similar type of mounting. In the days of equatorial mount telescopes it was convenient to have dedicated instruments for altitude and azimuth measurements. Since the sixteenth century a large number of designs for sextants, quadrants and altazimuths have been produced, which all performed the function of a theodolite. A modern version of the altazimuth is the cine-theodolite which is used for the precise tracking of military missiles undergoing testing.

A new exciting technique available to astronomy is 'gravitational microlensing'. As early as 1936, Albert Einstein proposed the idea of the gravitational lens which is caused, according to the General Theory of Relativity, by light being slightly bent when passing a star, or even an entire galaxy. This, in effect, produces a lens of astronomical proportions. In a case of perfect alignment a distant star can appear as a ring of light around a nearer star. Gravitational lenses are a comparatively rare phenomenon and do not remain in place for very long due to the movement of the stars. In gravitational microlensing the distortion of the light from a source star can result in a magnified image of the star so that the oblateness of the star can actually be measured and it is also possible to detect orbiting planets. The magnification achievable can be as much as 1000. When describing gravitational microlensing Einstein commented: *'Of course, there is no hope of observing this phenomenon directly'.* In February 2008, a solar system with planets similar to Jupiter and Saturn was discovered. On 2 June 2008 the smallest yet extrasolar planet was discovered orbiting a brown dwarf or red dwarf and which had a size of about three times that of the Earth. This was discovered by the microlensing group at the Mount John Observatory in New Zealand. Despite the feeble light and heat from the star it has been suggested that the planet could well have liquid water on the surface if there was an adequate internal heat source.

One may well ask why it is necessary to make such huge telescopes. The magnification power of telescopes is determined by the focal lengths of their lenses, not by their size, but this is not the whole problem. The resolving power of a telescope is a measure of its ability to distinguish between two closely spaced points and a single point. The resolving power is determined by the wavelength of light and the diameter of the objective lens hence the need to use lenses as large as is economically and practically possible. There is a practical limit to the size of a one-piece glass mirror – about ten metres. For larger mirrors it becomes necessary to assemble the mirror from a large number of usually hexagonal segments but in this case diffraction from the edges of the segments must be taken into account. In the past, segmented mirrors were polished to a spherical surface so that all the segments would have the same curvature but with advances in computer aided manufacturing aspherical surfaces are now possible. A new exciting development in this area is to use telescopes working together by computer control using an interferometric technique thereby having effective lenses of gigantic proportions. This means that it is possible to have resolution in the milliarcsecond range or even less. A milliarcsecond is a spread of less than 5 mm at a range of 1 000 km. This astonishing feat allows nearby stars in the Milky Way to be seen as discs instead of point sources and can even provide direct observation of their larger planets. Huge advances in astronomical science and engineering have made the detection of planets orbiting other stars in our galaxy a practical reality. There are several ways in which this can be done:

1. *Doppler shift.* The slight wobble of a star caused by a large orbiting planet can be detected as a shift in the colour of the starlight as the star moves towards or away from the Earth. This method is the most successful so far.

2. *Astronomical measurement.* This method relies on extremely high resolution observation to detect the sideways movement of the star due to its orbiting planets. In this case the distance of the star is limited by the resolving power of the telescope.

3. *Transit method.* Here extremely sensitive photometric measurements are used to detect the slight dimming as a planet transits in front of a star.

4. *Gravitational microlensing.* This method, as already mentioned, can only be

used when the target star and the lensing star are correctly configured. The gravitational bending of light around the lensing star is used to provide a magnified view of the target star and allow detection of its planets.

5. *Pulsar timing.* Pulsars rotate at a very constant angular velocity so that radiation sweeping the earth will have a very precise pulsation. Anomalies in the pulsation are indicative of planets orbiting the pulsar. The first extrasolar planets detected were discovered in this way.

6. *Spectrographic measurement.* The transit of a star by a planet with an atmosphere could result in selective absorption of certain wavelengths slightly altering the spectrographic signature of the star.

7. *Direct imaging.* This can be achieved by a coronagraph method where the star view is obstructed to reduce the huge contrast between the starlight and the reflected light from the planet. This is a rather complicated method as the shape of the baffle and aperture must be carefully selected and rotated so that the planet can peep through the diffraction bands of the star. Nulling interferometry is also used to reduce the huge contrast between light from the star and that from the planet.

A great advance in direct imaging is expected from the proposed New Worlds Mission. This is a spacecraft with a large baffle or occulter, of several tens of metres diameter, which can block the starlight and allow a space telescope to observe the planets of the obscured star. The starshade would typically be positioned 80 000 km in front of the telescope. It may seem likely that the shade should be of circular shape but this is not the case. A circular shade would diffract the starlight which would constructively interfere along the optical axis making detection of the planets impossible. The envisioned shade which would minimize diffraction problems is in the form of a sunflower with pointed petals. Another proposal is to use a large shade with a central hole which could be aligned with planets orbiting the star. It would be useful for at least two shades to be used so that telescope time would not be wasted while waiting for the shade to be aligned.

Some comments by University of Colorado Professor Webster Cash of the centre for Astrophysics and Space Astronomy:

"In its most advanced form, the New Worlds Imager would be able to capture actual pictures of planets as far away as 100 light-years, showing oceans, continents, polar caps and cloud banks. If extra-terrestrial rainforests exist, they might be

distinguishable from deserts. To me, one of the most interesting challenges in space astronomy today is the detection of exo-solar planets. We have created an affordable concept with very practical technology that would allow us to conduct planet imaging in visible and other wavelengths of light."

For telescopes, computers have changed everything. The size of mirrors was to a certain extent limited by the mass of the glass. Distortion of a mirror due to its own weight as well as distortion of the mount could be a serious problem with a mirror polished to within nanometres. Large mirrors can now be made to a surprisingly small thickness or in segments and the whole system kept in perfect optical geometry by means of computer controlled active optics. Another huge computer benefit is the use of adaptive optics to control the twinkling of stars due to currents of varying density air. Adaptive optics is achieved by means of adjustable mirrors which compensate for optical disturbances in the atmosphere. To achieve this it is necessary to select a bright reference star in the vicinity of observation so that the system can compute the correction required. If there is no convenient reference star in the vicinity the solution is quite brutal – make one! A reference star can be made by beaming a laser kilometres high into the upper atmosphere where it can excite sodium atoms looking to all intents and purposes like a star. Sodium is conveniently released by meteors passing through the upper atmosphere. Adaptive optics is also a prerequisite when aiming for milliarcsecond image resolution – something which would be quite impossible to attain with a static optical system. This is not the final word on modern marvels – there is a newcomer – advanced adaptive optics which can apply compensation over the telescope's entire field of view. Advanced adaptive optics uses a large secondary flexible mirror with a thickness of only about 2 mm which hovers in a magnetic field and which can change shape in milliseconds to provide the ultimate in sharp images. The mirror is shaped by hundreds of electromagnetic actuators which can adjust the mirror to within nanometres. A computer software technique known as 'Lucky Imaging' has recently become available. This provides further enhancement of the output from adaptive optics by selecting images at about 20 frames per second and retaining 'lucky' images undistorted by twinkle and discarding images which appear to be distorted. This is particularly useful for visible light imaging as adaptive optics is most effective in infrared wavelengths but less so in visible light. The results have been spectacular giving resolution comparable to that obtained by space telescopes.

With computer control, precise tracking of the sky using an altazimuth mounting is no problem at all. Astronomical images are digitally captured by charge coupled devices (CCDs) kept at extremely low temperature in order to improve signal to noise ratio. A similar device is used in digital cameras. The image is then available for various types of software enhancement after which the astronomer can study the results at his own convenience on his own computer. There is no need for the astronomer to go to the observatories which in many cases are situated in remote inhospitable places or even in space. Many telescopes do still, of course, capture images photographically. Astronomical images are also captured in light wavelengths extending well beyond the visible spectrum.

There are three basic types of optical telescope available: Refractors, Reflectors and Catadioptrics. Refractors are telescopes comprising lenses without the use of focusing mirrors. Hand held telescopes, binoculars and optical survey instruments are of this type. There are also many large old refractor telescopes still in operation. Newton Reflectors, designed and made around 1670 use a concave paraboloid mirror which reflects the light back up the tube where it is reflected by a small mirror to an eyepiece or camera at the side of the tube. This design was pre-dated by the Gregorian telescope described by Scottish mathematician and astronomer James Gregory (1638–1675) in his 1663 publication *Optica Promota*. The Gregorian telescope uses a concave paraboloid mirror which reflects the light to a concave ellipsoid secondary mirror which in turn reflects the light back through a hole in the primary mirror to an eyepiece. The secondary mirror is positioned beyond the focus point of the primary mirror which results in a much shorter tube than would be required by a Newton telescope with the same focal length. The Gregorian is still in use for selected applications but has largely been superseded by the Cassegrain design. For the amateur astronomer the mirrors of a Newtonian telescope are far easier to manufacture and test than those of a Cassegrain. Using a technique developed by Leon Foucault in 1858, spherical concave mirrors can easily be tested for accuracy of curvature. Other mirrors require more specialised methods. Having a hole in the centre of the object mirror does cause a slight loss of receiving area but there is also an advantage in that it slightly increases the resolving power of the telescope. It should not be supposed that James Gregory was responsible for the Gregorian Calendar! This was introduced by Calabrian doctor Aloysius Lilius and decreed by Pope Gregory on 24 February 1582. The introduction of this calendar worldwide was an extremely complex process which cannot be dealt with here. An interesting calendar anomaly was that Wednesday 2 September 1752 was followed by Thursday 14 September 1752. This shedding of eleven days caused much confusion and inconvenience and many people felt that they had been cheated out of eleven days of their lives.

Catadioptric telescopes use a combination of mirrors and lenses. As these telescopes reflect the light from the back to the front and then back again they are mounted in short fat tubes giving them a chubby appearance. There are two common types – the Schmidt Cassegrain and the Maksutov Cassegrain. These telescopes have excellent qualities and are widely available with computer control and sky database for use by the affluent amateur. Cassegrain telescopes appear very similar to the Gregorian except for the positioning of the secondary mirror which is placed closer to the primary mirror than its focal point. The Cassegrain telescope was developed by Laurent Cassegrain in 1672. The 'classic' Cassegrain consists of a paraboloid concave primary mirror which reflects light to a smaller hyperboloid convex secondary mirror which in turn focuses the light through a hole in the primary mirror to a focal plane. The Schmidt Cassegrain has the secondary mirror ground into the centre of a plate closing the tube which corrects any spherical abberation caused by the primary mirror. The Maksutov Cassegrin has a spherical primary mirror which reflects the light to a spherical corrector mirror at the centre of a spherical lens closing the tube. The Ritchey-Chrétien Cassegrain has primary and secondary hyperboloid mirrors. This telescope is suited to wide field photographic observations as it has a flat focal plane free of coma and spherical aberration. This telescope is in widespread use.

In addition to placing the telescope's detection instruments at the cassegrain focus, additional bulky instruments can be placed on platforms at the nasmyth focus points. These are points along the horizontal axis of the telescope's vertical movement. A tertiary mirror is used to reflect light along the centres of the axis trunnions to the instruments. The nasmyth focus points are named after Scottish engineer James Nasmyth (1808–1890) who is best known for his invention of the steam hammer.

Sir Isaac Newton proposed that a paraboloid mirror could be produced by rotating a container of mercury. Several attempts have been made to produce a liquid mirror telescope since Newton's time but the idea has only caught on seriously since 1982. The key components of a liquid mirror are: a rigid mercury container with nearly the same curvature as the required mercury surface in order to keep the quantity of mercury to a minimum, a vibration free direct drive motor, an air bearing to support the mirror, and observatory foundations designed to eliminate seismic vibrations. The liquid mirror will rotate at about 10 r.p.m. The rotational speed must be kept constant to typically one part in a million. The liquid mirror telescope can, of course, only point vertically upwards but there are niche applications, such as asteroid astronomy, where a zenith telescope can be used to advantage. The main advantage of a liquid mirror telescope is that it can be manufactured at a very small fraction of the cost of a full function tracking telescope. With a liquid mirror special consideration must be given to the safe storage of mercury and adequate ventilation is required to deal with extremely toxic mercury vapour.

The Earth's atmosphere provides essential protection from harmful electro-magnetic radiation and cosmic rays. Gamma rays and X-rays are fully filtered by the atmosphere. There is however a most convenient window in the filtering which allows the narrow band of visible light and heat to pass through practically undiminished. This is quite an astonishing phenomenon, as light wavelengths immediately above and below the visible and heat are heavily filtered. There is a small band of near infrared which can be utilised for astronomy but the site for the telescopes must be carefully chosen. Infrared is absorbed by water vapour and carbon dioxide in the atmosphere so the telescopes must be positioned at high altitude in very dry climates. Another complication for infrared astronomy is the emission of infrared radiation by the atmosphere as well as the observatory building, telescope mounting structure and even the telescope mirror itself. A mirror coating of silver is useful in reducing infrared radiation by the mirror. There is a window seemingly made to order for radio astronomy in the shorter wavelengths. Longwave radio energy is completely absorbed by the atmosphere. There is much valuable information to be gathered in the radio, infrared, ultraviolet, X-ray and gamma ray bands. Space observatories are ideally suited for observation in these wavelengths. There are more astronomical bodies in the universe radiating in the infrared spectrum than in the visible.

The wavelengths received by radio telescopes are typically a million times longer than those of light observed by optical telescopes. In order to obtain resolution comparable to that of optical instruments, gigantic radio antennae would be required. High resolution radio astronomy is achieved by using interferometry techniques on widely spaced observatories or even networks of observatories spread over several continents. For even greater resolution the baseline can be made larger than the earth by having part of the network as a space observatory orbiting the earth. This is known as Very Long Baseline Interferometry (VLBI). Radio observatories with very long baselines are actually capable of resolution in the milliarcsecond range rivalling that of the large optical telescopes. For VLBI radio astronomy the interferometry technique is quite different to that of optical telescopes. The received signals cannot be combined in real time by a central synthesis facility but must be captured on computer storage and later combined using exceptionally accurate timing information obtained by the use of atomic clocks. Due to the rotation of the earth, the various receivers will not all remain at a static distance from the object of observation and this must also be taken into account. Another technique that can be used for huge interferometric baselines is Earth Rotation Aperture Synthesis. This utilizes the rotation of the Earth to obtain baselines of thousands of kilometres. VLBI can also be used for a down to earth matter. By observing quasars, which are so distant that they can be assumed to be fixed in space, the distances between the telescopes of the network can be computed to within centimetres. This permits calculation of the drift of the various tectonic plates of the surface of the earth and also the continuously changing tilt of the earth. Not all radio telescopes are steerable. Some huge antennae are permanently fixed and rely on the rotation of the earth to scan the heavens.

Images produced by radio astronomy can be dramatically different to those of optical telescopes. Objects appearing bright to optical telescopes can be invisible to radio telescopes which in turn can produce images of objects with no visible output of light at all. Vast expanses of interstellar and intergalactic hydrogen are visible to radio telescopes and completely transparent to optical astronomy. An obvious question to ask is: why do astronomical bodies produce radio waves? There are several ways in which this can happen.

a. Low temperature thermal radiation (heat) from solid bodies such as planets.

b. Pulsed radiation from rapidly rotating neutron stars.

c. *Bremsstrahlung* from interstellar gas. Bremsstrahlung (braking radiation) occurs when charged particles undergo deceleration while passing through a field of atomic nuclei.

d. Spectral line radiation from atomic and molecular transitions in interstellar gas.

e. Synchrotron radiation from relativistic electrons or other charged particles moving through magnetic fields.

f. Cosmic background radiation. This is the remnant radiation from the Big Bang which has cooled to only a few degrees above absolute zero. In recent years this radiation has come under intense study using orbiting satellites built for the purpose. This radiation is easily received on earth but measuring the extremely slight unevenness in the radiation requires precise measurement from space. Even ordinary TV and radio receivers are able to detect this radiation as noise.

Radio astronomy is a relatively new science becoming established only in 1932. As early as 1899 Nikola Tesla became the first radio astronomer when he received radio waves from the cosmos, but his results were regarded with disbelief by the scientific community and ignored. Tesla thought that parts of the signals were from extra-terrestrial civilisations and made many attempts over the following years to establish communications, becoming the first scientist to launch a SETI project. Radiation from space spans a very wide range of wavelengths. In order to make meaningful studies of this radiation, it is necessary to selectively study small ranges of wavelengths.

The computer age and the space age have brought undreamt of advances in astronomy but this is only the beginning. Even greater advances are envisaged for the second decade of the twenty first century using technologies yet to be developed. The latest advance in astronomy is in the new science of Gravitational Astronomy. This is based on the detection of relativistic gravitational waves radiated by immense objects in close binary orbit. The unique feature of this type of observation is that it does not involve electromagnetic radiation at all and that gravitational waves are not filtered by anything in their path providing information of the early universe which would not be obtainable by any other means. At present it is not known for certain at what velocity the waves are propagated. The velocity of electromagnetic waves is determined by the electric permittivity and magnetic permeability of space. There is no reason to suppose that gravitational waves are propagated at the same velocity but it is currently generally accepted that light and gravity are indeed propagated at the same velocity.

We give here brief descriptions of some of the most well known telescopes as well as the lives and times of prominent people involved. Please do not take umbrage if your favourite telescope has not been mentioned - there are simply too many to describe them all.

Galileo

In any discussion on telescopes, we can only start with the work of one of the founders of modern science, Galileo Galilei (1564–1642). He did pioneering work on objects falling, or rolling down inclined planes and correctly discovered the formula for gravitational acceleration as well as the parabolic path of projectiles. By observing the swing of a chandelier in Pisa Cathedral he determined that the period of pendulum swing is determined by the length of the pendulum and not by its mass or amplitude. This made the construction of longcase (grandfather) clocks possible. The period of a pendulum is actually slightly dependant on the swing amplitude but this effect can usually be disregarded. A pendulum bob will swing in a circular arc but for the period to be independent of amplitude the bob would have to move in a cycloidal path. This is known as the tautochrone problem. Galileo devised an experiment to measure the velocity of light by means of two men with lanterns. The one lantern would be uncovered and the moment the second man saw the light he would uncover the second lantern and the time taken for the light to traverse both distances would be measured. This had no hope of working as the lantern operators would have had to react within nanoseconds. A mechanised version of this method of measuring the velocity of light was used by Fizeau in 1849 by means of a rapidly rotating toothed wheel. The wheel would be run with increasing speed until the light passing through the gap between two teeth would return through the next gap. The velocity of light could then be calculated from the length of the light path and the time taken for the wheel to turn through the angle of one tooth. Galileo's great claim to fame and downfall came with the invention of the telescope by Flemish spectacle makers. Galileo immediately realised the usefulness of this for astronomical observations and soon began to grind his own lenses. Telescopes of the time were simply two plano convex lenses at either end of a tube. At best these telescopes had a magnifying power of 8 but Galileo managed to produce his own instrument with a magnifying power of 30 and which also had an eyepiece to give an erect image. The lens material was of poor quality glass with a greenish tint from iron content and there was no possibility of chromatic abberation correction. The objective had to be severely stopped down to improve image clarity. The field of view was very limited, not even wide enough to see the full face of the Moon.

(Modern astronomical telescopes have a much smaller field of view but these are in a quite different class of instrument). Compared to an ordinary pair of modern binoculars this was truly miserable optics but was sufficient to Galileo's purposes. Galileo noticed that the planets were illuminated in phases similar to the Moon confirming the Copernican theory that all the planets, including Earth, moved in orbits around the Sun. Galileo was also able to observe the four large moons of Jupiter. He was able to calculate the orbit periods of the Jovian moons, with much difficulty, as it was not possible to optically distinguish one moon from the other. His observation of Saturn led him to believe that this was three separate bodies as it was not possible to see the rings with any clarity. Galileo also noticed the vastly increased number of stars which became visible when observed through a telescope and that the Milky Way was in fact composed entirely of stars. Democritus of Abdera (470–400 BC) also noticed that the Milky Way was a band of stars.

Mikolaj Koppernigk (1473–1543) a Polish cleric, medical doctor, mathematician and astronomer preferred to be known by the Latin version of his name Nicolaus Copernicus. His interest in astronomy grew from the astrology and horoscope determination which was part of medical training of the time. He wrote a most astonishing book *De Revolutionibus Orbium Coelestium* on a theory of the universe with the Sun at the centre. This was published close to the end of his life so his ideas did not land him in trouble with the religious authorities. The book contained the following seven axioms:

1   *There is no one centre in the universe.*

2   *The Earth's centre is not the centre of the universe.*

3   *The centre of the universe is near the Sun.*

4   *The distance from the Earth to the Sun is imperceptible compared with the distance to the stars.*

5   *The rotation of the Earth accounts for the apparent daily rotation of the stars.*

6   *The apparent annual cycle of movement of the Sun is caused by the Earth revolving around it.*

7   *The apparent retrograde motion of the planets is caused by the motion of the Earth from which one observes.*

These ideas were not all new. Anaxagoras of Clazomenae (500–428 BC) held similar ideas explaining the illumination of the Moon and the causes of solar and lunar eclipses. He also claimed that the Sun was a red hot stone. These ideas displeased the authorities who sentenced him to death for his impiety. The sentence was later commuted to banishment from Athens. Poor Anaxagoras could never have guessed that the same ideas about the universe would result in the same intolerance and despicable persecution by authority two thousand years later. Anaxagoras was probably the first mathematician to attempt the problem of squaring the circle – which means geometrically constructing a square with the same area as a given circle. This has subsequently been proved impossible due to π being a transcendental number. It is a trivial arithmetic task to calculate the size of a square with the same area as a given circle – the geometrical approach must be done with a ruler and compasses. Aristarchus of Samos (c. 310–230 BC) also had ideas very similar to those of Copernicus. He even attempted to calculate the distances of the Sun and Moon as well as their diameters. His figures were inaccurate due to the difficulty of accurately measuring angles and estimating the exact time of half illumination of the Moon. Leonardo da Vinci also claimed that the Sun was at the centre of the solar system but he did not publish his work so that his astronomical ideas went unnoticed. Galileo's support of the Copernican view got him into very serious trouble. Imprudently, Galileo wrote in a way which could be construed to suggest that he was making game of the pontiff. He certainly would not have been so foolish as to do this intentionally. He was summoned to appear before the Inquisition in Rome. He pleaded for his hearing to be held in Florence due to ill health but this met with no sympathy. He was warned that if he did not appear he would be taken to Rome in chains. The elderly Galileo had to face his examination under threat of torture and had no choice but to acquiesce to whatever demands were made of him. Galileo spent the last years of his life afflicted with blindness, failing health and permanent house arrest. He died in 1642. In this same year the next giant to walk the stage was born – Isaac Newton. Galileo was exonerated of his heresy in 1992.

Possibly the most advanced early thinker on the cosmos, far ahead of his time, was Giordano Bruno (1548–1600), Dominican friar, philosopher, mathematician and astronomer. He not only endorsed the solar system of Copernicus, he also conceived an infinite universe where the stars were similar to the Sun with orbiting planets. He even declared that the universe being infinite, there was no particular point which could be regarded as the centre.

These ideas got him into very serious trouble.

Another remarkable astronomer who we should mention here is Jeremiah Horrocks (1617–1640), a young clergyman who was the first astronomer to observe the transit of Venus across the face of the Sun and also determined that the Moon was in an elliptic orbit around the Earth. He managed to observe the transit of Venus by passing sunlight through a pinhole in a card and projecting the light on a screen. This was a remarkable achievement requiring meticulous calculations as it is impossible to see Venus before and after the transit.

Royal Greenwich Observatory (RGO)

Royal Greenwich is an observatory with rich historical associations like no other. Greenwich is the home of Greenwich Mean Time and the Greenwich Meridian. In 1675 Charles II directed none other than Sir Christopher Wren to build the observatory for establishing longitudes and perfecting navigation and astronomy. Wren built the observatory on a budget of £500 and with recycled materials purloined from the Tower of London and Tilbury Fort. Wren was himself an astronomer as well as an architect. He designed Flamsteed House of red brick with stone dressing and with cupolas and balustrade for 'the habitation of the observator and a little for pompe'. The lofty octagon room was equipped with 'the choicest instrument'.

English polymath Robert Hooke (1635–1703) collaborated with Wren in building the RGO, the Monument to the Great Fire and St Paul's Cathedral and conceived the method of constructing the dome. Hooke was also responsible for several other major architectural works. Hooke achieved fame as Surveyor to the City of London and chief assistant to Christopher Wren in rebuilding London after the Great Fire. He proposed a redesigned London street pattern with wide boulevards, arteries and affording a splendid prospect of St. Pauls. This wonderful plan was thwarted by petty property disputes. Hooke was particularly fascinated by biology and we have him to thank for the biological term 'cell' because observation of plant cells reminded him of monks' cells. Hooke explained his law of celestial gravity with:

All objects are pulled towards the Sun with a force proportional to their mass and inversely proportional to the square of their distance to the Sun.

Hooke's wide range of interest and expertise caused him to become known as 'London's Leonardo'. Robert Hooke and Giovanni Domenico Cassini (1625–1712) are both credited with the discovery of the Great Red Spot on Jupiter (ca. 1665).

The RGO Meridian building was later added to house the growing collection of instruments – notably the Harrison clock. Harrison won the substantial prize in 1773 for constructing a pendulum-free seaworthy clock which could be used at sea for determining longitude. The observatory building was named Flamsteed House after the first Astronomer Royal appointed in 1675, The Rev. John Flamsteed. It was from Flamsteed that Newton obtained vital astronomical data, despite the rivalry and acrimony of this relationship. Flamsteed installed his equatorial quadrant in 1677 for astronomical measurements. Scottish mathematician and astronomer James Gregory (1638–1675), originator of the Gregorian Telescope, was commissioned by St. Andrew's University in 1673 to proceed to London to purchase 'such instruments and utensils as he shall judge most necessary and useful' to establish an observatory. Gregory went to Greenwich where he sought Flamsteed's advice on the matter. Gregory ordered an astronomical clock from Joseph Knibb, a well known London clockmaker at the time.

A prominent personality in Newton's time was Edmond Halley (1656–1742) who established that comets moved in elongated elliptical orbits around the Sun. Halley's father was a soap manufacturer at a time when the use of soap was coming into fashion. We have Halley to thank for Newton's *Principia*. Not only did Halley proof read Newton's work, he even published the *Principia* at his own expense. The *Principia* was translated from Latin into French by brilliant mathematician and scientist Gabrielle-Emilie Marquise du Châtelet (1706–1749). Emilie was on intimately friendly terms with philosopher Voltaire and others. Halley sided with Newton in his infamous dispute with Leibniz over the discovery of calculus. In his later years, Newton was appointed as Warden of the Royal Mint, where he launched a successful campaign against counterfeiters, even sending men to the gallows. The counterfeiters risked their lives for a mere pittance. In desperation, a counterfeiter would make his coins into a roll and hand it to a girl to hide, but even this intimate secrecy did not elude the inspectors. Newton also successfully organised the introduction of a new currency. Halley was to become Astronomer Royal after Flamsteed's death. Flamsteed's widow was so enraged by Halley's appointment that she removed and sold all of her husband's instruments so that Halley would not have the use of them. Halley had a brilliant astronomical career. In 1698 he was appointed by William III as a sea captain, master of a warship with the delicious name *Paramore Pink* which he could use to further his work on navigation and southern hemisphere charting. Having the son of a soap manufacturer as captain did nothing to ease the problem for the sailors of keeping their clothes clean. Soap is of no use at all in sea water and the supply of potable water was certainly not available for doing the laundry. So what did a good sailor use to wash his clothes? Urine! This was also seen in ancient Rome as a convenient and economical ammonia based laundry lye. In 1692 Halley proposed a theory which, considering his brilliant astronomical work, seems utterly incomprehensible. He proposed that the Earth was hollow, being a shell of some 800 km thickness. This shell contained two inner shells and a central core, the diameters of the shells and core being similar to the diameters of Venus, Mars and Mercury and with atmospheres separating the shells. Even a person with no astronomical or mathematical knowledge at all would have found this ludicrous.

George Biddell Airy (1801–1892) is said by many to be the greatest ever Astronomer Royal. He was a mathematician who, by the age of 27, held both the prestigious Lucasian and Plumian professorships at Cambridge University. The Plumian Chair in astronomy was founded by Thomas Plume in 1704. The first incumbent of the Plumian Chair was Roger Cotes, a former student of Isaac Newton. Sir Arthur Eddington held the Plumian professorship from 1913 to 1944. Airy's dedication to his position is summed up by:

> "I am not a mere Superintendent of current observations, but a Trustee for the honour of Greenwich Observatory generally and for its general utility in the world."

When parliament decided in 1844 that the new buildings for the Houses of Parliament should incorporate a tower and clock, Airy was appointed to draft a specification. One of his requirements was that:

> "... the first stroke of the hour should register the time, correct to within one second per day, and furthermore that it should telegraph its performance twice a day to Greenwich Observatory, where a record would be kept."

Airy was instrumental in the standardisation of time throughout the British Isles by means of the railway telegraph system. The clock tower also needed a bell. This was ordered from the Whitechapel Bell Foundry in 1858. The enormous bell was the largest ever cast at the foundry weighing 13 760 kg. The profile of the bell remains to this day on display in the foundry to the astonishment of visitors. Four smaller quarter chime bells, tuned to the familiar B, G, A and D, were also cast. The large bell, affectionately known as 'Big Ben', was manufactured at a cost of £2401, a huge sum at the time. It was a festive occasion in the streets of London when a team of sixteen brightly beribboned horses drew the bell on its trolley from the foundry to the clock tower using a South Bank route. The bells are mechanically struck by hammers and not swung as would be the case with change-ringing bells.

Airy's magnificent altazimuth instrument was installed in 1847. This was effectively a non-portable theodolite and used for accurately measuring lunar positions. Airy placed an order with Ransomes and May for the iron castings and an order with Troughton & Simms of London for the optics. The author has sentimental attachments to these manufacturers: the magnificent Victorian machines of iron, flywheels and hissing steam were popular items for model building by Meccano enthusiasts, and in some years doing land surveying work, I became attached to the superb instruments manufactured by Cooks, Troughton & Simms. In 1848 Airy designed his Transit Instrument which would in 1884, at an International Meridian Conference held in Washington DC, be voted by twenty two nations to be the zero meridian of the world. The transit instrument remained in use for 103 years and took 600 000 observations.

Airy's Great Equatorial Refractor was installed in 1859 for which funding was facilitated by public sentiment of the time. In 1845, both Adams of Cambridge and Urbain Le Verrier of Paris predicted mathematically the existence of the planet Neptune. The first optical detection of this planet was made by Johann Gottfried Galle of Berlin Observatory. The Victorian English public was outraged that the planet had been discovered by a 'damned foreigner' and not in England. Airy was able to use this sentiment to his advantage in obtaining approval for a new instrument of his own design. The telescope had a 12,8 inch objective lens manufactured by Mertz of Munich. (Those damned foreigners again!) The telescope was upgraded to a 28 inch object glass in 1891. In 1896 the 12,8 inch lens was used for a guide telescope on the 26 inch Thompson photographic refractor.

In February 1894, the quiet routine of Greenwich observatory was disturbed in a late afternoon when a Continental anarchist approached the observatory building, bomb in hand. The device detonated prematurely causing no damage other than to dispatch the anarchist to a realm which was hopefully more to his satisfaction. To this day it has never been established why anyone should choose to attack so unlikely a target. Greenwich town has for several centuries had rich associations with the British Monarchy and the Royal Navy. Henry VIII was born here in 1491 and not in Pembrokeshire as one might suppose considering his Welsh Tudor ancestry.

The world renowned astrophysicist Sir Arthur Eddington held a post at the

RGO from 1906 to 1913. The 70 cm refractor lens telescope on equatorial mount was installed in 1893 and remained in use until 1960 after which it was retired to the Greenwich Onion Dome. In 1958 the RGO moved to the Herstmonceux castle near the coast of Sussex which became the home of the Isaac Newton 2,5 metre reflector telescope. All observational activities of the RGO ceased in October 1998 ending a 323 year role as a pioneer of British astronomical research. A sad end to an illustrious history but such is the path of progress. Modern telescopes are engineering marvels which must be installed in the crisp clear air of remote mountaintops in far away places.

Lick Observatory – Mount Hamilton

It would take nearly 300 years since Galileo's telescope before the building of the first large telescope on a mountain top. This was the 91 cm refractor lens instrument on Mount Hamilton in northern California. The first telescope domes were built between 1876 and 1888 funded by a bequest from James Lick, an astonishing and eccentric millionaire personality of nineteenth century California. This was the world's largest telescope and housed in an observatory at an altitude of 1284 metres. The manufacture of the glass blanks, a formidable task at the time, was entrusted to Charles Feil & Sons in Paris. It would take eighteen attempts and five years to make the glass discs. The shaping and polishing of the lens surfaces was undertaken by Alvan Clark and Sons of Cambridge, Massachusetts – a task which took another year. A smaller 30 cm telescope was also made. The option of building a reflector telescope was also considered but was thought too innovative for the time. Later additions were made – the Shane 120 cm reflector in 1959 and the Nickel telescope replacing the original 30 cm refractor in 1980. Soon after the original telescope was installed the first ever interferometric measurements of astronomical objects were made by Albert Michelson (1852–1931). Using a rather small baseline of slits in a mask he successfully measured the sizes of the large moons of Jupiter.

Mount Wilson

The Mount Wilson Observatory has a most distinguished history of astronomical achievements. The Observatory is located at an altitude of 1742 metres in the San Gabriel mountains northeast of Los Angeles. The first instrument installed, in 1904, was the Solar Snow telescope for solar research. 1908 saw the completion of the

18 metre solar tower. The 152 cm Hale telescope saw first light in 1908. This was to be the world's largest for 9 years. The 190 mm thick, 860 kg glass blank was cast by Saint-Gobain in 1896 in France. The Hale is no longer used for astronomical research – it has been fitted with eyepieces and made available to visiting members of the public. The Mount Wilson Hale telescope must not be confused with the enormous Hale telescope on Mount Palomar. In 1912 the 46 metre solar tower was built. The huge 2,54 metre Hooker telescope, funded by John D. Hooker, saw first light in 1917 and which was now the world's largest and had a resolving power of 50 milliarcseconds. This instrument was used by Michelson in 1919 to measure star diameters by means of a large interferometer mounted on the telescope. In 1923 Edwin Hubble, using the Hooker, discovered the vastly increased size of the universe when he found that the Andromeda nebula was in fact a galaxy of stars similar to the Milky Way. He also discovered a cepheid variable star in the Andromeda galaxy which was to provide a means of measuring the vast distances between Earth and the galaxies. In 1929 Hubble made his other famous discovery. He deduced that the red shift of distant galaxies increased with distance proving the Lemaître theory that the universe was continuously expanding.

An astonishing development at Mount Wilson was the establishment of the CHARA facility (Centre for High Angular Resolution Astronomy). In 2001 the first starlight fringes were achieved on its 331 metre baseline. The complex consists of six 1-metre telescopes with the resolving capability of a mirror of 331 metres. Light from the telescopes is conveyed through evacuated tubes to the Central Beam Synthesis Facility where the six beams are combined. The paths of the beams are matched to an accuracy of less than a micron. Depending on the wavelength of light being observed, the array is capable of resolving details, believe it or not, as small as 500 microarcseconds.

Mount Palomar

The Hale telescope on Mount Palomar has ruled supreme as the largest telescope on Earth for some forty years in the second half of the twentieth century. The huge 5,08 metre 20 ton Pyrex glass blank was cast by Corning in 1934. The disc took 8 months to cool and after polishing weighed an impressive 14,5 tons. The completion of the observatory was delayed by WWII resulting in the telescope only becoming operational in 1948. This was well before the computer age so it was necessary to provide the telescope with an equatorial mount – a vast structure of 530 tons of steel. To accommodate this a 1 000 ton rotary dome was required.

Despite its age, the Hale telescope has remained at the forefront of modern astronomy. The telescope has operational adaptive optics and a sodium laser facility for producing a reference star kilometres high in the upper atmosphere. The adaptive optics have been further enhanced by the introduction of the 'Lucky Camera' which uses computer software to select 'lucky' images and discard those apparently distorted by twinkling. The resolution results achieved compare favourably with those of space telescopes. The lucky camera is more formally known as PHARO (Palomar High Angular Resolution Observer). A spectacular image of the IW Tau binary stars has been produced. These stars have a separation of only 300 milliarcseconds.

An interesting, if not scary, project being run at Palomar on the 1,2 metre Samuel Oschin telescope is the Near-Earth Asteroid Tracking program (NEAT), which monitors the orbits of errant and possibly dangerous asteroids.

The primary telescope is named after George Ellery Hale and the observatory is owned and operated by Caltech.

Keck – Mauna Kea, Hawaii

Everything about the twin Keck telescopes can only be told in superlatives – modern marvels of optics and computer science. The telescopes are named in honour of William M. Keck. Gifts from the W.M. Keck foundation made the building of the telescopes possible. Keck 1 began observations in 1995. The design of the objective mirrors is awesome. The mirrors are each in 36 hexagonal segments having effective diameters of 10 metres. The segments have a thickness of only 75 mm spaced with a gap of 3 mm and made from Zerodur low-expansion glass-ceramic. The segments are adjusted twice per second under computer control to keep them in perfect optical adjustment as a single hyperboloid to within 4 nanometres. An ordinary sheet of copier paper has a thickness of 100 000 nanometres.

The Keck telescopes are provided with a magnificent suite of detection instruments:

DEIMOS   The Deep Extragalactic Imaging Multi-Object Spectrograph is capable of gathering spectra from 130 galaxies or more in a single exposure.

HIRES   The High Resolution Echelle Spectrometer breaks up incoming starlight into its component colours to measure the precise intensity of each of thousands of colour channels. This instrument has detected more extrasolar planets than any other in the world. The radial velocity precision is up to one metre per second.

LRIS   The Low Resolution Imaging Spectrograph is a faint-light instrument capable of taking spectra and images of the most distant known objects in the universe.

NIRC   The Near Infrared Camera for the Keck I telescope is so sensitive it could detect the equivalent of a single candle flame on the Moon.

NIRC-2   The second generation Near Infrared Camera works with the Keck Adaptive Optics system to produce the highest-resolution ground-based images and spectroscopy in the 1–5 micron range.

NIRSPEC   The Near Infrared Spectrometer studies very high redshift radio galaxies, the motions and types of stars located near the Galactic Centre, the nature of brown dwarfs, the nuclear regions of dusty starburst galaxies, active galactic nuclei, interstellar chemistry, stellar physics, and Solar System science.

OSIRIS   The OH-Suppressing Infrared Imaging Spectrograph is a near-infrared spectrograph for use with the Keck II adaptive optics system. OSIRIS takes spectra in a small field of view to provide a series of images at different wavelengths. The instrument allows astronomers to ignore wavelengths where the Earth's atmosphere shines brightly due to emission from OH (hydroxyl) molecules, thus allowing the detection of objects 10 times fainter than previously available.

The telescopes have, of course, altazimuth mounts. An equatorial mount for a telescope of this size would be unthinkable and in any case of no advantage with computer tracking available. The two telescopes can be used together using a modern interferometric method to give the resolving power of an objective of 85 metres! This will give the telescopes the ability to work in the milliarcsecond range making the measurement of the sizes of nearby stars and the detection of their planets possible.

Studies of the Gliese 581 system over the past 11 years have revealed that this red dwarf star, 20 light years away, has seven orbiting planets. The planets were detected by means of the HIRES spectrometer which could detect perturbations of the star by its orbiting planets. On 29 September 2010 the latest planet discovery announced, known as Gliese 581g, was a rocky planet with a diameter of about 1,5 times that of the Earth and was in the 'habitable' zone – where the planet would be able to sustain liquid water on its surface – one of the key requirements for life to exist. This planet orbits at a distance of only 22,5 million km and has a year length of only 37 earth days. The planet is tidally locked to its star so that the one side will be permanently extremely hot and the other very cold. The zone between the hot and cold sides is where life could be expected to exist.

The telescopes are installed at an altitude of 4 200 metres on Hawaii's dormant volcano Mauna Kea. The telescopes are owned and operated by Caltech and the University of California.

Subaru – Mauna Kea, Hawaii

This superlative addition to the cluster of Mauna Kea telescopes saw first light in 1999. Subaru is the Japanese name for the Pleiades star group. An eyepiece was fitted to the telescope for the inauguration ceremony so that Princess Sayako could look through the instrument. The telescope is owned and operated by the National Astronomical Observatory of Japan (NAOJ) and is at an altitude of 4 139 m. The primary mirror has a diameter of 8,2 m, a mass of 22,8 tons and is made of ultra-low thermal expansion glass. The mirror is polished to a mean surface error of only 14 nm. The mirror has a focal length of 15 m, the optics are Ritchey Chrétien and a resolution of 200 milliarcseconds is achievable. The adaptive optics have been upgraded to work with a sodium laser facility greatly enhancing image quality. The telescope has been provided with platforms at the nasmyth points so that bulky instruments can be conveniently used there. In all, the telescope has four positions for fitting detection instruments: in addition to the Cassegrain and Nasmyth focus points, instruments can also be placed at the prime focus instead of using the secondary mirror. The telescope has a selection of seven sophisticated instruments operating in the visible and infrared wavelengths. Several cameras and spectrographs can be mounted.

At Cassegrain focus:

Multi-Object Infrared Camera and Spectrograph (MOIRCS).

Cooled Mid Infrared Camera and Spectrometer (COMICS) with the ability to study cool interstellar dust.

Faint Object Camera And Spectrograph (FOCAS) with the ability to take spectra of up to 100 objects simultaneously.

At Nasmyth focus:

Infrared Camera and Spectrograph (IRCS). Used in conjunction with the new 188-element adaptive optics unit (AO188).

High Dispersion visible light Spectrograph (HDS). High-Contrast Coronographic Imager for Adaptive Optics (HiCIAO).

Infrared camera for hunting planets around other stars. Used with AO188.

At Prime focus:

Subaru Prime Focus Camera (Suprime-Cam) 80-megapixel wide-field visible-light camera.

Fibre Multi Object Spectrograph (FMOS)

Infrared spectrograph using movable fiber optics to take spectra of up to 400 objects simultaneously.

The list of achievements and discoveries made by this telescope is spectacular indeed. On 3 December 2009 Subaru was able to directly image a very large planet orbiting a star.

Provision is made for groups of visitors to view the observatory. This is not for the faint hearted or people with any kind of respiratory complaint. The observatory is reached by a three hour drive in a four wheel drive vehicle and a halfway stop is provided for visitors to acclimatize to the increased altitude. At the summit the atmospheric pressure and oxygen availability is reduced to 60%. Visitors are reminded that the observatory is a scientific instrument and is not designed for human habitation.

Gemini

The Gemini twin telescope observatory is the result of an international seven country partnership of the US, UK, Canada, Australia, Argentina, Brazil and Chile. The observatory boasts two identical 8,1 metre telescopes which have silver coatings instead of aluminium for optimisation of infrared astronomy. The mirrors have a thickness of 20 cm and a mass of 22 tons. The secondary mirrors have diameters of one metre. Nine metre wide coating chambers manufactured by the Royal Greenwich Observatory are used to re-coat the mirrors using a magnetron sputtering process. A significant benefit of the silver coating is that it can reduce infrared radiation from the mirror itself thus reducing unwanted signal noise. The silver is applied in four layers of about 100 nanometres thickness. The mirrors have a surface area of 50 square metres with a reflectivity of 98,75% The total quantity of silver used on a mirror is about 50 grams – enough metal to make a large athletics medal. The active optics of the mirrors have 120 actuators for maintaining perfect optical geometry of the surfaces. The telescopes are also equipped with adaptive optics mirrors for correction of atmospheric disturbances. The Gemini North telescope is mounted atop the Mauna Kea volcano in Hawaii at an altitude of 4214 metres and the Gemini South is on Cerro Pachon in Chile at an altitude of 2737 metres. The geographical positioning of these telescopes gives them full coverage of both northern and southern hemispheres.

JCMT – James Clerk Maxwell Telescope

This unusual telescope situated at an altitude of 4092 metres on Mauna Kea, Hawaii has a 15 metre microwave antenna and observes in wavelengths from 0,3 to 2 mm providing a view of the universe in the submillimetre range of wavelengths. The antenna consists of 276 individually adjustable panels with a surface accuracy of less than 50 microns. The telescope is mounted in a rotating carousel which has a transparent membrane stretched across the opening to protect the instruments and mounting from the elements. The telescope was funded by a partnership between the UK, Canada and Netherlands and is operated from the Joint Astronomy Centre in Hilo, Hawaii. The wavelengths observed by this telescope lie between those of infra red light and microwaves and receiving instruments designed for this purpose are employed – broadband continuum bolometers and spectral-line receivers. The Submillimetre Common Users Bolometer Array (SCUBA) provides unparalleled sensitivity with extensive wavelength range and field-of-view. The telescope is able to view an area of space as large as the face of the moon. The SCUBA is able to detect thermal emission of interstellar dust and can provide information of star and planetary system formation and the evolutionary history of galaxies. On 6th July

2005 the JCMT was able to measure the effects of NASA's deep impact mission to comet Tempel.

SPT – South Pole Telescope

The geographical south pole presents an ideal site for a telescope observing in the microwave to submillimetre wavelengths – a high altitude of 2800 m and an extremely dry atmosphere. The SPT is located at the Amundsen-Scott South Pole Station.

The SPT has a 10 metre offset Gregorian dish antenna which saw 'first light' on 16 February 2007 with a 960 element bolometer camera with an array of superconducting transition edge sensors. The primary antenna is accurate to within 25 microns and has a large field of view of 1 square degree. The secondary antenna is cooled to 10 K. A new camera with greater sensitivity and the capability to measure polarization was installed early in 2012. The first major survey was to find distant, massive, clusters of galaxies through their interaction with the cosmic microwave background, with the goal of constraining the Dark Energy equation of state. This was completed in October 2011. The SPT collaboration is made up of over a dozen (mostly North American) institutions. The SPO is funded by the National Science Foundation. An interesting aspect of a telescope at the geographical poles is that there is no difference between altazimuth and equatorial mounts.

European Southern Observatory (ESO) Cerro Paranal, Chile.

The ESO is owned by a consortium of ten member states: Belgium, Demnark, France, Germany, Italy, the Netherlands, Sweden, Switzerland, Portugal and the UK. The ESO headquarters are in Garching, Munich.

The observatory complex atop Cerro Paranal at an altitude of 2 632 metres is mind boggling. This desolate site 120 km from Antofagasto is one of the driest spots on Earth making it perfect for terrestrial astronomy. The giant 8,2 metre mirrors of the four main telescopes were cast by Schott Glaswerke in Mainz, Germany and subsequently polished near Paris. Shipping the twenty ton mirrors across the Atlantic and then hauling them to the top of Cerro Paranal was a major engineering mission in itself. The mirrors have a thickness of only 18 cm requiring permanent support to prevent them from fracturing under their own weight. After installation the mirrors had computer controlled active optics systems using 150 actuators to keep them in perfect optical geometry. The telescopes are connected by light tunnels to a central facility for interferometric combination of images. The telescopes are provided with adaptive optics deformable secondary mirrors which react 100 times per second to compensate for atmospheric conditions. The telescopes are also provided with laser facilities to create a reference 'star' in the upper atmosphere for use by the adaptive optics systems. The telescopes have an interferometric base line of 130 metres giving an optimum resolution of 500 microarcseconds! The telescope complex is further enhanced by three mobile 1,8 metre telescopes mounted on railway tracks which will provide additional interferometric base lines. The four very large telescopes are named Antu, Kueyen, Melipal and Yepun.

Two additional telescopes are under construction funded by a consortium of 18 UK universities. The VISTA (Visible and Infrared Survey Telescope for Astronomy) has a 4 metre mirror for highly detailed sky surveys. An additional 2,5 metre telescope will work in the visible spectrum. The High Accuracy Radial velocity Planet Searcher (HARPS) is a high precision echelle spectrograph installed in 2002 on ESO's 3,6 m telescope at La Silla. HARPS can attain a precision of 0,9 metres per second making it only one of two instruments worldwide with such accuracy. The instrument is contained within a vacuum vessel and the temperature controlled to within 0,01 K. This instrument has been used to discover 16 planetary objects in the southern hemisphere, including four multi-planet systems, as of May 2009. In October 2009, ESO announced the discovery of 32 additional extrasolar planets. In September 2011, it was announced that 50 planets were discovered including Super-Earths. On 2 April 2012, Mikko Tuomi of Hertfordshire University announced that HARPS had revealed that star HD 10180 had nine orbiting planets – the most ever detected orbiting a star.

In October 2012 HARPS detected an earth size planet orbiting Alpha Centauri B only 4,3 light years away. The Alpha Centauri system consists of binary stars Alpha Centauri A and B and further away Proxima Centauri which is the star closest to the solar system. The planet has a very fast orbit of only 3,2 days. This was a remarkable achievement as the planet causes the star to move towards and away from Earth at only 51 centimetres per second.

ESPRESSO, the Echelle Spectrograph for Rocky Exoplanet and Stable Spectroscopic Observations, is to be installed at ESO. This is hoped to achieve radial velocity precision of 0,35 km/hour or less which is comparable to the radial velocity induced by the Earth on the Sun. Operation is expected by late 2016 or early 2017.

On 13 December 2008, ESO astronomers using the SINFONI instrument on the VLT and adaptive optics, announced the results of scrutinizing the inner parts around the disc of a supermassive black hole 10 billion light years away. The observations were greatly improved by double gravitational lensing. The object of study was the famous cosmic mirage, the 'Einstein Cross' which appears as five light sources in the shape of a cross. The gravitational lensing was provided by a foreground galaxy a tenth of the distance of the quasar, and which caused it to appear as four objects. The cross was observed three times a month for three years monitoring all the brightness and colour changes of the five images. The macrolensing of the galaxy and microlensing of the stars provided a wealth of information.

The first ESO telescope complex in Chile was inaugurated in 1969 atop the 2 400 metre La Silla. In all 15 telescopes were installed over the years of which six are still in operation.

The Atacama Large Millimetre Array (ALMA) project will consist of up to 64 antennae of 12 metres in diameter at a site at an altitude of 5 000 m in Llano de Chajnantor, Chile. This is one of the largest ground based astronomy projects proposed for the next decade. ALMA is an international partnership between the US, Canada, Europe, East Asia and the Republic of Chile and is the most expensive ground-based telescope currently under construction. A contract has been awarded to General Dynamics for the construction of 25 antennae of 12 m diameter for delivery in 2007. The final configuration could be for 64 antennae and an additional compact array of 7 m and 12 m antennae. The antennae will be movable allowing an array of from 150 m to 14 km wide. ALMA began scientific observations in the second half of 2011 and is expected to be fully operational by the end of 2012. On 28 August 2012 a team of astronomers using the ALMA reported sugar related molecules, in this case glycolaldehyde, surrounding a Sun-like star, giving further evidence of the building blocks of life, in this case RNA, ready to be included in orbiting planets. Astronomers using ALMA have discovered an unexpected spiral structure as well as an outer spherical shell in the material of the star R Sculptoris. This is the first time that such a structure has been found around a red-giant star.

Transporting the 115 ton antennas from the Operations Support Facility at 2900 m altitude to the site at 5000 m presented huge problems – haulage was achieved by using two 28 wheel self-loading 130 ton heavy haulers made by Scheuerle Fahrzeugfabrik in Germany, and powered by twin 500 kW diesel engines.

The receivers will be designed for operation in frequencies from 30 GHz to 950 GHz. The ALMA Correlator, a specialised supercomputer that combines the information received from the antennae, will perform $1,6 \times 10^{16}$ operations per second (16 petaflops). This speed is humanly incomprehensible – let us put it another way: this is equivalent to 53,3 computer instructions being initiated each time a beam of light travels one millimetre!

Hubble and the NASA Great Observatories

The origins of the Hubble Space Telescope (HST) go back as far as 1946 when astronomer Lyman Spitzer proposed an extraterrestrial telescope and devoted much of his career pushing for a space telescope to be developed. The Hubble space telescope must surely be one of the greatest engineering achievements of the twentieth century. A huge 2,4 metre 11 110 kg telescope orbiting the Earth in space at an altitude of 599 km and moving at a speed of 28 000 km/hour circling the Earth every 97 minutes. The advantages of a telescope in space are considerable. The most obvious advantage is the absence of the atmosphere – no bad weather conditions or cloud cover, no filtering of starlight, no twinkling requiring adaptive optics, no dust and no illumination from lights. The other huge advantage is having perfect dark sky visibility 24 hours a day. The only obstructions to view are the presence of the Sun, Earth and moon which are hardly a problem. These benefits do of course come at a price – the huge cost of building the telescope and getting it into orbit. Sending up the repairman is also a major mission. The Hubble has produced many of the most astonishing photographs of deep space that have ever been seen. The amount of image data transferred to Earth is immense. The telescope can operate in light wavelengths of 115 nm in the ultraviolet to 2 500 nm in the infrared. The Hubble has made several hundred thousand observations of more that 25 000 astronomical targets and provided data for several thousand scientific papers.

The primary 2,4 metre mirror has a mass of 828 kg and the secondary 30 cm mirror 12,3 kg. The optics are Ritchey Chrétien which is the obvious choice for a Cassegrain instrument of this size. The mirror surfaces are polished to within 30 nanometres and are coated with a 75 nanometrer layer of aluminium and an even thinner layer of magnesium fluoride to improve reflectivity in ultraviolet wavelengths. The telescope is able to lock on to target by means of gyros to an accuracy of 7 milliarcseconds. This is a spread of 35 mm at a range of 1 000 km. The huge solar panels are able to supply the observatory with 5 680 W of power.

In December 1995 the Wide Field and Planetary Camera 2 was used to image the Hubble Deep Field. This was an area of only 2,5 arcminutes across chosen in an area containing very few Milky Way foreground stars, so that nearly all 3000 objects in the image were galaxies. Three years later a similar image was taken named Hubble Deep Field South. The Hubble Ultra Deep Field was imaged in September 2003. This field had an area of one tenth of the apparent size of the Moon and contained some 10 000 galaxies. This was an image looking back 10^9 years into the early universe. This field represents only one thirteen millionth of the entire sky. The number of galaxies in the universe is truly mind boggling.

In 2012 a new full colour image of the Ultra Deep Field was released which has been named the Extreme Deep Field. This is a composite image made from ten years of observations showing the development of galaxies from 13,2 billion years ago to the most recent. This represents exposure of 2 000 seconds by the two premier cameras – the Advanced Camera for Surveys and the Wide Field Camera 3. This is the deepest image of the sky ever obtained and reveals the most distant galaxies ever seen.

As part of the second maintenance mission in 1997, The NICMOS (Near Infrared Camera and Multi-Object Spectrometer) was installed. The NICMOS was originally cooled to a temperature of 77 K by means of a 104 kg block of nitrogen ice. This was later replaced by 'cryocooler' machine after the nitrogen ice had become exhausted. The NICMOS provided the Hubble with a spectacular additional near infrared capability with its multi-object spectrometer and three cameras.

In October 2006, a NASA news release announced the Hubble discovery of 16 extrasolar planets orbiting a variety of stars in the central region of the Milky Way in the course of a survey called 'Saggitarius Window Eclipsing Extrasolar Planet Search' (SWEEPS). Some 180 000 stars were examined 26000 light years away. The survey detected the planets by measuring the slight dimming of the stars by transits of the planets. This method is only effective with large Jupiter sized planets. An extrapolation of the results suggests that there should be about six thousand million Jupiter sized planets in the Milky Way. An interesting discovery was the existence of Ultra-Short-Period planets. These huge planets orbit very close to the parent star with an orbit period of less than a day. These planets orbit within the Roche limit but are prevented from disintegrating by their immense gravity.

On 18 July 2012 it was announced that the Hubble had detected the earliest spiral galaxy ever seen, 10,7 billion light years away. This was a rare phenomenon as early galaxies are not usually well formed like those less distant.

The Hubble, launched in 1990, continues to produce the most astonishing views of space but there is however a cloud on the horizon. The fourth scheduled maintenance mission was cancelled giving the telescope an uncertain future. This decision was primarily influenced by the high risk and huge expense of sending up a space shuttle. The telescope has much in-built redundancy so that it can accommodate failure of various components but a major failure in the gyros, batteries or instrumentation could render it inoperable and failure to correct orbit deviation could be catastrophic. To the jubilation of all concerned the Hubble received a reprieve when the SM5 maintenance mission was announced in October 2006. This final maintenance mission, launched on 11 May 2009, was carried aloft by the space shuttle Atlantis. Huge provision was made for the safety of the astronauts including a standby shuttle which could be used for a rescue mission in case of emergency. The upgrade included the installation of the Wide Field Camera 3 (WFC3) and the Cosmic Origins Spectrograph (COS) which greatly enhances the telescope's capabilities. The six ageing gyros and six batteries were replaced giving the Hubble a life expectancy until 2013 when it will be succeeded by a much larger and more sophisticated instrument. The Hubble is a joint project of NASA and the European Space Agency and is the first telescope of NASA's Great Observatories program.

The Hubble is named in honour of Edwin Powell Hubble (1889–1953) who is best known for his famous astronomical discoveries – that there are vast numbers of galaxies in addition to the Milky Way and also his optical confirmation of an expanding universe explaining the red shift of distant galaxies. Hubble was also proficient in athletics, fly-fishing and amateur boxing. He spent three years in England where he obtained his MA degree at Oxford majoring in Spanish. While in England he formed a life long identification with English mannerisms and dress. Back in the USA he would sometimes irritate his colleagues with his orotund accent and also his Inverness cloak and cape.

The successor to the Hubble is scheduled for launch in 2018. This is the James Webb Space Telescope. (Formerly known as the NGST – Next Generation Space Telescope). This is a monster version of Hubble with a 6,5 metre primary mirror and an orbit 1,5 million km high with a transit orbit time of 3 months. The huge mirror and much larger sunshield are too large to be housed in a launch vehicle so these will be folded and only deployed when the observatory is in space. This observatory will be optimised for observations in 0,5 micron to 20 micron visible and infrared wavelengths.

A recent poll has nominated ten images as the most spectacular ever taken of the cosmos by the HST. These are truly magnificent and are well worth seeking out on the Internet.

1. Sombrero Galaxy M104 (NGC 4594). This galaxy is 50 000 light years across at a distance of 28 million light years away and is estimated to contain 800 billion stars. The galaxy has a supermassive black hole at its centre with a mass of one billion suns. This galaxy is a strong source of synchrotron radiation in X-ray and radio wavelengths. Contemplating this galaxy is an almost mystical experience – it could well contain millions of planets inhabited by living creatures.

2. Ant Nebula Mz3. This is a young bipolar planetary nebula at 3 000 to 6 000 light years away in our own galaxy.

3. Eskimo Nebula NGC 2392. Also known as the Clown Face Nebula it looks somewhat like a face surrounded by a furry hood. This nebula was first discovered by William Herschel in 1787.

4. Cat's Eye Nebula NGC 6543. A planetary nebula 3 300 light years away in constellation Draco discovered by William Herschel in 1786. There is a spectacular composite image taken by the HST and the Chandra X-ray telescope.

5. Hour Glass Nebula MyCn 18. A young planetary nebula in constellation Musca.

6. Cone Nebula. A nebula 2,6 light years in length discovered by William Herschel in 1785.

7. A portion of the Swan Nebula (also named Omega Nebula). Messier 17 (NGC 6618) 5 500 light years away.

8. Starry Night (reminiscent of van Gogh's painting). A halo of light around a Milky Way star.

9. Swirling Cores of two merging galaxies (NGC 2007 and IC 2163) in the Canis Major constellation.

10. The Trifid Nebula – a stellar nursery 9 000 light years away where new stars are being formed.

The second of the Great Observatories was the Compton Gamma-Ray Observatory named in honour of Dr Arthur Holly Compton (1892–1962). At 17 tons this was the heaviest astrophysical payload ever flown at the time of launch in 1991. The four instruments on board could detect six decades of the em-spectrum from 30 keV to 30 GeV. The observatory detected some 2600 gamma ray bursts while in orbit. Astronomers have written 2000 papers from the vast amount of data gathered by the observatory. Due to a malfunctioning gyro, it was decided in June 2000, for safety reasons, to de-orbit the observatory and allow it to plunge into the Pacific Ocean rather than risk an uncontrolled descent.

The third item of the Great Observatories program is the Chandra X-Ray observatory which was launched into orbit in July 1999. Images in the X-ray wavelengths can be spectacularly different to those in visible light. The Chandra has already produced a vast wealth of astronomical data and has been particularly successful in observations of black holes. It has even been discovered that some black holes have a spin which is detectable by their effect on orbiting particles and distortion of space-time. In April 2006, the Chandra observed the eclipse of the supermassive black hole at the centre of galaxy NGC 1365 by a dense cloud of gas. This allowed the measurement of the size of the disc of material at the event horizon of the black hole. A spectacular NASA image, posted in September 2012, showed an accumulation of 270 hours of observation by the Chandra – the central regions of the Perseus galaxy cluster revealing the turmoil that has wracked the cluster for hundreds of millions of years. One of the most massive objects in the universe, the cluster contains thousands of galaxies immersed in a vast cloud of multimillion degree gas with the mass equivalent of trillions of suns. Using the Chandra, astronomers have also determined that the Milky Way galaxy, including the nearby Magellanic Cloud galaxies, is embedded in an enormous cloud of hot gas which extends for hundreds of thousands of light years.

A telescope working in the X-ray spectrum presents some unique problems. An ordinary Newton reflector is of no use at all as the X-rays would simply pass straight through. X-rays can be reflected by a mirror but only at a very oblique angle. The focusing mirror used is of a most unusual design. It consists of four pairs of slightly curved tubes, polished on the inside, and nested one inside the other with a combined mass of a ton of glass. Due to the extremely short wavelength the mirrors had to be polished to within less than a nanometre – the most precisely polished optical surfaces ever produced. The orbit of the observatory is also unusual in that it takes 2,6 days and is highly elliptical going a third of the way to the moon at aphelion in order to avoid the van Allen radiation belt. This observatory was named in honour of Subrahmanyan Chandrasekhar (1910–1995). There is an additional astronomical connection here – Chandra is the Sanskrit word for 'moon'.

The fourth of the Great Observatories is the Spitzer infrared space telescope launched in August 2003. This instrument has an 85 cm beryllium mirror and operates in the 3 to 180 micron infrared range. The results achieved are spectacular. Infrared astronomy is at a serious disadvantage for terrestrial telescopes due to filtering by the Earth's atmosphere. The telescope's CCD is cooled to an astonishing cryogenic temperature of only 5,5 K in order to obtain maximum signal to noise ratio. The observatory has been given an Earth trailing heliocentric orbit causing it to follow the earth in its orbit around the Sun. The telescope has been named in honour of Dr Lyman Spitzer Jr. (1914–1997).

SWIFT – Gamma Ray Burst Mission

The SWIFT mission is specifically designed for the study of Gamma Ray Bursts. This mission is the result of a collaboration between NASA, Italy and Britain. It was launched on 20 November 2004 and is controlled by the Goddard Space Flight Centre. This is an exciting addition to the growing number of space telescopes and is part of NASAs Mid-Sized Exploration program (MIDEX). SWIFT is not an acronym – it was named swift because it can swiftly react to gamma ray bursts and take appropriate action.

Gamma ray bursts were first observed in the 1960s in the course of taking military observations and it was thought that the strength of the bursts ruled out any possibility of them coming from beyond our own galaxy. It was subsequently found that the bursts were coming on a daily basis from all parts of the sky – millions of light years away from other galaxies. Energy bursts on this scale seemed unbelievable. These are the most powerful explosions in the universe. Stars with a mass of 10 times that of the Sun can produce supernovas at the end of their life cycle – stars with a mass of 40 Sun masses can produce hypernovas with an energy output of 100 times that of a supernova. The hypernova burst radiates more energy than the Sun in its entire life cycle. If a hypernova were to occur 300 light years away in our own galaxy it would appear a million times as bright as the Sun.

The main problem with observing a burst is a matter of timing. The gamma burst could last for a few milliseconds to a few minutes and the visible afterglow might only last a few hours. Speed is of the essence. The Swift mission is made up of three instruments:

- BAT – the Burst Alert Telescope. This is a very wide angle gamma ray detector with the function of detecting the bursts and their location in the sky. Within ten seconds the burst detector will transmit the information to ground based agencies for further observational action and also slew the observatory so that the narrow angle other instruments on board can make observations of the afterglow within 90 seconds of the burst observation.

- XRT – the X-Ray Telescope will then make detailed observations of the afterglow in the X-Ray spectrum.

- UVOT – this is the Ultra Violet and Optical Telescope which will also make detailed observations of the afterglow.

With these three instruments the em-spectrum from gamma rays down to visible light will be recorded. It would be possible to actually see a gamma burst if one happened to be looking in the right direction at the right time. The reflection of a gamma burst from the surface of the moon has actually been observed. The SWIFT mission is expected to provide comprehensive observational information of about 200 bursts in a two year period. On 9 May 2005, the Swift observed a burst which resulted from the collision of two neutron stars which then collapsed into a black hole. The X-ray telescope managed to see the afterglow but the visible light was too faint for observation. An alert from SWIFT in April 2009 was followed up by astronomers using two telescopes in Hawaii as well as a study of the afterglow using a telescope in Chile. The burst was from a star some 13 billion light years away which makes it one of the most distant astronomical object ever observed.

On 16 September 2012 Swift's Burst Alert Telescope was twice triggered and again the next day. This was caused by a rare X-ray nova in the Sagittarius constellation possibly near the galactic centre. This was caused by a flood of gas from a sun-like star plunging into a previously unknown stellar-mass black hole.

AGILE – Astro-rivelatore Gamma a Immagini L'Eggero

The Agile mission developed by the Italian Space Agency was launched in April 2007 on a three year mission by a Polar Satellite Launch Vehicle (PSLV) from the Satish Dhwawan Space Centre at Sriharikota on the Indian eastern coast. This is a small scientific mission dedicated to gamma ray astrophysics and will observe in the 30 MeV to 30 GeV gamma ray bands and in the 15 keV to 60 keV hard X-ray bands. Agile carries a gamma ray imager, an X-ray detector and a mini- calorimeter to hunt for transient events such as gamma ray bursts. The mission will investigate:

–   Active galactic nuclei, gamma ray bursts, pulsars, gamma ray unidentified sources, supernova remnants, compact objects and binary systems, gamma ray diffuse emission, TeV sources and fundamental physics. The science package will provide data continuity to build on discoveries made by NASA's Compton Gamma Ray Observatory. AGILE also helped astronomers lay the foundation for Fermi (formerly GLAST), a follow-on NASA mission.

Fermi Gamma-ray Space Telescope
Formerly: GLAST – Gamma ray Large Area Space Telescope

This major mission, built by General Dynamics, was launched by NASA on 11 June 2008. A high energy gamma ray observatory imaging in the 10 MeV to 100 GeV energy bands. For this mission NASA has teamed up with the US department of energy and a large number of institutions in France, Germany, Japan, Italy and Sweden. The entire project has been financed on a budget of $690 million.

The GLAST Large Area Telescope (LAT) has a huge field of view of more than 2,5 steradians. Its limit for source detection in an all sky survey is $1,6 \times 10^{-9}$ photons $cm^{-2} s^{-1}$ at energies over 100 MeV. The observatory is in a circular orbit 560 km high circling the Earth every 95 minutes and surveys the entire sky every three hours. Gamma ray bursts will be detected by the GBM (GLAST Burst Monitor). Fermi will explore the most extreme environments in the Universe and explain how black holes accelerate immense jets of material at nearly the speed of light. It will search for signs of new laws of physics and an explanation of Dark Matter. Fermi will help crack the mysteries of stupendously powerful explosions known as Gamma Ray Bursts and also answer long standing questions across a broad range of topics including solar flares, pulsars, the origins of cosmic rays, blazars and active galaxies and also studies of the early Universe. Within 40 days of being in orbit, to the delight of all concerned, Fermi detected 12 powerful gamma ray bursts. Fermi has also discovered a gamma-ray only pulsar which pulses three times per second and radiates 1 000 times the power of the Sun.

SOHO – SOlar and Heliospheric Observatory

The Solar and Heliospheric Observatory is a project of international co-operation between ESA and NASA launched on 2 December 1995 with a planned mission life of two years. The mission has since been extended to December 2012. The Observatory carries a payload of twelve specialised scientific instruments. This observatory is stationed 1,5 million km sunward of the Earth in a halo lissajous orbit at the L1 Lagrangian point of the Earth's orbit giving it permanent visibility of the Sun without ever being eclipsed by the Earth. A Lagrangian halo orbit is a small orbit at right angles to the plane of the main orbit. The halo orbit was necessary, as placing the craft in line with the Sun would have made communications difficult. The Lagrangian points of an orbit are not very stable so that objects placed there can easily be disturbed out of position. This happened on 25 June 1998 when the SOHO went missing. It was found again, using RADAR of the Arecibo Big Dish, on 4 August and subsequently rehabilitated. Needless to say, this telescope has produced a vast wealth of data on the Sun which would be unobtainable from terrestrial solar telescopes.

SOHO's ability to blank the Sun's glare allowed it to discover a large number of comets. Michal Kusiak of Polish Uniwersyet Jagielłñiski discovered SOHO's 1999th and 2000th comets on 26 Dec 2010.

SDO – Solar Dynamics Observatory

The SDO was launched by NASA on 11 February 2010 as part of its Living With a Star (LWS) program, with a planned life of 5 to 10 years.

SDO's goal is to understand the Sun's influence on Earth and near-Earth space by studying the solar atmosphere on small scales of space and time and in many wavelengths simultaneously.

A remarkable solar model was displayed over Federation Square, Melbourne, from 4 June to 4 July 2010. This was a huge helium filled balloon scaled to one 100 millionth the size of the Sun (about 14 metres). The balloon was illuminated and animated by five projectors using images animated by mathematical equations which never repeat. The project used the latest SOHO and SDO imaging overlaid by animations derived from Navier-Stokes, reaction diffusion, perlin and fractal flame equations. Spectators could disturb the animations in real-time using an iPhone, iPod touch or iPad.

Hinode (Sunrise) formerly Solar-B

This JAXA/NASA/PPARC mission was launched on 22 September 2006 from the Uchinoura Space Centre, Japan. Originally planned as a three year mission, after more than four years it remains fully operational. HINODE took high resolution photos of the 2012 Venus transit of the Sun.

The payload contains three specialized instruments to investigate the interaction between the Sun's magnetic field and its corona.

The SOT – Solar Optical Telescope is a 0,5m Gregorian optical telescope with a field of view of 400 arcseconds across and a resolution of 0,2 arcsecond. The focal plane package contains a Broadband Filter Imager, a Narrowband Filter Imager and the Spectropolarimeter which produces sensitive vector magnetograph maps of the photosphere.

The XRT – X-ray Telescope uses grazing incidence optics to image the corona's hottest components, 500 000K to 10^7K. The imaging field is 34 arcminutes across capable of imaging a full view of the Sun.

The EIS – Extreme-Ultraviolet Imaging Spectrometer obtains spatially resolved spectra in two wavelength bands: 17,0 – 21,2nm and 24,6 – 29,2nm. The temperature equivalent of these bands is from 50 000K to 20×10^6K. The EIS is used to identify the physical processes involved in heating the solar corona.

STEREO – Solar TErrestrial RElations Observatory

The Solar Terrestrial Relations Observatory was launched on 25 October 2006 as part of NASA's Solar Terrestrial Probes program. This two year NASA mission consists of two space solar observatories, the one leading the Earth in its orbit and the other trailing. In December 2007 the observatories were leading and trailing the Earth by approx. 0,4 AU. This configuration gives an excellent stereoscopic view of solar storms and the ejection of material from the Sun. The stereoscopic view is essential for following the progress of a CME (Coronal Mass Ejection) on its path from the Sun towards the Earth. Spectacular images of solar activity have been obtained. A sensational image of comet Encke was obtained when it passed through a solar storm while within the orbit of Mercury. The CME first deflected the tail of the comet and then detached it completely leaving the comet with the indignity of flying without a tail. The interaction between the CME and the comet's plasma tail were the result of opposite magnetic fields which later became connected with a consequent release of energy. There is some similarity here to the northern lights seen on Earth. The spacecraft will gradually spread further apart at about 44 degrees per year. In late 2009 they passed through the L4 and L5 Lagrangian points of the Earth's orbit. After passing behind the Sun they will not be recaptured into Earth orbit.

The largest solar superstorm ever recorded occurred in August/September 1859. This is known as the Carrington Event named after the discoverer and recorder Richard Carrington. The CME was so severe that aurorae were seen around the world and even over the Caribbean. Telegraph systems all over Europe and North America were disrupted, in some cases shocking the telegraph operators. The auroral light was greater than that of a full moon.

The spacecraft are each provided with four instruments:

1   SECCHI (Sun Earth Connection Coronal and Heliospheric Investigation) comprising: EUVI – Extreme UltraViolet Imager, COR1 – Inner CORonagraph, – Outer CORonagraph and HI – Heliospheric Imager.

2   SWAVES – Stereo-WAVES: an interplanetary radio burst tracker tracing the generation and evolution of travelling radio disturbances to the Earth.

3   IMPACT – In-situ Measurement of Particles And cmE Transients.

4   PLASTIC – PLAsma and SupraThermal Ion Composition. This instrument will provide plasma characteristics of protons, alpha particles and heavy ions.

The STEREO mission will provide a wealth of information on solar bursts and their effect on terrestrial communications.

Solar Probe Plus

The Solar Probe Plus NASA mission is due to be launched in 2018. This will use multiple gravity assists at Venus to incrementally decrease the orbital perihelion to achieve multiple passes at 6 million km from the Sun. A reinforced carbon composite shadow shield will be used to protect the craft from the incident solar intensity which will be 520 times that on Earth. When the probe is in orbit around the Sun it will achieve the highest velocity ever achieved by a man made object – 200 km/second. The scientific goals are:

- Determine the structure and dynamics of the magnetic fields at the source of the solar wind.

- Trace the flow of energy that heats the corona and accelerates the solar wind.

- Determine what mechanisms accelerate and transport energetic particles.

- Explore dusty plasma near the Sun and its influence on solar wind and energetic particle formation.

ULYSSES

The Ulysses space mission is a joint venture by NASA and ESA launched from space shuttle Discovery in October 1990 and used a solid fuel booster to accelerate it in an elliptical Hohmann transfer orbit to Jupiter. The unique feature of this mission is that it would take measurements and observations of the Sun's poles by taking an out-of-ecliptic orbit. The Sun's poles are nearly at right angles to the ecliptic and therefore not visible from Earth or spacecraft moving within the ecliptic. To launch a spacecraft directly into an out-of-ecliptic orbit would require a prohibitive amount of fuel; the orbit was achieved by means of a flyby past the south pole of Jupiter in February 1992 and then over the north pole of Jupiter and on to a swing over the north pole of the Sun. The result was an orbit at nearly right angles to the ecliptic with a perihelion just beyond the Earth's orbit and an aphelion just beyond Jupiter. The spacecraft is equipped with instruments to characterize fields, particles and dust. Due to the lengthy periods at a considerable distance from the Sun, the spacecraft could not be powered by solar panels and was therefore provided with a radioisotope thermoelectric generator (RTG) using plutonium-238 as a radioactive heat source.

In 1994–1995, Ulysses explored both the northern and southern solar poles and on 1 May 1996 it crossed the ion tail of comet Hyakutake and found the tail to be at least 3,8 AU in length. In 2000–2001, Ulysses explored the southern solar polar regions and found that the southern magnetic pole was widely dispersed and not clearly located as the northern magnetic pole. In the spacecraft's aphelion of 2003–2004 it made further observations of Jupiter. In 2004 the craft flew through the ion tailings of comet McNaught-Hartley. The Ulysses mission has been extended to at least march 2009 so that it could operate over the Sun's poles for a third time in 2007–2008. The spacecraft is equipped with a sophisticated selection of scientific instruments with a total mass of 55 kg.

On 10 June 2009 a new milestone was reached when the Ulysses Mission came to an end after becoming the longest running ESA spacecraft. The orbital path is now taking it further away from the Earth. On 9 June 2009 the Ulysses Mission team received the NASA Group Achievement Award.

DARWIN

This search for extraterrestrial life was planned by ESA (European Space Agency) for launch in 2015. This is not the usual type of SETI project which relies on the extraterrestrials being smart enough to send us radio broadcasts – the DARWIN would concentrate on 'biomarkers' which are subtle natural indicators of the presence of life which will include all forms of life. Oxygen will be a prime target. This is produced and used by life forms and will be a strong indicator. Without life, free oxygen will react with other chemicals, and eventually dwindle away. Other biomarkers are water and carbon dioxide. The oxygen will not be detected directly but by the presence of ozone which is another form of oxygen. The DARWIN would search several thousand star-planet systems and then concentrate on likely candidates for life. The DARWIN mission would consist of four free flying spacecraft stationed at the L2 Lagrangian point of the earth's orbit, three of which will be 3 metre infrared telescopes. The telescopes will direct their signals to the fourth craft which will combine the images to give the resolution of a telescope hundreds of metres in diameter. Needless to say, the telescopes will need to fly in extremely precise formation. Infrared light transmitted by a star is millions of times brighter than from its planets. A technique known as nulling interferometry will reduce the starlight and improve the signal to noise ratio of the light from the planets. The precise formation of the spacecraft would require precision to within nanometers – such precision between objects is hardly attainable using piezoelectric tripods in the laboratory. The craft could be disturbed out of position by the extremely slight pressures from solar radiation and the craft's own heat radiation. Work on the DARWIN mission has been discontinued.

PLANCK and HERSCHEL

Cosmic background radiation study and early universe evolution studies. Launched by ESA (European Space Agency) on 14 May 2009. The Planck mission is the third medium sized mission of ESA's Horizon 2000 Scientific Program (formerly COBRAS/ SAMBA) and is part of its Cosmic Visions Program. It will measure unevenness in the Cosmic Background Radiation with unprecedented sensitivity and resolution as a successor to the COBE mission. It will also test theories of the early universe and cosmic structure. The telescope has an aperture of 1,5 m diameter which feeds information to the LFI (Low Frequency Instrument) and the HFI (High Frequency Instrument). The LFI is an array of 22 tuned radio receivers which image the sky between 30 GHz and 70 GHz. HFI is an array of 52 bolometric detectors which image the sky in six frequencies between 100 GHz and 857 GHz. The detectors are cooled to an almost unbelievable cryogenic temperature of 0.1 K making them the coldest objects in the Universe. It has been mentioned that the measurements are like measuring the heat of a rabbit on the Moon from the Earth. The cryocoolers operate in three stages of 20 K, 4 K and 0.1 K. On 26 April 2010 the LFI completed 100% coverage of the sky. The HFI will also complete this coverage by 28 May 2010. The Planck mission was launched together with the Herschel satellite (formerly FIRST). The two were separated after launch and placed in different orbits at the L2 Lagrangian point of the earth's orbit. This places them 1,5 million km further from the Sun than the Earth. The Herschel satellite has a huge 3,5 metre mirror and collects data in the far infrared to submillimetre wavelengths from the coldest objects in the universe. The objectives are to study the formation of galaxies in the early universe, the evolution of stars, planetary systems, the molecular chemistry of the universe and the chemical composition of the surfaces of comets and planets. The Herschel was placed in a lissajous orbit at the L2 Lagrangian point. The mission is designed for a three year service life by which time the helium required for supercooling the instruments will be depleted. The three instruments on board are:

1   PACS – Infrared Photo conductor Array Camera and Spectrometer.

2   SPIRE – Spectral and Photometric Imaging Receiver.

3   HIFI – Heterodyne Instrument for the Far Infrared – a high resolution spectrometer which will observe spectra in thousands of frequencies simultaneously.

The Herschel achieved its first light observations using all three of its instruments by 10 July 2009. On 2 October 2009 the Herschel produced spectacular images of the Milky Way galaxy near the galactic plane using the SPIRE/PACS parallel mode of operation. At the time of launch the Herschel had the largest single mirror ever made for a space telescope. The Herschel has provided the most spectacular images of star formation and a vast wealth of scientific information.

In 2012 ESA's Herschel discovered that previously unseen galaxies are responsible for a cosmic fog of infrared radiation. Using two of ESA'a very large telescopes in Chile, and observing in two different wavelengths, astronomers were able to estimate the proportion of very distant galaxies which had hitherto remained unseen. The following is a revised estimate of the galaxies and stars in the visible universe:

Superclusters   10   million

Galaxy groups   25   billion

Large galaxies   350 billion

Dwarf galaxies   7   trillion

Stars   30   billion trillion (3×10^{22})

The Herschel has also provided astronomers with data, which indicated for the first time, cold water vapour enveloping a dusty disk around a young star. This suggests that the disk, which is poised to develop into a solar system, contains vast quantities of water. The quantity of water is estimated at thousands of times of that of the Earth's oceans. Water covered planets such as the Earth may be quite commonplace. Michiel Hogerheijde of Leiden University is the lead author of a paper describing these findings in the Oct. 21 issue of the journal Science.

SIM – Space Interferometric Mission (also known as SIM PlanetQuest)

This highly innovative and challenging mission was being developed by NASA and contractor Northrop Grumman. One of the main goals was to detect Earth sized extrasolar planets orbiting stars and also to construct a map of the Milky Way galaxy. Initial contracts for SIM were awarded in 1998. Work required scientists and engineers to move through eight specific new technology milestones and by Nov 2006 all eight had been completed. SIM was delayed several times by budget cuts and finally cancelled in 2010.

SAFIR – Single Aperture Far-InfraRed observatory

This observarory is planned for launch by NASA between 2015 and 2020 and is envisioned as a follow-on to the Spitzer Space telescope and the Herschel Space

Observatory. SAFIR will provide unprecedented sensitivity in the range between the infrared wavelengths probed by the James Webb space telescope and the microwave wavelengths observed by terrestrial radio telescopes. The telescope will have a 10 metre deployable antenna and operate in wavelengths from 20 microns to 1 mm. The orbit will either be at the L2 Lagrangian point or a 3 AU solar orbit.

FUSE – Far Ultraviolet Spectroscopic Explorer

This NASA Origins Program mission was launched on 24 June 1999, to explore the universe using high resolution spectroscopy in the far-ultraviolet spectrum from 90 to 120 nanometres which is unobservable by most other telescopes. This mission was developed and operated for NASA by the John Hopkins University. The development was in collaboration with the space agencies of Canada and France. FUSE is designed to investigate: conditions in the first few minutes after the Big Bang, dispersion of chemical elements throughout galaxies and the effect of this on the evolution of galaxies and also the properties of interstellar gas clouds from which stars and solar systems form. FUSE uses four mirror segments to reflect the light to focus. Two are coated with silicon carbide for good reflectivity in the shorter wavelengths and the other two with aluminium and lithium fluoride for reflectivity of the longer wavelengths. Each mirror reflects the light to one of four curved diffraction gratings to disperse the light into a spectrum. Fuse reached the end of its mission in January 2008.

GAIA – Global Astrometric Interferometer for Astrophysics

This is an ambitious ESA Cornerstone mission to create an extremely precise three-dimensional map of stars throughout our Milky Way galaxy and beyond, and map their motions which encode the origin and subsequent evolution of the Milky Way. Despite its name, GAIA will not use interferometry. GAIA will determine the positions, distances and proper motions of one billion stars (about 1% of the galaxy population) with an accuracy of 20 microarcseconds at 15 magnitude and 200 microarcseconds at 20 magnitude.   GAIA will also perform radial velocity measurements with an expected detection of tens of thousands of extrasolar planetary systems.   GAIA will have the capacity to discover apohele asteroids (with orbits between Earth and the Sun). It is difficult for Earth based instruments to detect these as most of the region is only accessible in daylight hours.   It is also hoped that GAIA will detect up to 500

000 quasars and accurately test Einsteins General Theory of Relativity.

The mission is scheduled for launch in 2013 from Kourou, French Guiana aboard a Soyuz Fregat rocket, to be placed in a lissajous orbit at the L2 Lagrangian point of the Earth's orbit for a planned mission life of five years.

GAIA is the successor to the Hipparcos mission.

The GAIA payload will consist of three instruments.

1. The astrometry instrument (ASTRO), which is dedicated to measuring the angular position of the stars of magnitude 5,7 to 20.

2. The photometric instrument, which allows the acquisition of spectra of stars over the 320-1000 nm spectral band.

3. The high resolution spectrometer to measure the radial velocity of the stars by acquiring high resolution spectra in the spectral band 847-874 nm (field lines for calcium ion) for objects up to magnitude 17.

APO - Apache Point Observatory

The Apache Point Observatory established at Sunspot, New Mexico in 1985 hosts a suite of four telescopes:

1. The 3,5 m Astrophysical Research Consortium (ARC) telescope.

2. The 2,5 m Sloan Digital Sky Survey (SDSS) telescope.

3. The 1,0 m New Mexico State University telescope.

4. The 0,5 m ARCSAT – small aperture telescope.

The ARC 3,5 m telescope is a Ritchey-Chrétien reflector with instruments attached at several focal points.

The ARC echelle spectrometer uses a 2048 x 2048 pixel CCD.

The Double Imaging Spectrometer is a low resolution optical spectrometer.

The Near Infrared Camera/Fabry Pérot Spectrometer is liquid nitrogen cooled and uses a 1024 x 1024 HgCdTe detector.

The Seaver prototype imaging camera is an optical imaging instrument with a 2048 x 2048 pixel CCD.

The TripleSpec is a near infrared spectrograph which provides continuous wavelength coverage over the 0,94-2,46 micron range.

Observations using the 3,5 m telescope can be carried out remotely by observers using the telescope users interface through the Internet.

The 2,5 m SDSS telescope is a Ritchey-Chrétien instrument with an unusually large 3° field of view to support its primary task of observing the entire sky. The camera is made up of thirty CCD chips each with a resolution of 2048 x 2048 pixels totaling approx. 120 megapixels and is fitted with an elaborate arrangement of filters. This telescope uses the drift scanning technique to scan small strips of the sky using the rotation of the Earth. The instrument can record 1000 spectra simultaneously by means of an elaborate system of optical fibres connected to holes drilled in an aluminium plate. A selection of drilled plates are used.

The Sloan Legacy Survey covers over 7 500 square degrees of the Northern Galactic Cap with data from nearly 2 million objects and spectra from over 800 000 galaxies and 100 000 quasars. The information on the position and distance of the objects has allowed the large-scale structure of the Universe, with its voids and filaments, to be investigated for the first time.

The 1,0 m NMSU telescope is a Ritchey-Chrétien instrument with a 2048 x 2048 pixel CCD mounted at a Nasmyth focus and provides a 15,7 arcminute view of the sky.

The 0,5 m ARCSAT is currently used for small research projects.

Corot Mission – COnvection, ROtation and planetary Transits

This ESA mission was launched on 27 December 2006. The mission is being led by CNES (Centre National d'Etudes Spatiales) in association with French Laboratories (CNRS) and partners in Belgium, Brazil, Spain Germany and Austria. The mission is in a circular polar orbit around the Earth. The mission uses stellar seismology to probe the inner structure of stars as well as detect extrasolar planets by observing the periodic micro-eclipses when a planet transits in front of its star. COROT aims to be a pioneer mission in the discovery of telluric extrasolar planets, bodies with properties comparable to the rocky planets of the Solar System. More than 12 000 stars with magnitudes from 12 to 15,5 will be surveyed. The 27 cm telescope optics includes two parabolic mirrors, 6 dioptric lenses and has a 1,2 m focal length. On

3 May 2007 COROT discovered its first extrasolar planet providing an image of a giant planet orbiting a star 1500 light years away. Some 9 extrasolar planets have thus far been discovered, the smallest having a diameter 1,7 times that of the Earth.

The Mission Centre and Control Centre are located in Toulouse. Communication and ground stations are in: Kiruna (Spain), Aussaguel (France), Hartbeeshoek (South Africa), Kouron (French Gyuana) and minor specific stations at Alcantara (Brazil) and Vienna (Austria).

OWL – OverWhelmingly Large telescope

This conceptual design by the ESO organization was originally envisioned to have a primary mirror of 100 m in diameter but the prohibitive cost of this resulted in the telescope being downscaled to a European Extremely Large Telescope with a 39 metre mirror and an expected cost of €1,2 billion. The site is expected to be in Chile and completion between 2017 and 2019. The primary mirror will be in 3048 hexagonal segments of 15 cm thick Zerodur ceramic. The focal length will be 175 m and the optics Cassegrain, Coudé. The proposed spectral range will be from 320 nm to 1200 nm in the ultraviolet to infrared.

Euro 50

This truly magnificent telescope for the future is being planned by a collaboration of scientists in Sweden, Finland, Ireland, Spain and the UK and with ESA participation. This monster with its 50 metre mirror has the potential to revolutionise astronomy. The primary mirror will comprise 618 two metre wide hexagonal segments and the secondary monolithic mirror will have 3 169 adaptive optics actuators. The telescope is planned for a Gregorian configuration with a primary focal ratio of f/0,85 and an exit focal ratio of f/13. Resolution is expected to be 2–3 milliarcseconds in the visual and 10 milliarcseconds in the infrared. The telescope will have a mass of 3 500 tons and soar 85 metres high.

GMT – The Giant Magellan Telescope

This gigantic addition to the astronomical scene is scheduled for completion by 2018. This instrument is like no other on earth. The GMT will have the resolving power of a 24,5 metre primary mirror using Gregorian optics. The mirror will consist of a central circular reflector of 8,4 metres in diameter surrounded by six more mirrors of the same size. The mirrors will be of lightweight borosilicate glass. The mirrors will be spherically ground to a radius of 36 m giving a focal ratio of f/0,7. The seven secondary adaptive mirrors will map directly on to the primary mirrors and will form a circle 3,2 m in diameter which will focus through a 1,78 m hole in the central primary mirror to 5,5 m below where large instruments requiring a minimum number of reflections will be located. Tertiary mirrors can also be deployed here to redirect beams to instruments mounted on a platform above the focal point. It is expected that an angular resolution of 15 to 30 milliarcseconds will be achieved in the near-infrared. The telescope will have a moving mass of 1 000 tons and will be housed in a cylindrical dome, 54 metres in diameter and 60 metres high, with moving shutters across the top and on one side to reveal the heavens to the instrument. The site for the GMT will probably be in central or northern Chile which is considered to be the best site on earth for large telescopes. The GMT is being developed by a partnership of: The Carnegie Institution of Washington, Harvard University, The Smithsonian Astrophysical Observatory, The University of Arizona, The University of Texas at Austin, The University of Michigan, The Massachusetts Institute of Technology, The Texas A&M University and The Australian National University.

XMM Newton Mission (X-ray Multi Mirror design)

This magnificent addition by ESA to the fleet of space observatories is the largest scientific satellite ever built in Europe. This is ESA's second 'Cornerstone' mission. The three telescopes are marvels of precision engineering. Each telescope lens consists of 58 high precision wafer thin concentric mirror tubes polished to unbelievable precision on the inside and provided with gold reflective coating. In order to reflect X-rays from the surfaces a grazing angle of only 30 arcminutes was used which would provide sufficient reflectivity in the 7 keV energy range. The observatory was launched by Ariane-5 rocket in December 1999 from Kourou, French Guiana. The mission life has been extended to 2012 but could technically remain operational until 2018. The observatory carries X-ray imaging cameras and spectrographs and an optical monitoring telescope. This three-axis stabilised spacecraft has a pointing accuracy of one arcminute. The prime conractor Dornier Satellitensysteme (A part of Daimler-Chrysler Aerospace) led an industrial consortium involving 46 companies from 14 European countries. The main scientific instruments carried aboard the craft are:

The three European Photon Imaging Cameras (EPIC) produced by a consortium of ten Institutes in four nations. One of the cameras uses a new type of CCD developed by the Max Planck Institute of extraterrestrial physics. The optics are designed to cover a spectral range of 12 keV to 0,1 keV. The total collecting area is 4300 cm^2 at 1,5 keV and 1 800 cm^2 at 8 keV. The two Reflection Grating Spectrometers (RGS). The Optical Monitor (OM) aligned with the X-ray telescope gives the mission a multi-wavelength capability. This is a 30 cm aperture Ritchey-Chrétien telescope with a 170–600 nanometre spectral range.

The telescopes were manufactured by Carl Zeiss (Oberkochen, Germany), Kayser-Threde (Munich, Germany) and Media Lario (Bosisio Parini, Italy). X-ray tests were performed in the PANTER test facility of the Max Planck Institute of extraterrestrial physics. The spacecraft is in a 48-hour elliptical Earth orbit with an apogee of 114 000 km and a perigee of 7 000 km.

The ESA XMM-Newton and NASA's Chandra telescope have much in common regarding design and capability. There is fortunately well established co-operation in the use of these observatories so that scientists seeking time on one of the observatories will have access to the other as well by making a single booking.

NuSTAR – NASA's Nuclear Spectroscopic Telescope Array

NuSTAR, a Small Explorer mission, was launched by aircraft over the Pacific ocean on 13 June 2012. The rocket was fired five seconds after being dropped from the aircraft, and the craft then took a 550 km high equatorial orbit. This mission is expected to provide unprecedented spatial and spectral resolution of the X-ray spectrum. This will complement data from the larger missions – Fermi, Chandra, Hubble and Spitzer. The unique telescope design includes a ten metre mast with a Wolter type telescope (ten metre focal length) working in the range of 5 to 80 keV. A Wolter telescope uses a grazing incidence angle of between 10 arc-minutes and 2 degrees and two tubular mirrors to give a usably wide field of view.

Its primary goals include a deep survey of black holes a billion times more massive than the Sun and to understand how particles are accelerated to within a fraction of a percent below the speed of light in active galaxies. Also to understand how the chemical elements are created in the explosions of massive stars by imaging the supernova remnants.

2MASS – Two Micron All Sky Survey

This near infrared mission was funded by NASA and the National Science Foundation. Northern hemisphere operations began at the Fred Lawrence Whipple observatory at Mt. Hopkins, Arizona in June 1977 and southern hemisphere operations at the Cerro Tololo Inter-American observatory in Chile in march 1998. The two matching 1,3 m telescopes were optimised for observations in the 1250 nm, 1650 nm and 2170nm bands. Operations were completed in 2001. The mission objectives were: Detection of galaxies obscured in visible light, Detection of brown dwarfs, Extensive survey of low mass stars. Data of some 300 million stars and a few million brown dwarfs were catalogued and an atlas of images of the entire sky was produced.

LSST – Large Synoptic Survey Telescope

Construction of this observatory is well under way and first light is hoped to be achieved by 2015. The telescope is designed for wide field survey work with a field of view of 3,5° – about seven times the apparent size of the moon. The primary mirror will be huge – 8,4 m and the spectral range will be from 320 nm to 1060 nm which includes both ultraviolet and infrared. The wide angle required will necessitate an unusual mirror arrangement of three mirrors. The secondary mirror will be 3,4 m in diameter and the tertiary 5,0 m. This will be located at the centre of the primary as a monolithic unit which will reduce the primary collecting area considerably. The focal length of this unusual telescope will be 9,9 m. The huge 3,2 gigapixel prime focus digital camera will take a fifteen second exposure every twenty seconds. The camera is expected to take 200 000 images (1,28 petabytes) per year. Vast computer resources will be required to process this data.

The site chosen for this observatory is on the peak El Peñón of Cerro Pachón, Chile, alongside the existing Gemini South and Southern Astrophysical Research Telescopes. The planned scientific goals of the survey are:

- Measuring weak gravitational lensing in the deep sky to detect dark energy and dark matter.

- Mapping small objects in the solar system, particularly Near-Earth asteroids and Kuiper belt objects.

- Detecting transient optical events such as Novae and Supernovae.

- Mapping the Milky Way.

In January 2008, software billionaires Charles Simonyi and Bill Gates pledged $20 million and $10 million respectively to the project. The project requires a further grant of $400 million.

Pan-STARRS – Panoramic Survey Telescope And Rapid Response System

This is a collaboration between the University of Hawaii Institute for Astronomy, MIT Lincoln Laboratory, Maui High Performance Computing Centre and Science Applications International Corporation. When in full operation the observatory will use four identical 1,8 metre telescopes with a wide 3° angle field of view and is expected to detect objects as dim as magnitude 24. The observatory will be located at Mauna Kea or Haleakala in Hawaii.

The unusual feature of this observatory is that it will conduct an astronomy and photometry survey of nearly the whole sky on a continuous basis. It will detect differences between present and previous observations of the same areas of sky and is expected to discover large numbers of new asteroids, comets, variable stars and other celestial objects. The observatory will utilize the largest digital camera ever built using 1,4 gigapixels per image. The exposure times will be from 30 to 60 seconds and allow an extra minute for computer processing. Some 10 terabytes of data will be captured every night. After processing the data for changes in positions and magnitudes of objects the data will be discarded. An area equivalent to the entire sky will be surveyed four times every month.

WISE – Wide-Field Infrared Survey Explorer

This NASA mission was launched on 14 December 2009. The craft with its 40 cm infrared telescope, was placed in a circular 525 km high polar synchronous orbit giving it an all-sky surveying capability, and can observe in the 3 to 25 micron range with exceptional sensitivity. During its seven month mission the telescope was expected to take 1,5 million images. A supply of hydrogen ice kept the telescope and detectors cooled to 15 K. As the telescope sweeps the sky a small mirror stabilised the image for eleven seconds for each exposure. Each image will have one megapixel of data for each of four wavelengths. The data would be downloaded four times a day for processing by Earth based computers. WISE studied study cool dim stars such as brown dwarfs, asteroids and infrared galaxies. WISE also provided an important source catalogue for the proposed JWST. The cost of the mission is estimated at $300 million.

During its active mission WISE found dozens of previously unknown asteroids every day. Over 33 500 new asteroids and comets, and over 154 000 solar system objects were discovered by October 2010 when the hydrogen coolant and original NASA funding ran out. Wise was decommissioned on 17 February 2011 after a four month 'warm mode' extension. Using data from WISE, astronomers have discovered packs of Trojan asteroids at the L4 and L5 Lagrangian points of Jupiter's orbit. So far 5 253 dark reddish Trojan rocks have been counted, many of them more than 1 km in size.

LINEAR – LIncoln Near-Earth Asteroid Research

This is a cooperative project between the United States Air Force, NASA and MIT's Lincoln Laboratory. The LINEAR robotic telescopes have made the majority of asteroid detections since 1998. By the end of 2007 LINEAR had detected 226 193 new objects which included 2019 near-Earth asteroids, tens of thousands of other asteroids and 236 comets. The original instrument installed in 1996 was the one metre Ground-based Electro-Optical Deep Space Surveillance telescope (GEODSS). A second one metre telescope was added in 1999 and later a 0,5 metre telescope as well. LINEAR observes each patch of sky five times in one evening mostly observing the ecliptic where most Near Earth Objects (NEOs) will be found. LINEAR uses a highly sensitive CCD of 1960 x 2560 pixels.

LBT – Large Binocular Telescope

The LBT is a joint venture of a large consortium of universities in the USA, Germany and Italy. This is of a most unusual design – two 8,4 metre mirrors on a common mounting, hence the name binocular. The mirrors are spaced at 15 metre centres giving an interferometric baseline of 22.8 metres and also a 15 metre baseline for nulling interferometry. The telescope saw first light with one mirror in January 2006 and second light with the other in September 2006. The telescope will operate in the visible and near infrared wavelengths. With the high resolution and also nulling interferometry it is expected that planets orbiting stars will be visible. The observatory is situated at 3 300 metres on Mount Graham in the Pinaleño Mountains of Arizona. The construction of the observatory went ahead after surviving forty lawsuits. The construction was opposed by the San Carlos Apache Tribe who regard the mountain as sacred, and also by environmentalists who claimed that the observatory would imperil the endangered Mount Graham Red Squirrel.

SALT – South African Large Telescope

This new project is the result of a co-operation between South Africa, Germany, Poland, the US and New Zealand. The observatory built at Sutherland, about four hours drive from Cape Town, was completed in 2005. The telescope saw first light on 1 September 2005 and the observatory was inaugurated on 10 November 2005. The telescope is of a most unusual design similar to the Hobby Eberly telescope in Texas. The mirror is a huge array, 10 by 11 metres in size, consisting of 91 hexagonal elements. The telescope is able to slew on its azimuth mounting but the inclination is fixed at 37° from the vertical. Light from the astronomical object is focused on an image plane, from where it proceeds to an optical instrument, while tracking the apparent movement of the sky. The observatory is well sited for southern hemisphere observations and will function in the visible and ultraviolet spectra. The primary instruments are SALTICAM, the telescope acquisition camera as well as an imager/photometer and PFIS, the Primary Focus Imaging Spectrograph.

GTC – Gran Telescopio Canarias – Large Telescope of the Canary Islands

This exciting addition to the telescope scene saw first light on 14 July 2007 using twelve of its planned thirty six hexagonal mirror segments. Scientific operations commenced in May 2009. The telescope is situated at an altitude of 2 400 m on the Roque de los Muchachos of the La Palma island. When fully assembled the primary mirror will have a diameter of 10,4 metres making the telescope well suited to deep space studies and also for searching for extra-solar planets. The ELMER optical instrument is optimised for observations in the 370 to 1 000 nanometre visible and infrared wavelengths. The CanariCam is a mid-infrared imager with spectroscopic, coronagraphic and polarimetric capabilities. The CanariCam will use a two-staged closed cycle cryocooler to cool the cold optics and cryostat interior down to 28 K and the detector to about 8 K.

The GTC is a Spanish initiative led by the IAC (Instituto de Astrofisica de Canarias).

The Liquid Mirror Telescopes

Although originally conceived by Sir Isaac Newton, the liquid mirror telescope had to wait until the 20th century for engineering expertise equal to the task. A liquid mirror telescope project was established by the University of British Columbia in 1994 for developing and testing liquid mirror technology. The Large Zenith Telescope was built in collaboration with scientists of Laval University and the Institut d'Astrophysique de Paris. The telescope was completed in 2003.

The rotating mirror is 6 metres in diameter with a mass of 3 tons, a focal ratio of f/1.5 and a focal length of 10 metres. The specially designed air bearing of stainless steel has a mass of two tons and its 56 cm spindle can support a load of 10 tons. The mirror is driven by a built-in brushless direct drive DC motor. A digital shaft encoder provides feedback to keep the rotational speed accurate to one part in a million. The corrector lens and detector are mounted on a hexapod with actuators in each leg which can be computer controlled to adjust the leg length to within 10 microns.

ALPACA (Advanced Liquid-mirror Probe for Astrophysics, Cosmology and Asteroids). This will be an 8 metre class liquid mirror located on Cerro Tololo in Chile.

The proposed LAMA (Large Aperture Mirror Array) telescope will consist of an array of 6,15 metre liquid mirrors simulating a mirror of 50 metres in diameter. The instrument will track within 4 degrees of the zenith with a tracking time of up to 30 minutes and will operate in the visible and infrared wavelengths. It is estimated that observations could be made in sub-nanoJansky flux levels in targeted areas.

Arecibo

The 305 metre big dish at Arecibo, Puerto Rico, is the largest single unit radio telescope on Earth. The giant spherical reflector is constructed from 40000 perforated aluminium panels and the downward facing movable receivers are suspended 168 metres above the surface. The dish is fixed in position pointing vertically upwards and relies on the rotation of the earth for scanning. The receivers are cooled by liquid helium for maximum signal clarity. The observatory operates from 50 MHz to 10 GHz (6 m to 3 cm). The telescope can transmit at up to a phenomenal 20 terawatts of power, and is sensitive enough to receive signals from a comparable telescope 1000 light years away. Waiting 2000 years for a reply is not however conducive to a lively conversation. A one MW planetary radar transmitter is used for the study of planets, comets and asteroids. With radar vision the telescope is able to detect the faint pulses from distant pulsars and quasars. The telescope, inaugurated in 1963, is owned and operated by Cornell University.

The Arecibo telescope is ideally suited to SETI projects. (Search for Extraterrestrial Intelligence). After a major upgrade in November 1974 a message was beamed in the direction of the M13 cluster of some 300 000 stars 25 000 light years away. A message of 1679 pixels forming a 23 by 73 array was chosen as these prime numbers are the only way of forming the data into a rectangle. The array encoded the following information as binary pixels (Ones and zeroes):

- The numbers 1 to 10 in binary code.
- Atomic numbers of hydrogen, carbon, nitrogen oxygen and phosphorus indicating the essential elements for human life.
- Formulas for sugars and bases in nucleotides of DNA. Graphical representation of DNA double helix.
- Stick figure representation of a human.
- Human population size of Earth at the time.
- Height of humans using the wavelength of the signal as a measure. Fourteen wavelengths gave a height of 176,4 cm.
- A simple representation of the Solar System showing the Sun and nine planets. Earth is offset by one pixel to show that this is where we live.
- The message ends with a depiction and info of the Arecibo radio dish.

In August 2001 a 'reply' was received. One would expect the reply to be sent in the same wavelength and direction from where it was received but in this case it was impressed on a wheat field, in the style of crop circles, near the radio telescope at Chilbolton in Hampshire, UK. Why would ET reply by vandalising planted crops? The reply was a replica of the original message but with the information edited to suit the circumstances of the sender. There was also a second image showing a portrait of the sender.

Some of the updates to the message were as follows:

- Silicon added to the list of life elements.
- Additional strand added to DNA and change in number of nucleotides. Figure of ET shown with large head and height of only 100,8 cm. Population shown as 21,3 thousand million.
- Solar system diagram modified to highlight fourth and fifth planets as well. This is highly suspect as it is extremely unlikely that more than one planet of a solar system could suit human life.

This 'reply' must surely be a very elaborate hoax but the most puzzling question is how could it possibly have been done in a single night under cover of darkness? One way of doing this would be to measure out the rectangle and drive in pegs to suit the rows and columns and then connect the pegs with strands of twine to mark all the squares. This could be done at the hoaxer's leisure on the days preceding the imprint. It would then be quite feasible for a team of stompers to form the image on a single pass through the rectangle at night. The wheat showed no sign of microwave damage and ET would have been hard pressed to beam microwaves in neat little squares. Pressing wheat down with planks would be much easier in squares than circles. To have replied so quickly (27 years) in such a physical manner would also imply that the recipient could not have been very far away from the Earth. The recipient probably replied by stomping a wheat field because he had no means of leaving the Earth to reply from space. The wheat field has long since been restored to its intended agricultural use.

ATA – Allen Telescope Array

A very exciting entry to the SETI scene occurred on 11 October 2007 when the first 42 radio dishes of the ATA observatory at Hat Creek near Mount Shasta, California, were activated. When the full compliment of 350 six metre dishes are operational this will be the largest radio telescope on Earth. The antennae are of offset Gregorian configuration with 2,4 m subreflectors. The first phase of this ambitious project was largely funded by a $25 million grant from the Paul G. Allen Family Foundation. Philanthropist Paul Allen is a co-founder of Microsoft. This is the first major observatory to be used primarily as a SETI instrument. The observatory is at an altitude of 1 000 metres and surrounded by volcanic peaks which will effectively shield it from unwanted signals from TV, radio and cellphone radiation. The heavens will be scanned in frequencies from 0,5 GHz to 11 GHz making this the first panchromatic, wide angled radio camera ever built. The observatory has a multi-tasking capability so that concurrently with SETI observations it will also be able to collect scientific data on supernovas, black holes and other exotic astronomical objects.

JBO – Jodrell Bank Observatory

The Jodrell Bank Observatory is part of the School of Physics and Astronomy at the

University of Manchester. The observatory is home to the Lovell Radio Telescope and the MERLIN / VLBI National Facility. The observatory was established by Dr Bernard Lovell in 1945. The Lovell radio telescope with its 76 metre dish, completed in 1957, is an awesome sight and was for many years the largest fully steerable radio telescope on Earth.

The MERLIN (Multi Element Radio Linked Interferometric Network) is an array of radio telescopes distributed around Great Britain with separations of up to 217 km and operating at 151 MHz to 24 GHz. (2 m to 1,25 cm) At 5 GHz the resolution is better than 50 milliarcseconds rivalling that of large optical telescopes. Even greater resolution can be achieved. This requirement is addressed by VLBI (Very Long Baseline Interferometry). There are plans to extend the UK network by an additional telescope in Ireland.

The proposed e-MERLIN Legacy Program is a major upgrade involving the installation of new receivers, analogue and digital electronics, optical fibre links to each telescope and a new correlator at Jodrell Bank. Full operations were scheduled to begin in 2010. This upgrade will provide considerably increased bandwidth and enhanced imaging using an array of seven radio telescopes spanning 217 km.

When fully implemented the e-MERLIN Legacy Program will have the following capabilities:

Observing bands at   1.3-1.8 GHz, 4-8 GHz, 22-24 GHz

Total bandwidth   4 GHz

Resolution   10 to 150 milliarcseconds

Sensitivity ~   1 microjansky

Astrometry, polarimetry, spectroscopy

The following projects have been approved for the e-MERLIN Legacy Program:

Astrophysics of Galaxy Transformation and Evolution (AGATE)

The e-MERLIN Cyg OB2 Radio survey: Massive and Young stars in the Galaxy (COBRaS)

Galaxy Evolution Survey (eMERGE)

Pulsar Interferometry Project (e-PI)

Feedback Processes in Massive Star Formation

Gravitational Lensing and galaxy evolution

Legacy e-MERLIN Multi-Band Imaging of Nearby Galaxies (LeMMINGs)

Luminous Infra-red Galaxy Inventory (LIRGI)

Morphology and Time Evolution of Thermal Jets Associated with Low Mass Young Stars

Planet Earth Building Blocks - a Legacy e-MERLIN Survey (PEBBLES)

Resolving Key Questions in Extragalctic Jet Physics

Super-CLASS: the Super Cluster Assisted Shear Survey – a weak lensing deep field survey using e-MERLIN.

The JBO generally operates with three telescopes for VLBI observation with the EVN (European VLBI Network) and/or global VLBI networks. The JBO also operates with VSOP observatories (Very long baseline interferometric Space Observatory Programs) which provide baselines larger than the Earth.

A most astonishing announcement was made in February 2005 – a Dark Galaxy. A galaxy composed entirely of matter emitting no visible light at all. This galaxy contains enough hydrogen to form more than a hundred million stars! The galaxy may also contain vast quantities of non baryonic dark matter. There may well be many more dark galaxies as dark matter is far more abundant than ordinary matter. The dark galaxy, named VIRGOH21, was first detected by the Lovell telescope and further confirmed at Arecibo.

A novel project has been launched with its centre at Jodrell Bank – The Spaced Out project. The models are situated as follows:

Mercury	Hermitage Primary School, Cheshire.
Venus	Alderley Edge Primary School, Cheshire.
Earth	Tytherington High School, Macclesfield.
Mars	Hartford High School, Cheshire.
Ceres (Asteroid)	Furness Vale Primary School.
Gaspra (Asteroid)	William Hulme's Grammar School, Manchester.

Jupiter	Techniquest@NEWI, Wrexham.
Saturn	Lancaster Girl's Grammar School.
Uranus	William Herschell museum, Bath.
Neptune	The Armagh Planetarium, Armagh.
Absolus (Centaur)	The Spaceguard Centre, Powys.
Pholus (Centaur)	Swanhurst School, Birmingham.
Chiron (Centaur)	The National Space centre, Leicester
Varuna (TNO)	Cambourne School, Cambourne.
Pluto	Robert Gordon's College, Aberdeen.
Halley's Comet	Forest Gate Community School, London.

This is a one in fifteen million scale model of the Solar System with the model of the Sun situated at Jodrell Bank. All of the models of the astronomical bodies are of the same size – one metre, but the spacing is to scale. The models are situated at various schools and other public places. The models include all the planets as well as some asteroids and centaurs. If the models were to scale the Sun would have a diameter of 93 metres and the Earth 85 cm at a distance of 10 km. Pluto would have a diameter of 20 centimetres and a distance from the Sun of 393 km. The Spaced Out project may not excite the professional astronomer but it does provide school pupils with a useful introduction, and appreciation of the Solar System.

NRAO – National Radio Astronomy Observatory

The National Radio Astronomy Observatory is a facility of the National Science Foundation, operated by Associated Universities Inc. The headquarters of the NRAO is in Charlottesville, Virginia, operating major radio telescope facilities in Green Bank, West Virginia and Socorro, New Mexico.

The Robert C. Byrd Green Bank Telescope (GBT) is the world's largest fully steerable radio telescope with a huge antenna of 100 by 110 metres and a moving mass of 7300 tons. The antenna consists of 2004 surface panels each with its own motorised surface actuator to compensate for distortions. The dish is of paraboloid geometry but tilted in such a way that the support structure for the receiving instrument does not obscure the incoming signals in any way. The focal length of the dish is 60 metres and the operational frequencies are from 1 GHz to 100 GHz. The telescope moves on a circular track with a diameter of 64 metres which is level to less than a tenth of a millimetre. The telescope is able to view the entire sky from above 5 degrees of elevation. The GBT has been equipped with a laser ranging system. Laser beams are reflected between the structure, the telescope and a series of ground stations surrounding the telescope. These beams are used to monitor deformations due to gavity, wind and temperature which allow the telescope's motors, subreflector and surface panel actuators to compensate for any distortions.

The Very Large Array (VLA) consists of 27 antennae in a Y-shaped configuration each of 25 metres diameter. This configuration gives the resolution of an antenna 36 km across. The VLA is located 80 km west of Socorro, new Mexico.

GMRT – Giant Metrewave Radio Telescope

The GMRT is situated in India, about 10 km east of Narayangaon town on the Pune-Nasik highway and will exploit six wavebands: 38, 153, 233, 325, 610 and 1420 MHz. This is a project of the National Centre for Radio Astrophysics (NCRA) of the TATA Institute of Fundamental Research (TIFR). The array consists of 30 gigantic fully steerable parabolic dishes of 45 m diameter. Fourteen of the dishes are located in a compact region of about 1 km square and the remaining 16 are arranged in a 'Y' configuration with the longest interferometric baseline of about 25 km. The angular resolution available will be from 1 to 60 arcseconds depending on the frequency. Aperture synthesis is used for producing images from radio sources. The long wavelengths received have permitted the antennae to be formed from wire rope stretched between struts to form the paraboloid dishes. The observatory is primarily used for observations of galaxies, pulsars and supernovae.

LOFAR – LOw Frequency ARray for radio astronomy

This innovative and ambitious project by the Netherlands Foundation for Research in Astronomy ASTRON (Stichting ASTRonomisch Onderzoek in Nederland), will survey the universe at radio frequencies from 10 MHz to 240 MHz. This observatory will not use the usual paraboloid dish antennae but will instead utilise a huge array of inexpensive omni-directional dipole antennae. The original feasibility study was carried out in 1999 and initial funding provided by the Dutch government in 2003. The first phase of the project will utilise 15 000 antennae distributed over 77 stations in the North East of the Netherlands and nearby parts of Germany reaching a baseline of 100 km. The final configuration will have 25 000 antennae extending further into Germany and also in the UK, France, Sweden, Poland and Italy. The first image made by LOFAR was produced in February 2007 using 16 microstations distributed over an area of about 400 metres in diameter. Twenty nine hours of information with a frequency of 50 MHz and a bandwidth of 0,5 MHz were used. The processing power of the supercomputer required to handle the vast input of data will be enormous. This is estimated in tens of teraFLOPS. (One teraFLOPS is 10^{12} FLoating point OPerations per Second) The central processing facility is located at the University of Groningen.

LOFAR will study the very distant universe and the formation of the first stars and galaxies and also probe the intergalactic gas. High energy cosmic rays will also be studied. Within our own galaxy LOFAR will study low frequency radiation from pulsars and also search for Jupiter-size extrasolar planets. Within the solar system coronal mass ejections from the Sun and the solar wind will be studied.

The climate and altitude of the array site may not be ideal but LOFAR can be expected to produce a vast wealth of information of the universe and also leave a valuable technological legacy for future radio astronomy. ASTRON will play a leading role in the development of the SKA, as well as the ESO Very Large Telescope and the James Webb Space Telescope. In collaboration with the British and Spanish, ASTRON runs two optical telescopes at La Palma: the 4,2 m William Herschel Telescope and the 2,5 m Isaac Newton Telescope, as well as the James Clerk Maxwell Telescope in Hawaii.

SKA – Square Kilometer Array radio telescope

This exciting addition to radio astronomy is expected to be operational by 2020 and will have a combined aperture of one million square metres. It will consist of central clusters and thousands of other antennae spreading out by as much as 3 000 kilometres. This is one of the most important international science projects of the 21st century. The SKA is a collaboration between 30 institutes in 20 countries. The cost is expected to be €1,5 billion. When fully developed the resolution of the images will be phenomenal, exceeding that of optical telescopes. The frequency range will be from 70 MHz to 10 GHz (4,3 m to 30 mm) and angular resolution less that 100 milliarcseconds. The volume of data from the dishes will be about ten times the global Internet traffic. The computer processing required to handle the immense volume of data from the dishes will also be phenomenal – possibly in the exaflops range. About the processing power of 100 million PCs. Several countries have submitted bids for hosting this array – Argentina, Australia, China, Brazil, the US and South Africa. The headquarters of the SKA will be at Jodrell Bank, Manchester. In Sept. 2006 the bids were shortlisted to Australia and South Africa. Important factors in the selection are locations more than 25° away from the equator and a low population density area with little microwave radiation from cellphones and TV transmissions. The Southern hemisphere is preferable for observing the Milky Way. An abridged timeline is as follows:

1991   Concept

2012   Final site selection – dual sites in South Africa and Australia/New Zealand

2013-15 Detailed design and pre-construction phase

2016-20 Phase one construction

2020   Full science operations with phase one

2020-24 Phase two construction

2024   Full science operations with phase two

2024 onwards – phase three

The exceptional angular resolution made possible with the far outlying antennae has enabled science projects which could not before be attempted.

Some of the projects envisioned are as follows:

1. Strong field tests of gravity

2. Cosmic magnetism

3. Cradle of life – imaging proto-planetary discs at cm wavelengths

4. Galaxy evolution, cosmology and dark energy

5. Probing the dark ages and epoch of ionization

6. Exploration of the unknown

7. Binary supermassive black holes

8. X-ray binary systems and relativistic jets

9. Small-scale structure and evolution in active galactic nucleus jets

10. Strong gravitational lensing

11. Absolute astronomy and geodesy

12. Relative astronomy: parallax and proper motions

13. Galactic masers

14. Mapping high mass star formation in nearby galaxies

15. Stellar winds/outflows

16. Imaging stellar atmospheres and resolving stellar radio flares

17. Spatial and temporal changes in the fundamental constants

18. Ultra high energy particle astronomy

19. Probing scattering in the intergalactic medium and extreme scattering events

20. Spacecraft tracking

This immense project is being developed by scientists in seventeen countries. The SKA will probe into the very early stages of the universe and study pulsars, colliding black holes, dark energy and the influence of magnetic fields on the development of stars and galaxies. The South African meerKAT (Karoo Array Telescope) has been launched as a precursor to the SKA. The full MeerKAT array of sixty four 13,5 m dishes will enable the technology required for the SKA to be developed and is expected to be ready by 2016. Meerkats are small mongoose-like creatures indigenous to the area. They have the cute habit of keeping a lookout by standing straight up with their front paws by their chests. Their preferred diet consists of scorpions and snakes which they are easily able to deal with despite the obvious risks. The KAT-7 seven dish testbed achieved a major milestone on 3 December 2009 when interference fringes were seen between two of the dishes.

The Australian government has approved funding of $100 million over four years for the thirty six dish ASKAP (Australian SKA Pathfinder). ASKAP received its first signals on 3 March 2010 from a satellite as part of a project to measure the shape of an antenna surface by means of holography. The antenna has a highly innovative design having moving axes of altitude, azimuth and polarization whereby the entire dish rotates in unison with the sky. ASKAP is expected to be completed by 2013.

On 25 May 2012 the SKA board chairman, John Womersley, announced at a press conference held at Schiphol Airport, Amsterdam, that both South Africa, and Australia/New Zealand had been chosen to host the SKA as a dual site.

Three types of antennae will be required to cover the wide range of wavelengths to be observed by SKA.

Low-band: A phased array of simple dipole antennae to cover the range of 70-200 MHz. These will be grouped in 100 m stations each containing about 100 elements.

Mid-band: This is likely to be a phased array of 'tiles' to cover the medium frequency range of 200 to 500 MHz. The 3 m tiles will be grouped into circular stations 60 m in diameter.

High-band: This will consist of thousands of offset Gregorian 12 m dish antennae standing up to 15 m in height. This will cover the frequency range of 500 MHz to 10 GHz.

Signals will be transferred to central supercomputers – in Perth for the Australian/New Zealand array and in Cape Town for the South African array. Data rates will be 420 Gb/sec per dish and 16 Tb/sec per aperture array.

Some other significant space missions are:

SWAS The Submillimetre Wave Astronomy Satellite was launched into low Earth orbit in December 1998 as one of NASA's Small Explorer Program missions. The mission was to make targeted observations of giant molecular clouds and dark cloud cores. SWAS was utilised from June to August 2005 to provide support for NASA's Deep Impact mission to comet Tempel.

INTEGRAL – INTErnational Gamma Ray Astrophysics Laboratory. Launched by ESA in October 2002 and after more than 9 years remains operational. This mission is in cooperation with the Russian Space Agency and NASA.

The observatory has a payload of four co-aligned instruments:

IBIS – Imager observes from 15 keV to 10 MeV.

SPI – Primary spectrometer observes from 20 keV to 20 MeV.

ACS – AntiCoincidence shield.

JEM-X – Observes soft and hard X-rays from 3 to 35 keV.

The craft is controlled from ESOC in Darmstadt, Germany, ESA's control centre, through ground stations in Belgium and California. This observatory is designed to image in the 15 keV to 10 MeV energy bands.

VSOP-2 Very long baseline interferometric Space Observatory Program 2 to be launched by ISAS (Japan) in 2013. The observatory will have a 9 metre offset parabolic microwave antenna observing in the 5 GHz, 22 GHz and 43 GHz bands. The expected resolution will be from 130 down to 40 microarcseconds. This is phenomenal resolution hardly obtainable from optical telescopes, requiring an exceptionally long interferometric baseline, which is achieved by having one observatory on Earth and the other in space. A resolution of 40 microarcseconds is equivalent to a spread of 73mm on the Moon as seen from Earth.

AKARI (Previously ASTRO-F) Japan's second satellite for infrared astronomy launched in February 2006 and placed in a Sun synchronous orbit at an altitude of 700 km. The 71 cm silicon carbide reflector is gold coated to increase infrared reflectance and cooled by liquid helium to 6 degrees above absolute zero. The all sky survey will be done by the FIS (Far Infrared Surveyor) in 50 to 180 micron wavelengths. The high sensitivity imaging and spectroscopic observation will be done by the IRCs (Infrared Cameras) in (1,7 – 5,5), (5,8 – 14,1) and (12,4 – 26,5) micron wavelengths. The objectives are: A search for primeval galaxies and investigation of star formation and evolution of planetary systems.

SOFIA Stratospheric Observatory For Infrared Astronomy. NASA and the DLR (German Aerospace Centre) are working together to create SOFIA. The 2,5 m reflector telescope is mounted in a Boeing 747SP aircraft. The first observations were made in April 2007. SOFIA is an observatory mission under NASA's Origins Program.

XEUS – X-Ray Evolving Universe Spectrometer is to be launched by ESA in 2014. This satellite is to be placed at the L2 Lagrangian point of the Earth's orbit and will be used for the study of matter under extreme conditions, the study of highly collapsed stars and also the study of gravity, space and time near massive black holes.

GALEX – GALaxy EXplorer. This is a NASA small explorer class ultraviolet imaging and spectroscopy mission for investigating galaxies actively creating stars.

The satellite, launched on 28 April 2003, is equipped with a 50 cm Ritchey-Chrétien telescope modified for a 1,2 degree wide field of view. This mission has provided a vast wealth of data, imaging hundreds of millions of galaxies. There is a spectacular colour composite image of galaxy M106 with the far UV shown in blue and the near UV shown in red.

LHC – Large Hadron Collider

This is neither a telescope nor a space mission but it must be included here as it is intimately involved with cosmological matters – the initial stages of the Big Bang and the ultimate nature of matter.

The LHC is the largest machine ever built. It will accelerate protons in circular paths of 27 km to an unprecedented energy of 7 TeV per particle (7 x 10^{12} electron volts). With this extreme energy level it will be possible to conduct experiments in conditions close to the initial stages of the Big Bang and possibly detect the existence of the Higgs boson. The tunnel housing the particle beam pipes is underground at a depth varying from 50 to 175 metres below the surface. The accelerator is housed at CERN, near Geneva, Switzerland with the accelerator path lying mostly in France and the rest in Switzerland. The tunnel was constructed between 1983 and 1988 and formerly used to accommodate the LEP (Large electron-positron collider). The construction of the LHC was approved in 1995. It has a life expectancy of ten years after which an upgrade or the next generation of research equipment will be required.

The particles will be grouped into nearly 3000 bunches which can complete a circuit in 90 microseconds. The bunches will therefore be separated by about 30 nanoseconds. The two beams of particles will move in opposite directions in two pipes with crossover points at six positions where observations in the gigantic detectors can take place. Before the particles are injected into the main accelerator they first pass through smaller accelerators which will successively raise the energy in stages of 50 MeV, 1,4 GeV, 26 GeV and 450 GeV.

It is not meaningful to convert the extreme particle energy of 7 TeV to joules as this will give a small misleading result. (About 1,1 μJ). In April 2012 the beam energy was raised to 8TeV. The total energy stored in the magnets is 10 GJ and each beam will carry 362 MJ of energy which is equivalent to 78 kg of high explosive, equivalent to the propellant charges of the giant WWII battleship batteries, so that the beam dump must be capable of absorbing a vast amount of energy.

A convenient way of getting a grasp of the enormity of the accelerator is to take a look at the cost. The estimated total cost is €6 billion which is eight times the cost of the Queen Mary 2 ocean liner. The costliest parts of the collider are the supercooled superconducting electromagnets. The beams are kept on their circular paths by 1296 dipole magnets, and 392 quadrupole magnets keep the beams focused. In all some 5000 magnets are needed to guide the beams. Most of the magnets each weigh over 27 tons and 96 tons of liquid helium are required to keep the magnets at an operating extreme cryogenic temperature of 1,9 K. The initial cooldown of 31 000 tons of material required 12 million litres of liquid nitrogen. CERN has the largest extreme refrigeration installation on Earth. During the excavation of one of the gigantic shafts required for lowering equipment to an underground cavern it was found that the shaft penetrated flowing underground water. This problem was overcome by freezing the earth around the excavation site. The underground caverns are the largest ever excavated and are of cathedral proportions. The magnets are made of niobium-titanium alloy operating at a field strength of 8,35 tesla. This is close to 10 tesla which is considered the upper limit for this alloy. The superconducting magnet coils are made of copper-clad niobium-titanium capable of carrying currents up to 15 kA and able to withstand forces of hundreds of tons weight per metre. The world record for magnetic field strength in a dipole magnet was shattered in 1997 at the Berkely National Laboratory when a field strength of 13,5 tesla was achieved. The previous record of 11,03 tesla was set by a Dutch group in 1995. The record has subsequently been set at 45 Tesla by the National High Magnetic Field Laboratory. The beam tubes are of highly complex design. The outer vacuum tube is circular, containing an inner beam screen of square cross section made from low magnetic permeability stainless steel, copper coated on the inside. The beam screen is designed to resist crushing magnetic forces and minimize atom desorption caused by synchrotron radiation. The beam tubes must also be capable of accommodating thermal shrinkage when cooled to cryogenic temperature. The beam path will shrink by 10 metres when cooled to cryogenic temperature. The tidal effect of the Moon raises the earth's crust by 25mm in the area and produces a variation of 1mm in the path length. The degree of vacuum achieved in the tubes is similar to that found in Solar System space.

The extreme particle energy of the experiments has caused several people, including some scientists, to voice the concern that there is a possibility that the collisions could result in a runaway black hole situation. If the Earth were to instantly shrink to the size of a single atom there will certainly be no possibility of complaints or lawsuits! People feeling jittery about this will find little comfort in the fact that the LHC is a research tool for experiments into little understood aspects of particle physics and some aspects which are understood not at all. This is of course frivolous tabloid chatter – there is no possibility that a runaway black hole could be produced by a terrestrial science experiment. When in full operation, some seven thousand scientists from eighty countries will have access to the collider. The LHC had its initial switch-on ceremony on 10 September 2008.

Like most other aspects of the LHC, the electrical power requirements are enormous. When in normal operation from 200 to 300 megawatts will be required which is twice the power used by the nearby city of Geneva. It is estimated that up to one terawatt-hour will be required annually. The experiments will be shut down for 22 days in winter when power costs become prohibitive. Reliable power supply is essential as magnet failure could result in an uncontrolled beam which could melt a ton of metal in a fraction of a second. CERN has guaranteed power from *Électricité de France* and in case of emergency can switch seamlessly to the Swiss power grid. For a final backup in case of catastrophic power failure, CERN has several massive diesel generators, designed to power submarines, which can be activated to allow an ordered and safe shutdown.

Russian mathematicians Irina Aref'eva and Igor Volovich have suggested that the extreme energy of the collisions could give rise to space-time wormholes linking present and past. Let us leave this matter to the mathematicians!

There are six detectors housed in the vast caverns along the path at the intersection points. The detectors are named:

ATLAS	A Toroidal LHC Apparatus
CMS	Compact Muon Solenoid
ALICE	A Large Ion Collider Experiment
LHCb	LHC – beauty
TOTEM	Total cross section, Elastic scattering and diffraction Dissociation
LHCf	LHC – forward

The ATLAS and CMS detectors are by far the largest.

To get a grasp of the enormity of these instruments it is best to view the excellent website photos. Analysing the results of the experiments, 600 million collisions per second, would be hopeless without vast computing resources. The Worldwide LHC Computing Grid, incorporating some 80 000 computers, has been set up for this purpose.

On 6 July 2012 the LHC received world wide publicity in the media when it was announced that results from the ATLAS and CMS detectors indicated that there was a very high probability that the existence of the Higgs boson had been proven. Most academics seem sure that this has been achieved and Prof. Stephen Hawking has conceded that he has lost his $100 wager that this would never be happen. The Higgs boson is not a component of ordinary physical matter. It is an essential hypothetical part of the Higgs mechanism which describes how matter is given the property of mass. The 'particle' has about 133 times the mass of the proton but a lifetime so short that it hardly exists. Only 1% of the mass of protons and neutrons is due to the Higgs mechanism acting to provide the mass of quarks. However without the Higgs mechanism, quarks and electrons, and consequentially all physical matter, would be without mass. This is work at the forefront of advanced theoretical physics and much remains to be done. It will be fascinating to see how bosons relate to the hyperdimensional manifolds of string theory.

On 10 September 2012 it was announced that a new underground particle collider was being considered – with a circumference of 80 km and to be constructed in 2025.

We have looked into many of the most sophisticated and technologically advanced instruments that the present state of science and engineering can produce, using the entire spectrum of electromagnetic radiation, from radio waves to gamma rays. The results have been spectacular beyond the wildest expectations giving a vast wealth of scientific information. The end is by no means in sight – new super-instruments are planned to push the boundaries even further. We have also taken a look at the Large Hadron Collider, a spectacularly advanced experiment into studying the fundamental nature of matter, but here too, this is also seen as a stepping stone to even greater things to come. In the following chapters our quest of perception of the universe takes us beyond electromagnetic radiation to the more abstract work of the theoretical physicists and mathematicians, the studies and experiments in gravitation, and ultimately in perception through pure consciousness.

PHOTON QUANTA AND RELATIVITY

In our quest of perception of the universe we now look into the imperceptible – that of which we have an awareness through the work of mathematicians and theoretical physicists. In addition to the work of the great and famous we will also look into their lives and times as well as their interest in music. Music and art, although essential to the human condition, persistently remain beyond the purview of mathematics and science. Speaking of 'absolute music' philosopher Arthur Schopenhauer said:

> *Music is the answer to the mystery of life. The most profound of all the arts, it expresses the deepest thoughts of life ... music is thus by no means like the other arts, the copy of the Ideas, but the copy of the will itself, whose objectivity these Ideas are. This is why the effect of music is much more powerful and penetrating than that of the other arts, for they speak only of shadows, but music speaks of the thing itself.*

Darwin had a more practical idea, describing music as a development from mating calls as an aid to procreation. As can be expected, there is a vast quantity of music which falls precisely into this category.

Let us now consider one of the most mysterious and enigmatic phenomena of quantised energy – photons. Photons are the jewels of the universe. Even in a dimly lit room we are in a sea of photons streaming in their countless millions in every direction without collision. Photons can travel for thousands of millions of years across vast tracts of intergalactic space only to end in a brief moment of glory on the retina of someone's eye or to donate its energy to some atom of matter. Photons come in a broad band of wavelengths but it is the spectrum of visible light that is of particular interest to us. Fortunately for us various materials can selectively absorb or reflect different wavelengths of photons and our eyes are able to signal the different wavelength information to our brains to give us the faculty of colour vision. C. F. Alexander's well-loved children's hymn *All things bright and beautiful* is an

Ode to Joy for the photon. A brief reflection: look about you wherever you happen to be. Photons are streaming into your eyes from every part of every object in sight and forming images on your retinas. Move about now – photons will continue to form images in your retinas wherever you happen to move. They are streaming in every direction from every point of every object without the slightest interference with each other. This is nothing less than miraculous. Equally astonishing, photons can come to us after travelling for thousands of years from another star in our galaxy and there is not the slightest deviation from the straight and narrow, forming a point image. Photons are invisible, unaffected by gravity, other photons, or electric or magnetic fields. There is no part of the space of the universe where starlight is not visible – this means that the entire universe is flooded with photons. In addition to visible light the universe is also flooded with longer wave radiation – the microwave background radiation which is a remnant of the energy of the Big Bang creation of the universe. The number of photons in the microwave background radiation is immense, outnumbering the physical atoms of the universe by about ten thousand million to one. The energy of the microwave background radiation exceeds that of ordinary physical matter by nearly twenty to one. The human eye is highly sensitive to photons – only half a dozen photons arriving in rapid succession are sufficient to be detected as a tiny flash of light.

Photons have some very special properties. They are quanta of electromagnetic energy which do not interact with the time vector. They are therefore confined to three dimensional space and cannot experience gravity or the passage of time. A photon can only exist in motion and must therefore constantly move in a straight line at the speed of photons or 'light'. The speed of light and speed of the time vector are two of the basic unchanging landmarks of the physical universe. The energy of photon quanta is not fixed but is proportional to the frequency or colour as determined by the famous Planck's Constant. There is a most intimate relationship between photons and electrons. Modern theory has it that electrons are surrounded by 'virtual photons' which come into and out of existence in less than a femtosecond! When the electron is ready to release energy this will pass to a virtual photon which will then become 'real'. This photon will then speed on its way in a perfectly straight line until it can annihilate itself by donating its energy to another electron. Photons go – and how they go! At the speed of photons or light which is almost thirty centimetres per nanosecond. This is about 7,5 Earth circumferences per second or to the Moon and back in less than three seconds. To the Moon and back is not just a fanciful idea; a corner-cube reflector has been installed on the surface of the Moon for this very purpose. The ejected photon can speed for millions of years across intergalactic space singing: *Because I did not stop for time, time kindly stopped for me!* (With apologies to Emily Dickinson). Photons cannot be observed in any way at all as they can only be captured by an atomic electron and the moment this happens they no longer exist. One can of course observe sunbeams shining through a cloud or the beam from a photographic projector but this is not the stream of photons that is visible – it is only some of the light reflected by dust or vapour and scattered from the path. Another curious property of photons is that they are readily captured by atomic electrons but are totally ignored by free electrons flowing in a cathode ray tube. It is most fortunate that photons can be created by atomic electrons – the frequency/wavelength of the photons ejected by an atomic electron is determined by the atomic element that the electron belongs to. This makes the vast science of spectroscopy possible. The photons of visible light are created by atomic electrons falling to a lower energy level but photons can also be created by accelerating electrons. Electrons moving in circles such as those in a magnetron are in constant acceleration and can be a powerful source of microwave radiation. Magnetrons are commonly used for powering microwave ovens and RADAR transmitters.

Thus far we have only looked at the extremely small particle nature of photons – this far from the whole story. Photons have very pronounced wave properties with a wide range of frequencies and wavelengths and also the properties of coherence and polarisation. The range of photon wavelengths is practically unlimited. The radiation of MW radio transmissions can have wavelengths of kilometres and at the other end of the scale we have gamma rays with wavelengths in the femtometre range. Radiation from low frequency powerlines is regarded as electromagnetic disturbance as the radiation would attenuate to zero long before a small fraction of a wavelength had been reached. Visible light forms an almost insignificant part of the spectrum with wavelengths from 380 to 700 nanometres. A detailed quantitative look at the electromagnetic spectrum is given in the appendix under the heading 'The Vibes of the Universe'.

How can a stream of particles moving at constant speed make waves and what is it that is waving? There are many experiences of wave motion encountered in everyday life. The ripples on the surface of water, waves across a field of grass caused by gusting wind, waves in an electric circuit, sound waves in the air or other media. In all these examples the medium of the wave only oscillates with a small if sometimes complex motion. The waves of photons are quite different to all of these – it is the photons themselves which are travelling at extreme constant speed. Another major difference between light waves and other waves is the matter of attenuation. A sound wave will gradually become weaker with increasing distance from the source until it practically disappears. Light waves also become weaker with distance but this is only due to the spreading out of photons. Individual photons retain their quanta of energy even after travelling for millions of years. This is a most convenient situation for astronomy. Stars which are so dim that they cannot be seen with even the largest telescope can still be photographed by using a sufficiently long exposure. Let us look at photons emitted from a longwave dipole radio antenna. Photons are ejected by accelerating electrons. A maximum emission of photons will occur when electron movement is changing direction. The oscillating signal in the antenna will therefore emit a similar signal of varying photon density even though the photons are all moving at the same constant speed. The wavelength in the case of longwave radio broadcasting can be as much as a few kilometres. This transmission will be polarised according to the alignment of the antenna and the receiving antenna will have to be aligned similarly

The particle/wave duality of light has baffled the most brilliant minds of science for centuries. The particle/wave duality is not a peculiarity of photons; this also applies to larger particles but to a lesser extent as the particles become larger and the uncertainty values become less. Even Einstein could not come up with a satisfactory explanation for wave/particle duality of fundamental particles so we can only wait patiently for the appearance of Mr Zweistein! The wave nature of photons permits them to be reflected by smooth shiny surfaces. Smooth can only have meaning for a surface of a vast number of atoms. At the atomic level the words 'smooth', 'polished', 'flat', and 'coloured' are meaningless. Electrons do of course emit photons with a particular colour but this simply implies a specific wavelength/frequency and does not mean that the photon is actually 'coloured'. A photon approaching a target of atoms (or molecules) may experience the target as reflective and retain its wave property and leave at the same incident angle. Metals, having free electrons, are particularly good at reflecting photons. If the photon is not reflected it will be captured by an electron increasing the energy of the associated atom increasing its temperature or could also contribute to the photo-electric effect. Photons also have the property of polarisation which is another aspect of their wave nature. A polarised filter is a material somewhat like a picket fence at the molecular level. Photons holding their 'sticks' in line with the picket fence will get through and those at right angles will not. It is actually much more complex than this. If the photons have their 'sticks' at 45° to the fence half will get blocked and those that get through will all leave with their 'sticks' aligned with the pickets. This gives a curious phenomenon. Take two polarised filters set them at right angles to each other and they will stop all light from passing through. Now take a third filter and place it between the other two with its polarisation aligned at 45° to the other filters. Light passing from the first filter will be aligned with it. Half of this light will pass through the second filter and then leave realigned. On reaching the third filter half of the light will again pass through and once more be realigned so that instead of all the light being blocked by two filters, some of the light will be allowed to pass right through by adding a third filter. The polarisation can also work in a corkscrew fashion either left handed or right handed.

A huge advance in the understanding of the nature of light was made when Scottish physicist James Clerk Maxwell (1831–1879) formulated his electromagnetic theory of light. He described light waves as electric and magnetic fields oscillating at right angles to each other. This theory also explained the nature of polarisation, and astonishingly, he was also able to calculate the speed of light from the electrical and magnetic properties of space. This was a prestigious achievement in theoretical physics ranking in stature with the work of Newton and must surely have seemed at the time to be the last word on the nature of light. Maxwell's partial differential equations for the electromagnetic field in free space are one of the great mathematical achievements of the nineteenth century. Maxwell has the distinction of taking the first colour photograph in 1861.

Maxwell's equations express:

- Electric charges produce electric fields (Gauss's law) Currents produce magnetic fields (Ampere's law)

- Changing magnetic fields produce electric fields (Faraday's law of induction)

In 1864 Maxwell was the first to notice that a correction was required to Ampere's law – changing electric fields act like currents producing magnetic fields. Maxwell showed that the equations, with his corrections, predicted waves of oscillating electric and magnetic fields that travel through space. In 1865 he wrote:

This velocity is so nearly that of light that it seems we have strong reason to conclude that light itself (including radiant heat and other radiation) is an electromagnetic disturbance in the form of waves propagated through the electromagnetic field according to electromagnetic laws.

This laid the foundation for many future developments in physics such as special relativity and its unification of electric and magnetic fields in a single tensor quantity. The Maxwell equations were reformulated (some would say butchered) in 1884 by Oliver Heaviside and Willard Gi bbs to a far simpler representation using vector calculus. Maxwell's 1865 formulation was in terms of 20 quaternion equations in 20 variables. The Heaviside-Gibbs simplification resulted in four equations in four variables. Tom Bearden commented: A higher group symmetry such as quaternions will contain and allow many more operators than a lower algebra such as tensors, which itself contains more than an even lower algebra such as vectors. Einstein described Maxwell's work as the most fruitful that physics had experienced since the time of Newton. When asked if he stood on the shoulders of Newton, Einstein replied: 'No, on Maxwell's'. Albert Einstein commented:

The precise formulation of the time-space laws was the work of Maxwell. Imagine his feelings when the differential equations he had formulated proved to him that electromagnetic fields spread in the form of polarised waves, and at the speed of light! To few men in the world has such an experience been vouchsafed ... it took physicists some decades to grasp the full significance of Maxwell's discovery, so bold was the leap that his genius forced upon the conceptions of his fellow-workers.

— (Science, May 24, 1940)

Quaternions are a four dimensional non-commutative extension of complex numbers first described by Irish mathematician Sir William Rowan Hamilton in 1843. It is interesting to note that when extending two dimensional complex numbers to higher dimensions, the next stage is four dimensions; three dimensional complex numbers are not possible. The idea of quaternions came to Hamilton while walking along the Royal Canal, Dublin, on his way to the Irish Academy. He was so pleased with his discovery that he scratched the fundamental quaternion multiplication identity into the stone of Brougham Bridge. No trace of Hamilton's graffito remains, however a plaque commemorating his famous discovery has been affixed to the bridge. The plaque reads as follows:

Here as he walked by
on the 16th of October 1843
William Rowan Hamilton
in a flash of genius discovered
the fundamental formula for
quaternion multiplication
$$i^2 = j^2 = k^2 = ijk = -1$$
& cut it on a stone of this bridge

Hamilton's immense contributions to mathematics are of inestimable value to modern physics. It has been claimed that Maxwell's original equations hold the key to tapping electromagnetic energy from the vacuum state and also to electrogravitation. These goals were passionately pursued by Nikola Tesla – see chapter eight.

Maxwell was an accomplished poet; let us take a look at a poem written in honour of multi-dimensional geometer Arthur Cayley, which mentions hyperdimensionality and matrix algebra:

Oh wretched race of men, to space confined!
What honour can ye pay to him, whose mind
To that which lies beyond hath penetrated?
The symbols he hath formed shall sound his praise.
And lead him on through unimagined ways
To conquests new, in worlds not yet created.

First, ye Determinants! In ordered row
And massive column ranged, before him go,
To form a phalanx for his safe protection.
Ye powers of the nth roots of -1!
Around his head in ceaseless cycles run,
As unembodied spirits of direction.

And you, ye undevelopable scrolls!
Above the host wave your emblazoned rolls,
Ruled for the record of his bright inventions.
Ye cubic surfaces! By threes and nines
Draw round his camp your seven-and-twenty lines –
The seal of Solomon in three dimensions.

March on, symbolic host! With step sublime,
Up to the flaming bounds of Space and Time!
There pause, until by Dickenson depicted.
In two dimensions, we the form may trace
Of him whose soul, too large for vulgar space,
In n dimensions flourished unrestricted.

The Dickenson referred to was a portrait artist of Maxwell's acquaintance.

This poem is written in a 60 syllable six line iambic pentameter, suggesting with its factors of 3, 4 and 5 a right angled triangle. See chapter eight for Tesla's interest in these figures. Interestingly, Maxwell's reference to the twenty seven lines of cubic surfaces relates to the higher dimensional manifolds of string theory of advanced particle physics – something of which Maxwell could have known nothing. It is even tempting to seek a link to poet William Blake's *twenty seven Heavens and Hells*.

Max Planck said of Maxwell's theory of light on the occasion of the centenary of Maxwell's birth in 1831:

... remains for all time one of the greatest triumphs of human intellectual endeavour.

Maxwell's theory was based on the assumption of the existence of a luminiferous ether. Maxwell used the idea of infinitesimal vortices in the ether to describe magnetic field effects. Alas, this theory did not account for the particulate nature of light and it fell to Albert Einstein to re-introduce the Newtonian corpuscular theory as quantised packets of energy which became known as photons. This idea was actually proposed two thousand years earlier by Roman poet Lucretius in 50 B.C. stating:

The light and heat of the Sun are composed of minute atoms which, when they are pushed off, lose no time in shooting right across the interspace of air in the direction imparted by the push.

(Lose no time? They don't lose any time at all!)

At the time, a prevalent idea was that sight was achieved by particles streaming out of the eyes and then bouncing back much like RADAR. If this was true then people would be able to see in total darkness but this inconvenience to the theory was brushed aside.

Max Karl Ernst Ludwig Planck (1858-1947), the originator of quantum physics, came from a family of distinguished academics. His father, grandfather and paternal great-grandfather were all professors and his paternal uncle was a judge. The family moved to Munich in 1867 where Max graduated at high school. Max was musically gifted, taking voice lessons as well as piano, organ and cello. Before he began his studies at the University of Munich he discussed the possibility of a musical career with a musician. He was told that if he needed to ask for advice then he should study something else. When Max decided to study physics he received miserable advice from physics professor Philipp von Jolly who said: "In this field, almost everything is already discovered, and all that remains is to fill in a few holes."

The young Max had a different idea:

The outside world is something independent from man, something absolute, and the quest for the laws which apply to this absolute appeared to me as the most sublime scientific pursuit in life.

In 1877 Planck went to Berlin for a year of study with the famous physicists von Helmholtz and Kirchhoff and also mathematician Weierstrass. In 1879 Planck presented his dissertation 'On the second fundamental theorem of the mechanical heat theory' (entropy) and in 1880 his thesis 'Equilibrium states of isotropic bodies at different temperatures'. He became associate professor at Kiel University in 1885 where he furthered his work on entropy and its treatment, especially as applied to physical chemistry. He became full professor in Berlin in 1892 and also joined the local Physical Society. Of this time he wrote:

In those days I was essentially the only theoretical physicist there, whence things were not so easy for me, because I started mentioning entropy, but this was not quite fashionable, since it was regarded as a mathematical spook.

Planck started a six semester course of lectures on theoretical physics.

Lise Meitner commented: "dry, somewhat impersonal."

English participant James Partington said: "using no notes, never making mistakes, never faltering; the best lecturer I ever heard."

An amusing incident occurred when Planck went to give a lecture on radiant heat. When he asked in which room the lecture was to be held he was told: "You are much too young to be attending the lecture of the esteemed professor Planck."

In 1894 Planck devoted himself to the problem which would make him famous – black body radiation. How does the intensity of electromagnetic radiation emitted by a black body (a perfect absorber) depend on the frequency of the radiation? There were two laws to describe this – the Rayleigh-Jeans law which gave correct results only at low frequencies and the Wien law which gave correct results only at high frequencies. Planck found that he could only derive a satisfactory equation if light was emitted in 'packets'. He did not for many years think that these 'packets' corresponded with reality but were merely a mathematical trick. By 1900, he was able to present a theoretical derivation of the law but this required him to use ideas from statistical mechanics. He had a strong aversion against any statistical interpretation of entropy which he regarded as of axiomatic nature. He was forced to conclude that the energy of the packets was proportional to the frequency and hence the famous "Planck's Constant".

> " ... the whole procedure was an act of despair because a theoretical
> interpretation had to be found at any price, no matter how high that may be."

The discovery of Planck's Constant led him to define a new universal set of natural physical units, Planck length, Planck time etc. The natural units were derived in terms of five natural constants and setting these constants to unity. The Planck units are described further in the appendix under the heading 'The Beautiful Natural Planck Units of the Universe'. Planck despaired: "My unavailing attempts to somehow reintegrate the action quantum into classical theory extended over several years and caused me much trouble." In the following years physicists Rayleigh, Jeans and Lorentz set Planck's constant to zero to align with classical physics in defiance of Planck's non-zero value. Planck lamented: "I am unable to understand Jeans' stubbornness – he is an example of a theoretician as should never be existing, the same as Hegel was for philosophy. So much the worse for the facts, if they are wrong."

Planck's quantum theory did not initially receive widespread acceptance. As late as 1911, Paul Ehrenfest used the term 'Ultraviolet Catastrophe' to describe the problem that black body radiation, where the radiation according to the

Rayleigh-Jeans law, would result in infinite energy output. This impossibility was independently considered by Einstein, Lord Rayleigh and Sir James Jeans in 1905. It was Einstein who pointed out that the difficulty could be avoided by making use of the hypothesis proposed by Max Planck five years before.

Physicist Max Born wrote: "He was by nature and by the tradition of his family conservative, averse to revolutionary novelties and skeptical towards speculations. But his belief in the imperative power of logical thinking based on facts was so strong that he did not hesitate to express a claim contradicting to all tradition, because he had convinced himself that no other resort was possible. "

Planck was awarded the 1918 Nobel Physics Prize for his epoch making discovery of quantum theory. The following is an extract from his Nobel Lecture:

> For many years my aim was to solve the problem of energy distribution in the normal spectrum of radiating heat. After Gustav Kirchhoff has shown that the state of the heat radiation which takes place in a cavity bounded by any emitting and absorbing material at uniform temperature is totally independent of the nature of the material, a universal function was demonstrated which was dependent only on temperature and wavelength, but not in any way on the properties of the material. The discovery of this remarkable function promised deeper insight into the connection between energy and temperature which is, in fact, the major problem in thermodynamics and so in all of molecular physics ...

In 1887 Planck married Marie Merck who bore him four children; Karl, the twins Emma and Grete and a second son Erwin. The Planck home became a social and cultural centre for jointly playing music and was visited by well-known scientists such as Albert Einstein, Lise Meitner and Otto Hahn. Lise Meitner once mentioned a musical evening at Planck's Berlin home when Planck, Einstein and a professional 'cellist played Beethoven's B-flat major piano trio:

> Listening to this was marvelously enjoyable, despite a few unimportant slips from Einstein ... Einstein was visibly filled with the joy of the music and smiled in a light hearted way that he was ashamed of his dreadful technique. Planck stood quietly by with a blissfully happy face and, hand on heart, said 'That wonderful second movement!'

After several happy years Planck had to endure a series of grievous family losses with stoic resignation. In 1909 his wife died, possibly from tuberculosis. In 1911 he married his second wife Marga who bore him his third son Hermann. His eldest son was killed in action at Verdun. Erwin was taken prisoner by the French in 1914. The twins died in childbirth within two years of each other. Their daughters survived and were named after their mothers. A terrible fate befell Planck's son Erwin. He was executed by the Nazis for complicity in the failed attempt to assassinate Hitler in July 1944. Niels Bohr commented on Planck's immense contribution to science with:

> Scarcely any other discovery in the history of science has produced such extraordinary results within the short span of our generation as those which have directly arisen from Max Planck's discovery of the elementary quantum of action. This discovery has been prolific, to a constantly increasing degree of progression, in furnishing means for the interpretation and harmonising of results obtainable from the study of atomic phenomena, which is a study that has made marvellous progress within the past thirty years.

Planck retired in 1926. He was succeeded in his professorship at Berlin University by Erwin Schrödinger. During the Second World War, Planck was forced to leave his home in Berlin and move to the relative safety of the countryside. An air raid destroyed his Berlin home together with all his scientific records and correspondence.

Turning to Albert Einstein (1879–1955), in December 1921 he was awarded the Nobel Physics prize for:

> His services to Theoretical Physics, and especially for his discovery of the law of the photoelectric effect.

In his presentation address, Professor S. Arrhenius, Chairman of the Nobel Committee for Physics stated:

> There is probably no physicist living today whose name has been so widely known as that of Albert Einstein. Most discussion centres on his theory of relativity. This pertains essentially to epistemology and has therefore been the subject of lively debate in philosophical circles. It will be no secret that the famous philosopher Bergson of Paris has challenged this theory, while other philosophers have acclaimed it wholeheartedly. The theory in question also has astrophysical implications which are being rigorously examined at the present time.

Einstein was unfortunately unable to attend the ceremony. In 1954 Einstein made several much quoted statements; let us have a look at four of them:

In the year nineteen hundred, in the course of purely theoretical investigation, Max Planck made a very remarkable discovery: the law of radiation of bodies as a function of temperature could not be derived solely .from the laws of Maxwellian electrodynamics. To arrive at results consistent with the relevant experiments, radiation of a given .frequency had to be treated as though it consisted of energy atoms of individual energy. This discovery became the basis of all twentieth-century research in physics and has almost entirely conditioned its development ever since. Without this discovery it would not have been possible to establish a workable theory of molecules and atoms and the energy processes that govern their transformations. Moreover, it has shattered the whole .framework of classical mechanics and electrodynamics and set science a .fresh task: that of finding a new conceptual basis for all physics. Despite remarkable partial gains, the problem is still far from a satisfactory solution.

All these fifty years of conscious brooding have brought me no nearer to the answer to the question, 'what are light quanta?' Nowadays every Tom, Dick and Harry thinks he knows it, but he is mistaken.

If, then, it is true that the axiomatic basis of theoretical physics cannot be extracted from experience but must be freely invented, can we ever hope to find the right way? I answer without hesitation that there is, in my opinion, a right way, and that we are capable of finding it. I hold it true that pure thought can grasp reality, as the ancients dreamed. I consider it quite possible that physics cannot be based on the field concept, i.e., on continuous structures. In that case, nothing remains of my entire castle in the air, gravitation theory included, and of the rest of modern physics.

Also in 1918:

The supreme task of the physicist is to arrive at those universal elementary laws. from which the cosmos can be built up by pure deduction. There is no logical path to these laws; only intuition, resting on sympathetic understanding of experience, can reach them.

At a lecture in Glasgow in 1933:

The years of searching in the dark for a truth that one feels but cannot express, the intense desire and the alternations and misgiving until one breaks through to clarity and understanding, are known only to him who has himself experienced them.

After a sea voyage from Europe to New York with Chaim Weizmann in 1921, Weizmann told reporters:

Einstein explained his theory to me every day, and on my arrival I was fully convinced that he understood it.

Another quote from Einstein:

I am happy because I want nothing from anyone. I do not care for money. Decorations, titles or distinctions mean nothing to me. I do not crave praise. The only thing that gives me pleasure, apart from my work, my violin, and my sailboat, is the appreciation of my fellow workers.

Einstein on the laws of the Universe:

Everyone who is seriously involved in the pursuit of science becomes convinced that a spirit is manifest in the laws of the Universe – a spirit vastly superior to that of man ... in the face of which we ... must feel humble.

Einstein said of Gandhi in 1939:

Generations to come, it may be, will scarce believe that such a one as this in flesh and blood walked upon the earth.

Einstein devoted the second half of his life to the search for a unified field theory which would mathematically unite the gravitation of the General Theory of Relativity with a theory of electromagnetism. This proved to be the impossible dream of theoretical physics, as was also the even more ambitious TOE (Theory of Everything). Einstein did manage to develop a theory which 'wrapped' electromagnetism and gravitation into a common metric tensor but this theory broke down at high field strengths. Einstein was supportive of Theodor Kaluza's attempt to unify General Relativity with Maxwell's electromagnetism in 1919 by adding a fifth dimension to space–time and which was taken further by Oskar Klein to provide a basis for modern string theory. Another theory developed by Einstein is known as the Einstein-Schrödinger theory. Many scientists have regarded a unified field theory as a Quixotic quest for the impossible. The impossible dream has defied resolution but Einstein resolutely kept to the task until the day before his death. A short time before he died Einstein remarked:

> I cannot finish this work, it will be forgotten, but it will be rediscovered in the future.

At the time of Einstein's death only a nurse was present. She said that just before he died he mumbled a few words in German which she could not understand. He did definitely not say, as a recent TV documentary would have us believe: "*Die Katze ist auf der Tisch.*"

In 1999 Einstein was named by TIME magazine as the greatest person of the Twentieth Century. 1905 has been called the *annus mirabilis* of Einstein's life when he wrote his most famous papers. Another significant year for Einstein was 1915 when he published his General Theory of Relativity. *Annus mirabilis* (year of wonders) is often used to describe a year of good times. This comes to us from a poem by John Dryden (1631–1700) written in 1667 describing sensational events in London in 1665–1666 including the Great Fire.

Einstein's *mirabilis* papers were:

1. *On a Heuristic Viewpoint Concerning the Production and Transformation of Light.*

This paper explained the photoelectric effect and was cited for his Nobel prize. The paper also provided a confirmation of Max Planck's hypothesis of light quanta.

2. *On the motion – Required by the Molecular Kinetic Theory of Heat – of Small Particles Suspended in a stationary Liquid.*

This paper dealt with the study of Brownian motion and provided empirical evidence for the existence of atoms. Professor Wilhelm Ostwald confessed to professor Arnold Sommerfeld that Einstein's explanation of Brownian motion had converted him to a belief in atoms. In 1901 Einstein's father wrote to professor Ostwald in an unsuccessful attempt to get his son a position as an assistant at the University of Leipzig.

3.   *On the Electrodynamics of Moving Bodies.*

This paper introduced the Special Theory of Relativity. This theory embraced time, distance, energy, mass and electromagnetism. This work was possibly done with the assistance of his first wife Mileva, a mathematician, and certainly requiring the pioneering work of mathematical giants, Henri Poincaré and Hendrik Lorentz.

4.   *Does the Inertia of a Body Depend upon its Energy Content.*

In this paper Einstein showed that from the axioms of relativity it could be shown that there is an equivalence between mass and energy and is expressed in what was to become the most famous equation in physics.

Curiously, Newton discovered 'Newton Rings', the coloured fringes produced when a convex glass surface is brought into contact with a flat one or when a thin film of oil floats on water, but he did not realise that these were in fact interference fringes demonstrating the wave nature of light. Newton was not however adamant that light did not have a wave nature. Newton decided to sidestep the fringe issue:

I forbore to treat of the colours, because they seemed of a more difficult consideration, and were not necessary for establishing the properties of light there discoursed of.

The world was left with the problem of how can a wave with a wavelength millions of times larger than an atom suddenly donate all of its energy to a single electron and then instantly disappear.

The ultimate nature of light also fascinated polymath Thomas Young (1773–1829) English physician and physicist:

The nature of light is a subject of no material importance to the concerns of life or the practice of the arts, but it is in many other respects extremely interesting.

Young had a phenomenally wide range of interests and expertise which caused him to become known as: 'The man who knew everything'. Young will be known to every student of applied mathematics for: 'Young's modulus of elasticity'.

Dutch physicist Dr. Hendrik Antoon Lorentz (1853–1928) made huge contributions to relativity and electromagnetic theory. He was fluent in English, French and German. The following extract is from an obituary which appeared in

'The Times':

> In an early memoir, which became famous, Lorentz applied for the first time considerations relating to discrete molecules, to electric propagation in material bodies, and incidentally arrived at a rational reflection-equivalent for each substance independent of its density. In 1884 he began to study the effect which magnetization exerts on the polarization of reflected light. His "Theorie Electromagnetique de Maxwell et son application auz Corps Mouvants" and his "Versuch einer Theorie der Elektrischen und Optischen Erscheinungen in bewegten Körpern" were published in 1892 and 1895 respectively. They embodied the first systematic appearance of the electrodynamic principle of relativity, and in 1920 he brought out "The Einstein Theory of Relativity: A Concise Statement". In 1909 he published his "Theory of Electrons", based on a series of lectures at Columbia University, and in 1916 he published in French at Leipzig an account of statistical thermodynamic theories, based on lectures delivered at the *College de France* in 1912. An edition of his University lectures, entitled "Lessons on Theoretical Physics", began to appear, under his supervision, in 1919. He was also the author of a textbook of the differential and integral calculus; "Visible and Invisible Movements", 1901; and "Clerk Maxwell's Electromagnetic Theory", 1924.

Lorentz is also famed for his work on the FitzGerald-Lorentz contraction, which is a contraction in the length of an object at relativistic speeds. Lorentz transformations, which he introduced in 1904, form the basis of Einstein's Special Theory of Relativity. They describe the increase of mass, the shortening of length, and the time dilation of a body moving at speeds close to the velocity of light. The length shortening and time dilation of objects moving at relativistic speed did much to publicly sensationalise the Special Theory of Relativity. In 1902, Lorentz and Pieter Zeeman, were jointly awarded the Nobel physics prize:

> *In recognition of the extraordinary services they rendered by their researches into the influence of magnetism upon radiation phenomena.*

Einstein remarked that without the work of Lorentz he could never have discovered Special Relativity, and:

He meant more to me personally than anybody else I have met in my lifetime.

One of the most influential and universal mathematicians on the 19th and 20th centuries, David Hilbert (1862–1943) also made profound contributions to mathematics and physics and did groundbreaking work on the General Theory of Relativity. At the International Congress of Mathematicians held in Paris in 1900, Hilbert presented a list of 23 unsolved problems. After more than a century some of these remain unsolved.

French mathematician Jules Henri Poincaré (1854–1912) was another giant to walk the stage of the time. His cousin Raymond was the prime minister of France for five terms of office and also served as President of the Republic from 1913 to 1920. Henri's range of interests was phenomenal. He made huge contributions to mathematics, physics, celestial mechanics, fluid mechanics, and is regarded as the originator of algebraic topology and chaos theory. Poincaré was a tutor and later an associate of Norwegian physicist Kristian Birkeland. In 1887 King Oscar II of Sweden announced a competition for the solution of the three body orbiting problem. On awarding the prize to Poincaré, one of the judges, mathematician Weierstrass declared:

> *This work cannot indeed be considered as furnishing the complete solution of the question proposed, but that it is nevertheless of such importance that its publication will inaugurate a new era in the history of celestial mechanics.*

This work was also of great value in particle physics. When studying the three body orbiting problem he discovered a chaotic aspect as well as aperiodic orbits. Poincaré is regarded with Einstein and Lorentz as one of the originators of the Special Theory of Relativity. Practical examples of Poincaré's algebraic topology can sometimes be seen in TV adverts when a shape is morphed into something quite different. In topology, the entities under consideration are known as manifolds; these can be lines, areas, solids or even abstract objects in higher dimensions. The distinguishing feature between manifolds is the number of holes penetrating right through – sizes or angles play no part in algebraic topology. Different manifolds (non homeomorphic) are distinguished by their Betti numbers and torsion coefficients. A cup and saucer are simple examples – the saucer, not having any holes, can be morphed into a sphere or other lump. The cup however has a hole through its ear so that it can be morphed into a lump with a hole through it – the part of the cup that can contain a liquid is not significant as it can be morphed into the rest of the lump. A teapot is a more complex example – the spout and hole at the top are not important as these can be morphed away – the significant features are the hole through the handle and the holes for straining tealeaves that might enter the spout. One, two and three dimensional manifolds may seem trivial but topology is not simply a matter of playing with modelling clay. Manifolds can be in any number of higher dimensions which are beyond human visualisation and the mathematics involved is fiendishly difficult. A practical application of algebraic topology is in the Calabi-Yau manifold of string theory of subatomic particles. In 1904 Poincaré proposed his famous conjecture which he left to the mathematicians of this planet to find a proof which would make his conjecture a theorem – a task which would defy the efforts of mathematicians for a century. This was something of a Lorelei siren luring mathematicians like Rhine sailors to their doom. Mathematicians who attempted the problem were said to be afflicted with Poincaritis. The eventual proof was posted on the Internet by Grigori (Grisha) Yakovelevich Perelman in 2002–2003. The original concise 68 page proof was re-written as a 473 page paper by mathematicians Morgan and Tian to make it more accessible to other mathematicians and students. To the astonishment of the mathematical world, Perelman declined the prestigious Fields Medal and million dollar prize, preferring to remain in relative obscurity and poverty. His adherence to a higher ideal is expressed in the ancient Irish hymn:

> *'Be Thou my vision …'*
> *Riches I heed not, nor man's empty praise,*
> *Thou mine inheritance, now and always: …'*

Perelman had a passion for living naturally which caused him to refrain from trimming his hair or shaving his beard. These habits were readily accepted by those of his acquaintance – his reluctance to trim his fingernails was not.

An interesting branch of topology is the study of mathematical knots. These are different to those known to every sailor and Boy Scout in that the strands and braids are always endless loops. A remarkable aspect of three dimensional knots is that they fall apart in higher dimensions. As a trivial example draw two circles on a piece of paper, the one inside the other. In two dimensions it is impossible to separate the circles without cutting the outer circle, however in three dimensions you can simply lift the inner circle and take it away. No need to deal with the Gordian

Knot with a sword slash like Alexander the Great.

Joseph-Louis Lagrange (1736–1813) is regarded as a French mathematician but was actually born in Italy and named Giuseppe Lodovico Lagrangia. Lagrange made huge contributions to mathematics and in particular to differential and integral calculus and also worked on the propagation of sound and the theory of vibrating strings. He also worked on the problems of the orbits of Jupiter and Saturn, the libration of the Moon (slight oscillation of the face of the Moon presented to the Earth), the three body orbiting problem and perturbations of the orbits of comets by the planets. While studying the three body orbiting problem and previous work by Euler, Lagrange discovered five points in the orbit of a planet orbiting the Sun where comparatively small bodies could remain with reasonable stability while moving as a whole with the planet. These points are known as the Lagrangian points of an orbit and are of vital importance to observatories in space as well as other space missions. Spacecraft utilizing the Lagrangian points are not simply parked there but are placed in small halo or lissajous orbits at right angles to the plane of the planetary orbit.

French chemistry pioneer Antoine Lavoisier (1743–1794) intervened on behalf of Lagrange during the French Revolution saving Lagrange from arrest but was himself arrested, charged with watering the soldiers' tobacco and accepting funding from the 'General Farm' for his chemistry research (a considerable fortune). The 'General Farm' had been established for 'farming' taxes and revenue for which Lavoisier was despised by the populace. On 8 May 1794, after a trial that lasted less than a day, a revolutionary tribunal condemned Lavoisier, his father in law and 26 other beneficiaries of the 'General Farm' to death. When Lavoisier pleaded for the tribunal to allow him to complete the task that he was busy with before his execution he was told: "The Republic has no need of scientists." Lagrange said on the death of Lavoisier, who was guillotined on the afternoon of the day of his trial:

> It took only a moment to cause this head to fall and a hundred years will not suffice to produce its like.

Antoine Lavoisier and Joseph Priestley did much pioneering research on the identification and properties of gases. Priestley discovered oxygen which he called 'dephlogisticated air'. Henry Cavendish (1731–1810) discovered hydrogen which he called 'inflammable air' and that the combustion of oxygen and hydrogen produced water. Lavoisier named the gases oxygen (acid maker) and hydrogen (water maker).

Both oxygen and hydrogen could reasonably be called 'water makers' but oxygen is poorly described as 'acid maker'. To interchange the names of these gases would now be unthinkable. There is a similar curiosity with the assignment of the electron charge. One would naturally think that the fundamental unit of electric charge, the electron, should be positive, but negative it is and negative it will stay. Joseph Priestley had an astonishing language ability learning Greek, Latin, Hebrew, French, Italian, German, Chaldean, Syriac and Arabic. We are indebted to Joseph Priestley for two inventions still in common use: the use of India rubber for erasing pencil writing and the use of carbon dioxide to make drinks fizzy. A profound discovery of Priestley states:

> The injury which is continually done to the atmosphere by the respiration of such a large number of animals... is, in part at least, repaired by the vegetable creation.

Jean Baptiste Joseph Fourier (1768–1830) will be well known to every student of electrical engineering. He determined that a complex wave form could be synthesized by a series of simple sinusoidal waves of higher frequencies and smaller amplitudes. He could have had no idea of the immense implications of his discovery for several branches of electrical engineering a century later. Fourier also did much work on the mathematical theory of heat. He coined the 'greenhouse' analogy to describe the similarity between the Earth's atmosphere and the glass of a greenhouse in trapping solar heat. Fourier was originally enthusiastically supportive of the French Revolution and joined the local revolutionary committee. Developments caused him to change his mind, but citizen Fourier found that attempting to withdraw from politics was a life threatening exercise. He was twice arrested, imprisoned and released and even thought that he would be sent to the guillotine. Pleas, possibly by Lagrange, colleagues and students resulted in his being able to walk free.

In 1798 Fourier joined Napoleon's army in its invasion of Egypt as scientific adviser. Fourier provided a useful service to Napoleon by determining the fastest rate at which cannons could be continuously fired without overheating. The mathematics of Fourier's *Théorie analytique de la chaleur* was not only useful to Napoleon's overheating cannons, this also provided the theory of Ricci flow (somewhat like heat flow) in mathematical manifolds, required in the proof of Poincaré's famous conjecture two centuries later. The invasion of Egypt was part of a bold plan to sever connections between England and India. Napoleon, on his arrival in Egypt, had the brazen audacity to declare: "I am a Muslim and have come to free this country from the Mamelukes!" Napoleon openly admitted that he preferred Islam to the religion of his own country. His ill considered idea that his troops should convert to Islam came to an abrupt end when they discovered that they would, in addition to a surgical procedure, never again be allowed to consume any alcohol. The winners of Napoleon's invasion of Egypt were the historians, artists, scientists and many other experts who contributed to the 23 volume *Description de l'Egypte* which inspired huge interest in Egyptian archaeology. They even discovered the famous Rosetta Stone. Napoleon suffered a humiliating defeat at the Battle of the Nile, where Horatio Nelson, suffering disproportionately less loss than his enemy, destroyed or captured almost the whole of Napoleon's warship fleet. The most spectacular and horrific event of the battle was the destruction of the magnificent French flagship *L'Orient* which was blown to smithereens when the powder magazines exploded. The combatants of both sides were so overawed by the ghastly spectacle that they stopped firing for twenty minutes. This explosion is depicted in bas-relief on the plinth of the memorial to Captain George Westcott in St. Paul's Cathedral. The flagstaff of L'Orient was salvaged and kept as a souvenir by Nelson at his home in Merton. After several horrific encounters with the Turks, Napoleon returned to France abandoning his army to its fate in Egypt. In this same year King Ferdinand of Naples conferred on Nelson the title 'Duke of Bronte' in gratitude for services rendered. Bronte is a small town in Sicily. After this, Nelson would sign his letters, even those to Emma, as Nelson & Bronte. This new title would result in the future father of the literary sisters of Haworth changing his name, as a young man, from the various Irish forms to Brontë with its curious accent. In Greek, 'bronte' means thunder; this was not lost on Charlotte who would playfully call herself 'Charles Thunder'.

The discovery of the Rosetta Stone provided a wonderful key to the decipherment of Egyptian hieroglyphics. The stone was inscribed in three registers of text written in hieroglyphics, demotic and Greek. We are very fortunate that the stone has survived at all; it had been used as recycled building material. No trace of the topmost register of figures of gods and priesthood remains. Of the hieroglyphic text only about a third remains; this was the language of the gods and priesthood. The demotic text, the everyday language of the time, is the most complete with only a small corner being lost. The Greek register, a requirement of the Ptolomaic dynasty, is the most mundane part of the stone; a simple alphabetic language with no religious symbols. Several of these granite stelae would have been made and erected at important temples. The Rosetta stone was exactly what was needed by

Jean-François Champollion (1790–1832). Champollion was a phenomenal linguist having learned Hebrew, Arabic, Syriac, Chaldean, Chinese, Coptic, Ethiopic, Sanskrit, Zend, Pahlevi and Persian. Champollion noticed that the Greek and demotic texts contained the name 'Ptolmys' which he was able to link to the cartouche in the hieroglyphics. He determined that the hieroglyphics are partly iconic and partly alphabetic. Studying an obelisk taken to England he found a cartouche containing the name 'Kliopadra'. This obelisk has become known as 'Cleopatra's Needle' and is a well known sight on the Thames Embankment. Champollion also noticed that an inscription at Abu Simbel contained the name 'Rameses'. A major contribution to the decipherment of the Rosetta texts was made by physicist Thomas Young who concentrated on the demotic language which is derived from hieroglyphics. To this end Young found it necessary to study the Coptic language as well, which, being derived from the demotic, gave an indication of the pronunciation of the hieroglyphic words and names. Young wrote an encyclopaedia article comparing 400 languages. The great decipherment success of Champollion and Young has resulted in modern Egyptologists being able to read hieroglyphics with relative ease. There remain however several other ancient languages which have defied decipherment. The final custody of the Rosetta Stone raises the question of ownership of priceless art treasures. This is not simply a matter of boys playing with, and capturing marbles. Art treasures belong to human civilisation and are best left with those who are best able to keep them in safe custody and also make them accessible to those able to appreciate them.

In any discussion on quantum physics it is almost *de rigueur* to mention the celebrated and paradoxical two-slit interference pattern experiment devised by physicist Thomas Young. This experiment has been widely written about but in case it is new to some readers it is briefly thus: The experiment consists of a monochromatic coherent light source, a barrier with slits, and a bit further a target screen which would probably be a photographic film.

We start with one slit. It will come as no surprise to see that light passing through the slit will produce a stripe of light on the screen which becomes dimmer away from the centre. We now use two closely spaced slits. Light reaching the screen from two slits will travel slightly different distances to parts of the screen producing light and dark bands because of interference between the two sources. At some points the wave troughs and crests will cancel each other out causing a dark band and at others the crests will reinforce each other causing bright bands. No big deal – this is exactly what one would expect from wave motion and could also be demonstrated by ripples on a pond of water. We now repeat the experiment by projecting photons one at a time. With one slit the result is the same as before. How about two slits – do we get a bright band for each of the slits? Not likely! Photons fired singly still produce the interference pattern! Part of the photon cannot get stuck on the barrier between the slits – it is a quantum of energy so it all gets through or nothing and it cannot split in two and go through both slits. This is not the end of the story – it gets stranger. The experiment is repeated using electrons instead of photons. Firing hard particles individually through the slits could not conceivably cause an interference pattern. A solid particle passing through a slit could not possibly know that there is another slit nearby and then hang around waiting for another particle to pass through the second slit in the hope of having some interference fun. What happens? You have guessed it: the electrons also produce the interference pattern. What is going on here? Electrons in flight have mass and must therefore be particles but they are sufficiently wavelike to behave in the same way as photons! It would be naïve to suppose that photons or electrons could divide in two and then proceed through both slits. There is an additional twist to this experiment. If detectors using light beams are installed to determine through which slit the individual electrons pass then there is no uncertainty regarding the electron path and no interference pattern is formed. It will be found that by increasing the wavelength of the light being used, a point will be reached when the wavelength becomes more than the separation of the slits. When this happens the resolution will be too poor to tell through which slit the electron passed and the interference pattern will be again restored. Many explanations have been offered for the paradoxical two slit experiment, some even more bizarre than the experiment itself, and no doubt many more explanations will be still forthcoming! We have here a situation which, if not spooky, is certainly bizarre. A modern approach to this problem is that the path taken by the photons or electrons is, like most other happenings at quantum level, a matter of probability and that the straight path is only the most likely route bearing in mind that nothing can be known about a photon until it is captured. We now have a situation where the light intensity of classical physics is replaced by a probability density function, and wave amplitude by probability amplitude, leaving us with the philosophical problem of having the distinction between a mathematical concept and a 'thing' becoming blurred. A mathematical probability function need have no difficulty in passing through any number of slits or holes. Can an abstract mathematical function become a 'thing' that you can drop on your toe? Of course not – but it is getting close! Other experiments have been devised for testing the single particle interference paradox. A photon is fired at a half silvered mirror (beam splitter) and can take one of two paths which are recombined. The photon has equal probabilities of taking either path. The probabilities are able to interfere with each other even though the physical photon could take only one of the paths.

Dr Louis Victor Pierre Raymond duc de Broglie (1892–1987) made a huge contribution to quantum theory and wave mechanics. In his doctoral thesis of 1924 he described electron waves based on work by Einstein and Planck. Broglie is a town in Normandy. At the 1929 Nobel Prize award ceremony C. W. Oseen, Chairman of the Nobel Committee for Physics of the Royal Swedish Academy of Sciences paid tribute to de Broglie ending his presentation address with:-

> Monsieur Louis de Broglie. When quite young you threw yourself into the controversy raging round the most profound problem in physics. You had the boldness to assert, without the support of any known fact, that matter had not only a corpuscular nature, but also a wave nature. Experiment came later and established the correctness of your view. You have covered in fresh glory a name already crowned for centuries with honour. The Royal Academy of Sciences has sought to reward your discovery with the highest recompense of which it is capable. I would ask you to receive from the hands of our King the Nobel Physics Prize for 1929.

De Broglie described himself as:

> ... having much more the state of mind of a pure theoretician than an experimenter or engineer, loving especially the general and philosophical view.

The unknowability of fundamental processes presents some serious philo-sophical problems. This leads on to the famous Schrödinger's Cat thought experiment introduced in 1935. This is of course a thought experiment and no one's furry pet has perished in a foolish cat-in-the-box demonstration. Crudely put, it comes down to the argument that it is impossible to know that something has occurred unless it has been observed. This is not the same as a policeman looking for a witness to confirm that something has happened. We are here dealing with fundamental uncertainty at quantum level. The idea that the cat can be both dead and alive is offered as an absurdity and not as a fact. In the experiment, observation is implied as that of a human observer, and does not consider the existence of a universal consciousness. If we bring a universal consciousness into the experiment then there is not the slightest uncertainty regarding the life of the cat or a Biblical sparrow for that matter. Readers wishing to pursue this further will find no shortage of published material dealing with the matter in great detail.

Erwin Rudolf Josef Alexander Schrödinger (1887–1961) made profound contributions to general relativity, radioactivity, wave mechanics and colour vision. Schrödinger was fascinated by the matter waves of Louis de Broglie. In November 1925 Schrödinger gave a seminar on De Broglie's work which was to have a profound influence on the future of physics. Dutch physicist Peter Debye, a student of Sommerfeld, suggested that there should be a wave equation. Within a few weeks Schrödinger had found his epoch making wave equation. He published his revolutionary work on wave mechanics and general relativity in 1926 which was received with great acclaim.

Planck described it as:	*Epoch making work*
Einstein wrote:	*... the idea of your work springs from true genius ...*
and again ten days later:	*I am convinced that you have made a decisive advance with your formulation of the quantum condition ...*
Paul Ehrenfest wrote:	*I am simply fascinated by your wave equation theory and the wonderful new viewpoint it brings. Every day for the past two weeks our little group has been standing for hours at a time in front of the blackboard in order to train itself in all the splendid ramifications.*

Schrödinger's obituary in the Times stated:

> *The introduction of wave mechanics stands ... as Schrödinger's monument.*

Ehrenfest is probably the most tragic figure in all of mathematical biography. Schrödinger's famous partial differential equation mathematically describing matter waves is interesting in that it contains the Bombelli complex number concept of $i = \sqrt{-1}$ Schrödinger said of his wave equation:

> *To each junction of the position and momentum coordinates in wave mechanics there may be related a matrix in such a way that these matrices, in every case satisfy the formal calculation rules of Born and Heisenberg... The solution of the*

natural boundary value problem of this differential equation in wave mechanics is completely equivalent to the solution of the Heisenberg algebraic problem.

By 1960 more than 100 000 papers had been written based on the application of Schrödinger's wave equation. Schrödinger's wave equation is regarded by many as the most brilliant mathematical achievement of the twentieth century. Not everyone was pleased with Schrödinger's Wave Mechanics. In a letter to physicist Wolfgang Ernst Pauli (1900–1958) in 1926, Heisenberg wrote:

The more I think about the physical portion of Schrödinger's theory, the more repulsive I find it ... What Schrödinger writes about the visualisability of his theory 'is probably not quite right,' ... This was followed by an expletive with no exact English equivalent.

Schrödinger was not enthusiastic about Heisenberg's Matrix Mechanics and wrote in 1926:

I knew of Heisenberg's theory of course, but I felt discouraged, not to say repelled, by the methods of transcendental algebra, which appeared difficult to me, and by the lack of visualisability.

There was a problem with Schrödinger's wave equation in the interpretation of the electron matter wave. Schrödinger saw this as a spreading out of matter which was rich in some regions but scarce in others. Max Born tackled this problem and deduced that this was a probability wave and that the matter was actually concentrated in points. This provided a fundamentally new approach to particle physics. Physicist John von Neumann (1903–1957) developed a mathematical theory showing that Schrödinger's Wave Mechanics and Heisenberg's Matrix Mechanics were mathematically equivalent. The proving of this equivalence caused mathematician David Hilbert to remark:

Physics is obviously far too difficult to be left to physicists and mathematicians still think they are God's gift to science.

Hilbert actually suggested to Heisenberg that he find the differential equation corresponding to his matrix mechanics. Had he done so he may well have discovered Schrödinger's famous equation before Schrödinger. Hilbert's work is of immense value to modern physics. He is famous for his 'Hilbert Space' – a space of infinite dimensionality. He discovered the correct field equations for General Relativity and also proposed 21 geometric axioms, the greatest development in geometry for two thousand years.

Schrödinger learned German and English together as a child as both languages were spoken at home. His father was Austrian and his mother had English and Austrian parents. He had a brilliant academic career and received his doctorate in 1910 for his dissertation *On the conduction of electricity on the surface of insulators in moist air.* During the war years Schrödinger was under much suspicion by the Nazi regime who regarded him as being politically unreliable. He fled from Germany in 1933 due to the gathering storm of political rumblings and was appointed as professor at Magdalen College Oxford. (Magdalen is pronounced Maudlin) Although not Jewish himself, Schrödinger felt that he could not live in a country where the persecution of Jewish people had become national policy. In this same year he was awarded the Nobel prize in physics jointly with Paul Dirac. He returned to Austria in 1936 to take up a position at Graz University. After the Austrian *Anschluss* he naïvely thought that he could appease the Nazis by declaring his support and loyalty to the Nazi cause. This was an act which he would regret for the rest of his life. The Nazis were not amused and Schrödinger had to flee to Rome in 1938. Schrödinger apologised to Einstein with:

> *I wanted to remain free – and could not do so without great duplicity.*

In 1943 Schrödinger wrote a paper on a unified field theory of which he was foolishly over enthusiastic. He was devastated when Einstein declared the work to be without merit.

When Schrödinger went on foreign lecture tours he would travel with two wives, the one formally wedded to him and the other not. This did not go down very well with his hosts at academic institutions in the USA and UK. When Schrödinger moved to Ireland in 1939 to take up a position at the Dublin Institute for Advanced Studies he went with an unusual *ménage:* his wife Annemarie, his daughter Ruth and Ruth's mother Hilde, wife of his research assistant Arthur March. Having two wives did not however deter him from scandalous encounters with female students. Matters were further complicated when our Teutonic version of Don Giovanni, now aged fifty two, fathered two more illegitimate daughters by two Irish women. The one woman, a typist young enough to be his daughter, was outraged when she discovered that Ruth was Schrödinger's illegitimate child. She left Dublin in high dudgeon, while still pregnant, and went to live in Liverpool.

On his retirement in 1956, Schrödinger returned to Vienna. He wrote innumerable articles and papers and the following books which give an indication of his huge range of interests:

1935	Science and the Human Temperament
1944	What is Life? This little book is one of the great scientific classics of the 20th century and an early hint at the function of the DNA molecule.
1946	Statistical Thermodynamics.
1949	Gedichte – a collection of poems.
1950	Space-Time Structures.
1954	Nature and the Greeks.
1958	Mind and Matter. Schrödinger asks what place consciousness occupies in the evolution of life.
1960	Meine Weltansicht. Expressing his own metaphysical outlook.

A few quotations from Schrödinger:

"The scientist imposes two things, namely truth and sincerity, imposes them upon himself and upon other scientists."

"If you cannot – in the long run – tell everyone what you have been doing, your doing has been worthless."

"Every man's world picture is and always remains a construct of his mind and cannot be proved to have any other existence."

"The idea of the continuum seems very simple to us. We have somehow lost sight of the difficulties that it implies ... We are told such a number as the square root of 2 worried Pythagoras and his school almost to exhaustion. Being used to such queer numbers from early childhood, we must be careful not to form a low idea of the mathematical intuition of these ancient sages; their worry was highly credible."

Schrödinger took a keen interest in ancient Vedic writings as can be seen from the following quotation dating from 1918:

"Nirvana is a state of pure blissful knowledge ... it has nothing to do with the individual. The ego or its separation is an illusion. Indeed in a certain sense two

*"I"s are identical namely when one disregards all special contents – their Karma.
The good of man is to preserve his Karma and develop it further ... when a man
dies his Karma lives and creates for itself another carrier."*

A profound experiment of physics is the entangled photon observation problem. Two photons are simultaneously created (in superposition) and ejected in opposite directions. The polarisation of the photons is not defined or knowable until observation by capture. When one of the photons is captured its polarisation can be observed but this will instantly cause the other photon to assume a matching polarisation even at a great distance away! The casual science reader can be excused for wondering how this could possibly be a problem for anyone, but a problem it is, and one with profound implications. This brings us to the inexplicable phenomenon of events being instantaneously linked over vast distances without the delay of the speed of light. In 1935 Albert Einstein, Boris Podolsky and Nathan Rosen introduced a thought experiment (the EPR Paradox) to illustrate the absurdity of quantum entanglement. Einstein was adamant that a physical event could not take place without a physical cause. The EPR Paradox would require that entangled pairs of electrons would somehow remain linked when separated by a great distance and would mysteriously continue to instantly influence each other without the delay of the speed of light. This introduced the idea of 'non-locality' and 'hidden variables' beyond the space-time continuum. Einstein described this as *Spukhafte Fernwirkung* (Spooky action at a distance). At this early stage it was impossible to perform an experiment to prove the EPR paradox. When it eventually became possible to experimentally test the paradox, using paired photons, the results proved the concept of entangled quanta, plunging quantum mechanics yet deeper into mystery. This was somewhat akin to another epoch making experiment of physics, the Michelson-Morley experiment (MMX) which proved the opposite of what the experimenters set out to prove. The EPR Paradox leaves us with the idea that everything in the universe is mysteriously linked to everything else, beyond the space-time continuum, suggesting that the entire universe was entangled at the time of the Big Bang and has remained so ever since! Is this not a case of experimental physics making an approach towards the universal consciousness?

Niels Bohr remarked:

*"How wonderful that we have met with a paradox.
Now we have some hope of making progress!"*

Physicist John Stewart Bell (1928–1990) is famous for his work on the EPR Paradox and is best known for his Bell's Theorem which can be simply stated as:

"No physical theory of local hidden variables can ever reproduce all of the predictions of quantum mechanics."

Physicists John von Neumann and Erwin Schrödinger have written in philosophical and mathematical depth on the EPR Paradox. When an atomic electron is ready to release energy it will instantly fall from the energy state of its orbital to an orbital of lower energy state and in the process eject a photon. This is called the 'quantum leap' but it is nothing like the swift and graceful leap of a ballet dancer! Leap is something of a misnomer – the electron can either be 'here' or 'there' but nowhere in between. Before the energy release it will have a high probability of being in the higher energy orbital and after the leap it will have a high probability of being in the lower energy orbital. The notion of movement does not apply here at all. To attempt an explanation of this process in terms of a billiard ball model of an atom would be ludicrous. What of the photon that is sent speeding on its way? This is a quantum of energy – and this energy, according to Planck's constant, will have a precise frequency, and in accordance with the velocity of light, a precise wavelength. The photon will also have a polarisation. This is a vector quantity with a precise value and direction. The photon will end its existence by donating its energy to another atomic electron orbital which could be beyond the extent of our galaxy. The photon cannot experience time so that the transfer will occur instantaneously even though to us it may appear to travel for millions of years. Unlike the entangled quanta effects, the photon quantum appears to us to travel between the electron orbitals at the speed of light. This is at the borderline between 'local' and 'non local'. The concept of the photon simply implies that a quantum of energy is transferred from one electron orbital to another – the existence of the photon cannot be verified in any other way! This puts us in a heavy involvement with Einstein's Special Theory of Relativity leaving us with the shocking question: 'do photons exist at all?' One could of course take a few mirrors and send a stream of photons bouncing from the one mirror to the next but this is only an illusion! Photons cannot bounce. When a photon strikes a mirror the quantum of energy will be taken up by an electron orbital and then ejected to the next electron target but it is not the same photon. The surface of the mirror plays a vital role in the reflection of photons yet at atomic level the idea of a surface is meaningless. The probability wave front of the photon plays a crucial role in the reflection of the photon from a surface of millions of atoms. When Max Planck thought that his discovery of light quanta was merely a mathematical trick with no physical existence he may well have been more accurate than he knew. Is the world now ready for the *Entanglement Theory of Light?*

Samuel Avery writes in his book *Transcendence of the Western Mind: Physics, Metaphysics and Life on Earth:* "Light is more fundamental than space and time. Things cannot go faster than or be smaller than light because that is what they are made of. There is no medium for light in the physical world because light is not in the physical world. The physical world is in light. Light is visual consciousness itself."

Relativistically, the concept of a photon is bizarre indeed: from the photon's frame of reference it cannot experience time or speed as the universe appears without size – it is simply an instantaneous transfer of energy from one electron orbital to another. It is only when viewed from within the space-time continuum that photons present the illusion of moving at vast speed from one point to another. As an illusory entity there is no contradiction in the photon appearing either as a particle or a wave depending on the method of observation.

A possible example of instantaneous influence over vast distance is the effect of gravitation. It is not at present known for certain whether gravitation effects are instantaneous or are propagated at a finite velocity. It is currently widely held that gravitation is propagated at the same velocity as that of electromagnetic radiation. The velocity of electromagnetic radiation is determined by the electric permittivity and magnetic permeability of space: there is at present no theory which links these properties of space to gravitation. In chapter nine we will deal with the clairvoyant examination of subatomic particles – this too could be done instantaneously at great distance. An interesting case occurred when the investigator was working in Australia but was unable to find a local source of a particular mineral. He could however remember the location of the mineral in a museum in Dresden and was able to investigate the required particle without difficulty. This leads on to practitioners of alternative medicine who are able to remotely diagnose the medical condition of their patients.

The entanglement effect of fundamental particles has given rise to a new, well funded and researched branch of computer science – quantum computing. At present, all computers have a fundamental architecture of binary numbers and logic. Binary numbers and arithmetic are very easily explained and understood without the aid of abstruse mathematics – quantum computing is not. Binary digits (bits) can have only two values, 0 and 1. Quantum binary digits (qubits) are of quaternary nature and have values of 0, 1, or a quantum superposition of these. A three bit binary number can have a single value of from 0 to 7, a sequence of three qubits can be in a quantum superposition of 8 states. As the sequence of qubits increases the numbers become astronomical. The great promise of quantum computing is the massively parallel processing of tasks millions of times faster than can be achieved by conventional computers. An obvious target application is in encryption cracking where an exceptionally large number comprising hundreds of digits, which is the product of two exceptionally large prime numbers, must be factorised into primes. Conventional computers are unequal to this task. There are no fundamental obstructions to the development of quantum computing. Another line of research in the quest for massively parallel computing is molecular computing which will utilize DNA molecules for processing. On 8 October 2008 a practical demonstration of quantum data transmission was given. The data was transmitted by means of individual photons in fibre optic cable. Any attempt to intercept this data would immediately result in its destruction exposing the attempted security violation.

How does one go about getting a good understanding of quantum theory? You don't! Quantum theory is an abstract philosophy dealing with the innermost secrets of nature at the borderline of the real and unreal. Our ordinary notions of space, time and movement are almost irrelevant at the level of quanta where 'here', 'there', 'yes' and 'no' are all matters of probability. Quantum theory is the domain of specialists with exceptional reasoning and mathematical ability. It is no shame for persons of non-genius intellect to recoil from quantum theory in confusion and disbelief. Quantum theory may seem an outrage to common sense but it must be remembered that common sense is developed from everyday experiences which have little relevance at the level of quanta.

The preceding paragraph was written in humility, not arrogance. The author makes no claim to being able to 'understand' quantum theory. Physicist Richard Feynman once stated that he never understood quantum mechanics. Does this not suggest that quantum mechanics is not humanly understandable?

Dirac felt less than satisfied:

"I feel that we do not have definite physical concepts at all if we just apply working mathematical rules; that's not what the physicist should be satisfied with."

Niels Bohr commented:

"I do not believe that Quantum Mechanics is understandable, at least for the usual meaning of the word 'understand'."

A comment from Heisenberg:

"I think that modern physics has definitely decided in favour of Plato. In fact the smallest units of matter are not physical objects in the ordinary sense; they are forms, ideas which can be expressed unambiguously only in mathematical language.

The physicist may be satisfied when he has the mathematical scheme and knows how to use it for the interpretation of experiments. But he has to speak about his results also to non-physicists who will not be satisfied unless some explanation is given in plain language. Even for the physicist the description in plain language will be the criterion of the degree of understanding that has been reached."

A remark of quantum despair from Schrödinger:

"I do not like it, and I'm sorry I ever had anything to do with it."

More from Schrödinger:

"If we think that we can picture what is going on in the quantum domain, that is one indication that we've got it wrong."

There is a common misconception that Einstein had a distaste for quantum theory – in fact he did much to develop the theory. It was the Heisenberg uncertainty principle and the random decay of atomic nuclei, which occurred without any physical reason, that caused Einstein to make his famous remark, writing to Max Born, in 1926:

"Quantum theory is very impressive but an inner voice tells me that it is not yet the real thing. The theory yields a lot, but it hardly brings us any closer to the secret of the Old One. In any case I am convinced that He does not throw dice."

Lucasian professor Stephen Hawking responded to this remark by adding:

"Not only does God play dice, He sometimes does not let us see where He has thrown them."

Niels Bohr added:

"Einstein, stop telling God what to do!"

In 1920, Werner Karl Heisenberg (1901–1976), began with fellow student Pauli to study theoretical physics and mathematics under the renowned professor Arnold Sommerfeld. Sommerfeld tutored many of the most eminent physicists of the early twentieth century. Heisenberg also attended lectures by Niels Bohr. In 1924 he visited Niels Bohr at the Institute for Theoretical Physics in Copenhagen where he met Einstein for the first time. Heisenberg and Pauli were both destined to become Nobel laureates for the discovery of closely related fundamental principles which would bear their names and would profoundly influence the future of physics. In 1925 Heisenberg invented matrix mechanics, the first version of quantum mechanics. He was awarded the Nobel Prize in physics in 1932 for:

> *The creation of quantum mechanics, the application of which has, inter alia, led to the discovery of the allotropic forms of hydrogen.*

Despite this puzzling statement, the chairman, in his presentation address, gave an excellent account of the significance of Heisenberg's work. Heisenberg used part of his prize money to purchase a Blüthner grand piano. This instrument had come from the production series which had won Blüthner the first prize in the 1910 world exhibition in Brussels. Heisenberg's introduction of his Matrix Mechanics in 1925 was the first formulation of Quantum Mechanics. This did not at first receive general acceptance due to the abstruse mathematics and lack of visualisation of the particles. Schrödinger's Wave Mechanics which came later in 1926 was generally more favoured. In matrix mechanics the physical quantities are not ordinary variables but mathematical matrices. The electron is interpreted as a particle with quantum behaviour using sophisticated matrix computations which introduce discontinuities and quantum jumps. The matrix formulation was built on the premise that all physical observables must be represented by matrices. The set of eigenvalues of the matrix representing an observable is the set of all possible values that could arise as outcomes of experiments conducted on a system to measure the observable. Paul Dirac in Cambridge and Pascual Jordan in Göttingen created unified equations known as 'transformation theory' which was to become Quantum Mechanics. While studying the papers of Dirac and Jordan, Heisenberg discovered a problem in the way one could measure basic physical variables appearing in the equations. While in frequent correspondence with Pauli, Heisenberg discovered his epoch making Uncertainty Principle which he described in a 1927 paper as:

> *The more precisely the position is determined, the less precisely the momentum is known in this instant, and vice versa.*

Another quotation from the same paper:

> *I believe that the existence of the classical "path" can be pregnantly formulated as follows:*

> *The "path" comes into existence only when we observe it. In the sharp formulation of the law of causality – "if we know the present exactly, we can calculate the future" – it is not the conclusion that is wrong but the premise.*

'Uncertainty' is a good word for describing Heisenberg's principle but there is a slight discrepancy in translation. Heisenberg used the word *Unbestimmtheid*. Heisenberg's uncertainty principle constituted an essential component of the broader interpretation of quantum mechanics, including Bohr's complementarity, and the statistical interpretation of Schrödinger's wave function, known as the Copenhagen Interpretation.

In a lecture on Reality, physicist-philosopher Peter Russell raised a chuckle from his audience when he banged Heisenberg and Schrödinger's heads together by declaring: *"Particles are the eigenvalues of a wave equation".*

It was in 1927 that Heisenberg attended the Solvay conference in Brussels. He wrote in 1969:–

> *"To those of us who participated in the development of atomic theory, the five years following the Solvay Conference in Brussels in 1927 looked so wonderful that we often spoke of them as the golden age of atomic physics. The great obstacles that had occupied all our efforts in the preceding years had been cleared out of the way, the gate to an entirely new field, the quantum mechanics of the atomic shells stood wide open, and fresh fruits seemed ready for the picking."*

Heisenberg and Born delivered a paper at the 1927 Solvay Congress which stated:

"We regard quantum mechanics as a complete theory for which the fundamental physical and mathematical hypotheses are no longer susceptible of modification."

The fifth Solvay conference, held in 1927 and devoted to the theme 'Electrons and Photons' was a glittering affair. Some of the participants were: Einstein, Lorentz, Mme Curie, Lemaître, Planck, Bohr, de Broglie, Compton, Dirac, Heisenberg, Pauli, Schrödinger and Ehrenfest. Seventeen of the participants were, or were to become, Nobel laureates. After this conference the term *photon* proposed by Gilbert Lewis of Berkeley came into general use. The first Solvay conference, on the theme 'Radiation and the Quanta', was held in 1911 and chaired by Hendrik Lorentz. The Solvay conferences, which are held every three years in Brussels, were founded by Belgian industrialist Ernest Solvay.

Heisenberg, an accomplished pianist, met his wife Elisabeth Schumacher at a concert at which he was performing in 1937. On this occasion he played the Beethoven G Major trio and described his playing of the slow movement as a continuation of his conversation with his ardent listener. They married in the same year in Berlin. They had seven children; three sons and four daughters. Elisabeth took an active role in the musical education of their children and Heisenberg guided them in their choice of instruments so that the family could perform together as an ensemble. Heisenberg had an excellent repertoire with a particular interest in Beethoven and Chopin. He once referred, in a letter, to Schubert's B Major trio as the most beautiful trio on earth. He could play the Beethoven concertos from memory and also played Mozart and Beethoven concertos with Max Born, on two pianos, the one scientist playing the piano part and the other the orchestral. Max Born played Beethoven piano and violin sonatas with his friend Albert Einstein. A famous quotation of Max Born:

"I am now convinced that theoretical physics is actually philosophy. ... I believe there is no philosophical high-road in science, with epistemological signposts. No, we are in a jungle and find our way by trial and error, building our road behind us as we proceed."

In a letter to his parents from Copenhagen, Heisenberg wrote:

"Tonight I am going to play Beethoven 'cello and piano sonatas with a young physicist. That will be great. It is really impossible to live without music. When one listens to music, one sometimes arrives at the absurd idea that life might have meaning."

Heisenberg was not able to identify with modern music:

"In music, I find that the compositions of recent years are no longer as compelling as those of earlier times. During the 17th century music was still largely orienting itself on the religious core of life at the time; in the 18th century there was a transition into the emotional realm of the individual, and the romantic music of the 19th century has penetrated the innermost depths of the human soul. But in recent years music seems to be caught in a noticeably restless and rather feeble stage of experimentation, in which theoretical notions play a more important role than the secure awareness of progressing along a predetermined course."

In his autobiography he mentions the sadness as when he and his friends were thinking of the great epoch of European music as gone forever. On his 70th birthday, Heisenberg was invited by the Bavarian Radio Orchestra to play a Mozart concerto with full orchestra in a broadcast program.

Albert Einstein also had a profound interest in music as can be seen from a quotation:

"If I were not a physicist, I would probably be a musician. I often think of music, I live my daydreams in music. I see my life in terms of music."

Einstein was a regular concert-goer. A memorable concert that he attended was the *début* of the thirteen year old violinist Yehudi Menuhin in 1929 with the Berlin Philharmonic orchestra conducted by Bruno Walter. The mammoth program comprised a Bach concerto and the celebrated Beethoven and Brahms concertos. Einstein was so moved by the brilliant performance that he embraced the young teenager and exclaimed: "Now I know that there is a God in heaven!" Lord Menuhin died in Berlin in 1999 after one of the most brilliant musical careers of the twentieth century. One of his most famous violins was the 'Lord Wilton' Guarneri del Gesu made in 1742.

Heisenberg was interested in the philosophy of physics writing two books on the subject. In 1950 He proposed a theory of elementary particles involving a non-linear spinor field.

During the war years, Heisenberg found himself dangerously caught between two political fires. In the West he was under much suspicion for his involvement in the Nazi nuclear weapons program and in Germany he was under suspicion for his deep involvement in relativity and quantum theory which were regarded by the Nazis as 'Jewish science'. 'Jewish science' was not an unreasonable description. Most of the pioneering scientists responsible for the development of quantum theory, relativity and nuclear physics were indeed of Jewish ancestry and the flight of many of them to escape Nazi persecution caused a brain drain from Germany which the Third Reich could ill afford. Heisenberg was vilified in a 1937 SS propaganda newspaper which labelled him as a 'White Jew' and included the chilling fanatical comment:

Heisenberg is only one example of many others ... They are all representatives of Judaism in German spiritual life who must all be eliminated, just as the Jews themselves.

Heisenberg was in mortal danger and his attempts to put an end to the mindless attacks were to no avail. The problem was eventually dealt with when Heisenberg's mother spoke to Anna, a friend whom she had known for some time, the mother of Reichsführer-SS Heinrich Himmler. A respectfully worded letter from Himmler assured 'the esteemed Herr Professor' that there would be no more attacks. Himmler, controller of the SS (Schutzstaffel) and the Gestapo (Geheime Staatspolizei) would become one of the most feared and despised officers of Nazi Germany. Himmler had a passionate interest in occult matters and also in the breeding of a master 'Aryan' race to rule the world. He also became grandmaster of occult secret societies. He died in detention shortly after the war quite likely from a self administered dose of potassium cyanide. Heisenberg was also under threat from the West; an assassin was sent to eliminate him in Zurich. The thug, armed with a pistol, actually sat through a lecture by Heisenberg on S-matrix theory in elementary particle physics, but fortunately the gunman goofed and Heisenberg lived. This plot was ill advised, ill conceived and inept on a dizzying scale. In May 1945 (after the war) Heisenberg and nine other German scientists were arrested by US forces and held at Farm Hall, near Cambridge, for six months. They were given excellent accommodation and meals, but in his letters, Heisenberg repeatedly mentioned his longing to return home where his family was living in reduced circumstances. In one letter he mentioned that the meals were as good as one might expect on an ocean liner. While in detention, Heisenberg was heard (under surveillance) to say, on hearing of the American deployment of the atomic bomb in Japan:

> *"I believe the reason why we didn't do it was because all the physicists didn't want to do it on principle. If we had all wanted Germany to win the war we could have succeeded."*

Otto Hahn responded with:

"I don't believe that, but I am thankful that we didn't succeed."

Heisenberg headed the Nazi nuclear weapons program and it was his recommendation that the Vemork hydro electric plant in Norway be upgraded in its production of heavy water for use as a nuclear reaction moderator. The Norsk Hydro-Elektrisk was powered by a massive 144 metre waterfall from the Hardangervidda. Heavy water was originally produced as a byproduct of fertiliser production developed in 1900 by physicist Kristian Birkeland. Ironically, graphite could have been used instead of heavy water as a neutron moderator. The heavy water production was seen by the Allies as an extreme military threat requiring drastic action. The first attack against the plant ended in total disaster. Operation Freshman was launched in November 1942. One of the two bomber tow planes and two large gliders crash landed in poor visibility and mountainous terrain. All thirty four Royal Engineers of the 1st British Airborne Division and the air crews either died in the crash landings or were shot. A second raid, this time by Norwegian commandos, was launched in February 1943. This raid was more successful resulting in the loss of 500 kg of heavy water but the damage to the plant was quickly repaired. A massive but disastrous American bombing raid in November 1943 missed the target but caused considerable loss of civilian life. A final sabotage attempt in February 1944 was the most successful, when the ferry *Hydro* carrying a shipment of heavy water was sunk but unfortunately with loss of civilian life. The ferry sank in the Tinnsjö (lake Tinn) in 400 metres of water and beyond any means of salvage at the time. After the war, Heisenberg had a highly successful career in theoretical physics, lecturing in several countries and also received many awards and academic honours.

Nuclear fission was discovered in 1939 by Lise Meitner, Robert Frisch, Otto Hahn and Fritz Strassmann. The proof of fission required the physical insights of Meitner and Frisch and the chemical findings of Hahn and Strassmann. Meitner became a target for anti-Jewish discrimination and credit for her brilliant achievements was shamefully claimed by Hahn. Meitner was posthumously honoured in 1992 when element 109 was named meitnerium. Meitnerium cannot occur naturally as it has a half life of a few milliseconds and only a few individual atoms have been created by bombarding a bismuth target with iron atoms. It is astonishing that a shy Austrian girl, and a pianist of modest accomplishment, could have made epoch making discoveries influencing the future of science and human history. Meitner had a working relationship with her physicist nephew Otto Robert Frisch (1904–1979) and they would on occasion play piano duets together. They had a standing joke that *allegro ma non tanto* meant 'fast but not auntie'.

Meitner's parents placed great value on education and gave her a private tutor as secondary schooling for girls was not fashionable at the time in Vienna. She entered Vienna University in 1901 where she could study her two passions, mathematics and physics, under the inspiring and famous (albeit tragic) professor Ludwig Boltzmann. She received her doctorate in 1907. When Lise went to work at the Kaiser Wilhelm Institute in Berlin she had to initially work in a basement room as women were not permitted to set foot in the institute. The Kaiser Wilhelm Institute would later be renamed the Max Planck Institute.

After working together for thirty years, Hahn, as a matter of political expediency, requested Meitner's dismissal from the Kaiser Wilhelm Institute. In order to avoid arrest, Meitner had to leave Germany quickly without travel documents. She was spirited away by train, assisted by Dirk Kostner, to Holland using a lightly travelled route. Wolfgang Pauli remarked: *"You have made yourself as famous for the abduction of Lise Meitner as for the discovery of hafnium."* Hahn could not continue without Meitner's assistance and continued to seek her advice after she had fled to Stockholm in 1938. In December of 1939 Meitner and her nephew Frisch explained and named 'nuclear fission' using Bohr's 'liquid drop' model of the atomic nucleus. When Frisch rushed to Copenhagen to tell Bohr of his aunt's discovery Bohr exclaimed: *"Oh! What idiots we have all been! Oh but this is wonderful."* Frisch and Bohr proved Meitner's ideas experimentally in Copenhagen after which Frisch and Meitner published a paper explaining newly titled 'fission'. These results were immediately confirmed around the world. Meitner was invited to join the Manhattan Project but declined saying that she would have nothing to do with the development of a bomb. One of the modest Meitner's famous remarks: *"I am not important: Why is everybody making such a fuss over me?"*

In 1944 Otto Hahn was awarded the Nobel Prize in chemistry for the 'discovery' of the fission of heavy metals. In the presentation address, Prof A. Westgren, chairman of the Nobel Committee for Chemistry of the Royal Swedish Academy of Sciences, made mention of Hahn's long association with Meitner. Hahn did not attend the award ceremony but it was remarkable that the ceremony was held at all. Germany was in the convulsive throes of its Armageddon and a few months from capitulation. Hahn received his award at the 1946 ceremony and in his lecture also made mention of his long association with Meitner. Hahn offered to share a portion of the prize money with Fritz Strassmann, but Strassmann would have no part in this unworthy claim to fame. Another story told about Meitner was a proposal to make a movie about the development of nuclear fission and the bomb. She declined stating: *"I would rather walk the length of Broadway in the nude than see myself in a movie."* Meitner became a visiting professor at the Catholic University in Washington D.C. and also lectured at the Bryn Mawr women's college near Philadelphia. Meitner moved, after her retirement, to a cottage in Cambridge, England, in 1958 where she remained until her death in 1968. Her gravestone bears an inscription composed by her nephew Frisch which reads: *"Meitner – a physicist who never lost her humanity."* Meitner never married but possibly a great comfort during her life was the constant friendship of such great people as Niels Bohr and his wife Margarethe, Max Born, Max Planck, Wolfgang Pauli, James Chadwick and Albert Einstein. The great loves in her life were music and walking in the Austrian mountains.

Heisenberg was clearly troubled by the thought of placing nuclear weapons in the hands of a military regime of doubtful morality and permanence. In September 1941, possibly putting his life at risk, he went to Copenhagen for a meeting with his erstwhile mentor Niels Bohr. Denmark was Nazi occupied territory at this time. No transcript of the meeting was made as the scientists spoke alone while walking where they could not be overheard. A great deal has been written about this meeting but it has never been determined what Heisenberg hoped that it might achieve. This incident has the powerful emotional tension of a Greek tragedy. Bohr is said to have been greatly troubled by the meeting. After reading the Robert Jungk book "Stærkere end tusind sole" (Brighter than a thousand suns) Bohr wrote, but never posted, a letter to Heisenberg which gives the impression that Heisenberg was convinced that Germany would win the war and would have access to nuclear weapons. Germany had access to supplies of uranium and plenty of expertise making it the most favourable country for the development of nuclear weapons yet the nuclear program came to nothing. Possible explanations were political indifference or skepticism, lack of funding or even Heisenberg slow pedalling the project in order to prevent calamity. Fortunately London remains today as part of the physical objective universe and we can experience it without having to refer to old books and movies to know what it was like.

After the war, a search was made for Nazi nuclear facilities and only one experimental reactor was found and which had not yet been put into operation. The reactor was found in a cave under a church in Haigerloch, Bavaria. This church provided Heisenberg with an opportunity for occasional organ playing. He had a particular interest in Bach fugues and even attempted to write a fugue himself. A salvage operation on the sunken Norwegian ferry *Hydro*, using modern remote controlled salvage equipment, was carried out sixty years after the sinking. The low concentration of heavy water in the barrels indicated about only half a ton of heavy water – an insignificant amount for a nuclear weapons program. The contents and numbering of the barrels indicated that the cargo manifest was indeed correct. The barrels were filled according to the concentration of heavy water so that some of the barrels were only partly filled. These floated after the sinking and were recovered by Nazi forces. The ultimate facts, when viewed with hindsight, serve only to heighten the level of tragedy and futility of the heavy water episode. The Nazis had concluded that to win the war it would have to end quickly and correctly predicted that nuclear weapons would not be ready in time to make any difference to the outcome. Germany was in any case unable to launch a weapons program on the vast scale of the Manhattan project. As with so many military operations, whether successful or otherwise, it was the civilian population who had to pay the most dearly.

In 1940 Japan launched a nuclear weapons program headed by world renowned physicist Dr Yoshido Nishina, a friend of Niels Bohr and an associate of Einstein. Japan collaborated with Germany in weapons development and was able to obtain supplies of uranium from Korea. On the 14th May 1945 U-boat U-234 surrendered to US forces on the capitulation of Germany. The submarine, a large mine-laying vessel, had a most interesting cargo. Amongst three crated Messerschmidt aircraft, spare engines and other items were 10 canisters containing 560 kg of uranium oxide bound for Japan. The unrefined material contained a rather small quantity of uranium U-235. This consignment would, in any case, have been too late to be of any use in a nuclear fission weapon for the war effort. It would have been possible however to use the uranium in powdered form in a conventional bomb to cause widespread lethal radioactive contamination over a city. The two Japanese officers on board the submarine died with honour and were buried at sea.

Niels Hendrik David Bohr (1885–1962) was born in Copenhagen into an influential and academic family. He received his doctorate from Copenhagen University in 1911 for his thesis on the electron theory of metals. In 1912 he moved to Manchester where he joined Ernest Rutherford's group studying atomic structure and built on the atomic model proposed by Rutherford. Ernest Rutherford (1871–1937) 1st Baron Rutherford of Nelson was born in New Zealand. He was awarded the Nobel Prize in Chemistry in 1908. He tutored a large number of renowned scientists, ten of whom were to become Nobel laureates. Bohr returned to Copenhagen in the summer of 1912 and married Margarethe Nørlund. They had six sons. His first major scientific discovery, in 1913, was his model of the atom consisting of a nucleus with orbiting electrons. Bohr was aware that the Rutherford atom was mechanically and electromagnetically unstable but was able to explain stability by means of the new quantum theory being developed by Planck and Einstein. This model also explained the chemical properties of atoms by the number of electrons in the outer orbits and also the ejection of light quanta by electrons falling to lower orbits. After serving as a lecturer in Manchester and then in Copenhagen, Bohr was appointed to a professorship in 1916. The university created the Institute of Theoretical Physics for Bohr in 1921 where he served as director for the rest of his life. The Institute became an international centre for work on atomic physics and quantum theory. After meeting with Bohr in 1920 Einstein wrote in a letter: *"Not often in my life has a human being caused me such joy by his mere presence as he did."*

Bohr used his Principle of Complementarity to explain the difference between the two forms of helium: parhelium and orthohelium. These are not isotopes but differ only in their spectral lines. Bohr deduced that the difference lay in the configuration of their circular and elliptical electron orbits. Niels Bohr was awarded the 1922 Nobel Prize in Physics. Professor Arrhenius concluded his award presentation address with:

> *"Professor Bohr. You have carried to a successful solution the problems that have presented themselves to investigators of spectra. In doing so you have been compelled to make use of theoretical ideas which substantially diverge from those which are based on the classical doctrines of Maxwell. Your great success has shown that you have found the right roads to fundamental truths, and in so doing you have laid down principles which have led to the most splendid advances, and*

promise abundant fruit for the work of the future. May it be vouchsafed to you to cultivate for yet a long time to come, to the advantage of research, the wide field of work that you have opened up to Science."

In 1943 Bohr felt that his arrest by Nazi forces was imminent due to his Jewish ancestry and outspoken anti-Nazi views. He was assisted by the Danish resistance movement who took him with his family to Sweden at night in a fishing vessel. A few days later he was flown in an unarmed British bomber on a hazardous flight to England. Bohr and his son Aage worked for a few months in England and then moved to the US to work on the Manhattan Project. In 1947 the Danish government awarded Bohr the prestigious Order of the Elephant. This award allowed Bohr to place a carving of his family coat of arms on a wall of fame. Bohr did not have a coat of arms so he designed one himself. As a centrepiece to the design Bohr decided to depict his famous Principle of Complementarity. He chose the ancient Chinese Taoist Yin-Yang (female-male) symbol which represents Taoist pairs of opposites which also finds expression in the black and white chequered floor tiles or tessellated pavements of religious buildings. The Yin-Yang consists of a circle divided in half by two semicircles and the two halves are depicted in black and white. Bohr's inscription reads 'CONTRARIA SUNT COMPLEMENTA' which means 'Opposites are Complements'. Bohr pointed out ways in which the idea of complementarity could throw light on many aspects of human life and thought. He died in Copenhagen in 1962.

Wolfgang Ernst Pauli (1900–1958) was born in Vienna. (Pauli must not be confused with German physicist Wolfgang Paul). His second name was given in honour of his godfather, physicist Ernst Mach. Pauli is most famous for his Exclusion Principle which is fundamental to chemistry and also explains the formation of white dwarfs and neutron stars in the cosmos. It also explains why fundamental charged particles do not clump themselves together. The Exclusion Principle is closely linked to Heisenberg's Uncertainty Principle. It is speculated that the Pauli Exclusion principle may also be fundamental to the structure of the atomic nucleus all the way down to quark and superstring level. The Exclusion Principle is given as a fundamental fact of nature without explanation. Pauli graduated with distinction at the Döblinger Gymnasium in Vienna in 1918 and shortly afterwards published a paper on Einstein's General Relativity. He studied under Sommerfeld in Munich receiving his doctorate in 1921. Sommerfeld asked Pauli to write an article on relativity for an encyclopedia. This prestigious document was praised by Einstein. He spent a year as assistant to Max Born in Göttingen, a year with Niels Bohr in Copenhagen and six years as lecturer at the University of Hamburg. In 1928 Pauli was appointed as professor at the Federal Institute of Technology in Zurich and also held visiting professorships in Michigan, Princeton and Purdue Universities.

Pauli married in 1929 but this marriage to a cabaret performer unfortunately lasted less than a year. In 1931 Pauli had a severe breakdown and consulted psychotherapist Carl Jung. Pauli and Jung both lived near Zurich and both held professorships at the Federal Institute. Jung began to interpret Pauli's deeply archetypal dreams and Pauli became one of Jung's best students. Pauli felt that he had a great deal of trouble with women. Jung sent him to his student Erna Rosenbaum for psychotherapy. After Jung had declared Pauli fully cured and released from analysis, Pauli continued to have archetypal dreams, not concerning emotional problems, but with assumptions in physics and natural science.

In 1934 he married Franca Bertram, a marriage which would last for the rest of his life. Pauli moved to the US in 1940, on the invitation of Einstein, where he became professor of theoretical physics at Princeton. In 1945 he was nominated by Einstein, and awarded, the Nobel Prize in physics for:

His decisive contribution through his discovery in 1925 of a new law of nature, the Exclusion Principle.

Pauli was unfortunately not able to be present at the award ceremony.

In 1930 Pauli proposed a new particle, the neutrino, in order to explain the small loss of energy when a neutron decays into a proton and an electron (Beta decay). The neutrino would not be observed experimentally until 1959. It has subsequently been determined that there are three types of neutrino. Pauli collaborated with Eugene Wigner in the development of Quantum Electrodynamics. Pauli published little, preferring to write of his discoveries in letters to Bohr and Heisenberg. Pauli was a prolific letter writer. In addition to a large number of letters written to Carl Jung, he also wrote frequently to Einstein, Bohr and Heisenberg. These letters have been published in several volumes.

Physicist and molecular biologist Dr. Leo Szilard (1898–1964) was born in Budapest where he completed his engineering studies specialising in thermodynamics. He moved to Berlin in 1920 where he enrolled as a physics student at the University of Berlin taking physics classes from Einstein, Planck and von Laue. In 1922 he wrote his dissertation on phenomenological thermodynamics and received his doctorate in physics. Szilard collaborated with Einstein in the development of a home refrigerator with no moving parts. In 1928 he held teaching seminars on quantum theory with John von Neumann. Szilard filed German patents for the linear accelerator and the cyclotron and later the electron microscope. He taught at seminars on theoretical physics, nuclear physics and chemistry with Schrödinger, John von Neumann and Lise Meitner. In 1933 Szilard fled to Britain to escape Nazi persecution. Szilard read an article in 'The Times' quoting Lord Rutherford as saying, "anyone who looked for a source of power in the transformation of atoms was talking moonshine." This prompted Szilard, while walking in London, to conceive the idea of a neutron chain reaction for which he filed a British patent the following year. Astonishingly, this was a few years before the discovery of nuclear fission. Not surprisingly, his request for laboratory facilities at Cambridge to investigate chain reactions was rejected by Rutherford. He continued his research, first at St. Bartholomew's Hospital in London, and then at Oxford. Szilard at first thought that beryllium, and then indium could be used for demonstrating chain reactions – it was only a few years later discovered that uranium could be used for this purpose.

The immensity of Szilard's contributions to scientific research is awesome. Linear and cyclotron accelerators are fundamental to research in nuclear physics and the electron microscope has become indispensable in many branches of science and engineering. Whether for good or evil, the nuclear chain reaction is a discovery of epoch making proportions. Ironically, particle accelerators are the largest machines ever manufactured and are used for research into the smallest of sub atomic particles. The underground particle accelerator complex at CERN, partly in France and the remainder in Switzerland, is awesome. (The Large Hadron Collider is described in chapter six). Szilard moved to the US in 1938 where he continued with his research. He was later to become firmly opposed to the use of nuclear weapons and also became involved in international politics. In 1960 he attended a meeting with Nikita Khrushchev in New York and proposed methods of reducing US-USSR tensions including the establishment of the Washington-Moscow hotline. Szilard organised the founding of the EMBL (European Molecular Biology Laboratory) in Heidelberg. The Szliard Library on the EMBL campus boasts a collection of some 25 000 items. In 1962 Szilard founded the Council for a Liveable Earth which has a dolphin as its emblem. Szilard was a scientist inclined to brilliant ideas of genius but would leave it to others to take his ideas through the industrial development model stages and consequently has been described as 'the genius in the shadows'.

A very small but important matter concerning research into the atomic nucleus is the neutron. In 1919 Rutherford had succeeded in disintegrating nitrogen atoms by bombarding them with alpha particles. Using alpha particles was hardly very useful as it would be very difficult to approach a strongly charged positive nucleus with a positive particle. Things would be dramatically changed by the discovery of the neutron by James Chadwick (1891–1974) in 1932. The neutron, being without any electrical charge, was the perfect go-anywhere particle. After graduating at the Honours School of Physics at Manchester University in 1911, Chadwick spent two years working under professor Rutherford on various radioactivity problems, obtaining his M.Sc in 1913. On being awarded a scholarship he went to work in the Berlin Physikalisch Technische Reichsanstalt at Charlottenburg under professor Geiger. His accommodation was probably not to his satisfaction – he was interned as an enemy alien for the duration of the war in the Zivilgefangenenlager at Ruhleben. Chadwick returned to England in 1919 to accept the Wollaston Studentship at Gonville and Caius College, Cambridge. (Caius is pronounced 'keys') He joined Rutherford at the Cavendish laboratory in studies of atomic nuclei. He was elected a fellow of Gonville and Caius College in 1921 and became Assistant Director of the Cavendish Laboratory in 1923. He was elected a fellow of the Royal Society in 1927.

Chadwick was awarded the Nobel Prize for Physics in 1935 for the discovery of the neutron. A German physicist, Hans Falkenhagen, had discovered the neutron at about the same time as Chadwick, and shocked by the immensity of his discovery and fearful of the consequences decided not to publish his work at the time. Chadwick generously offered to share the Nobel Prize with Falkenhagen but he declined the offer. Chadwick was elected in 1935 to the Lyon Jones Chair of Physics at the University of Liverpool. From 1943 to 1945 he worked in the USA as head of the British Mission attached to the Manhattan Project. When he realised that the atomic bomb was not only possible but inevitable he had to resort to the regular use of sleeping pills. He was Master of Gonville and Caius College, Cambridge from 1948 to 1958. He was knighted in 1945.

Dr Paul Adrien Maurice Dirac (1902–1984) began his studies at Cambridge in 1923 where he wanted to undertake research into general relativity. He also became interested in quantum theory and completed his first paper dealing with quantum problems in 1924. At the time a new theory was required as classical physics could not explain the behaviour of atoms and electrons. Dirac was given proofs of a paper by Heisenberg to read in the summer of 1925. The significance of the algebraic properties of Heisenberg's commutators struck Dirac when he was out for a walk in the country. He realized that Heisenberg's uncertainty principle was a statement of the non-commutativity of the quantum mechanical observables. He realised the analogy with Poisson brackets in Hamiltonian mechanics. This similarity provided the clue which led him to formulate for the first time a mathematically consistent general theory of quantum mechanics in correspondence with Hamiltonian mechanics. Mathematician Sir William Rowan Hamilton (1805–1865) developed his Hamiltonian function in 1835 which would become indispensable in quantum mechanics nearly a century later. In 1852 Hamilton used the term 'Occam's Razor' referring to the law of succinctness *lex parsimoniae* proposed by 14th century English Franciscan Friar William of Ockham. This simply means: shave away all non-essential material from your theorem. Dirac's widely acclaimed theory included the wave mechanics of Schrödinger and the matrix mechanics of Heisenberg. Dirac's work was held in very high regard by Heisenberg.

The non-commutativity of variables in matrix algebra implies that the multiplications A×B and B×A do not give the same result. There is nothing unusual in multiplications giving more than one result. As a simple practical example take a stick and multiply it by three – the result will usually be expected to be three sticks but it could also give a stick three times as long or one three times as wide. In vector algebra there are two methods of multiplication involving quite different operations and giving quite different results. I can remember a very practical application of vector algebra when detailing large diameter steel pipes in a civil engineering drawing office. The bends and elbows in the pipes were drawn with precise dimensions and angles but when two pipes met coming from different angles in the drawing as well as from different angles to the plane of the drawing, calculating the angle between the pipes was regarded as an impossible mathematical problem and it was left to the boilermaker to cut and weld the pipes on site as best he could. When I showed the other draughtsmen that determining the angle between the pipes, using vector multiplication, was a quick and trivial calculation, they looked on in open mouthed disbelief. I doubt if any of them ever attempted the calculation.

Dirac was awarded a Ph.D. in 1926 for his doctoral thesis *Quantum mechanics* after which he went to Copenhagen to work with Niels Bohr. Accepting an invitation from Ehrenfest, he spent a few weeks in Leiden on his way back to Cambridge. Dirac made frequent visits to the Soviet Union. He laid the foundations for Quantum

Electrodynamics 'QED'. In 1928 he found a connection between relativity and quantum mechanics, his famous spin-1/2 Dirac equation which could explain the mysterious magnetic and 'spin' properties of the electron. In 1929 he made his first visit to the United States, lecturing at the Universities of Wisconsin and Michigan. After the visit, along with Heisenberg, he crossed the Pacific, lectured in Japan and returned home via the trans-Siberian railway. Dirac used his famous equation to predict the existence of the antiparticle to the electron, the 'positron'. All charged particles are now known to have antiparticles.

In 1930 Dirac published his classic book *The Principles of Quantum Mechanics* and for this work he was awarded, jointly with Schrödinger, the Nobel Prize for Physics in 1933. He at first thought to turn down the prize due to his dislike of publicity, but when it was pointed out to him by Lord Rutherford that he would receive far more publicity if he declined this great honour and large sum of prize money, he decided to accept it.

Dirac was appointed Lucasian Professor of Mathematics at the University of Cambridge in 1932, a post he held for 37 years. In 1933 he published a pioneering paper on Lagrangian quantum mechanics. A memorial meeting was held at the University of Cambridge on 19 April 1985 and P. Achuthan, reviewing the papers presented wrote:

> ... we vividly see everywhere the brilliant imprints of Dirac, unifier of quantum mechanics and relativity theory. Each of the pieces not only is in praise of an exceptionally gifted intellect but also places on record how deeply and abidingly the human mind can delve into the realms of mathematical insight and modelling, keeping intact the spirit of beauty and clarity of a creative genius. Only a few Nobel laureates ever can compare as well with this giant of mathematical sciences in whose demise the world of original thinking certainly has lost one of the most precious souls retaining fortunately still the glory for others to sing and emulate for a long time to come.

In November 1995, a plaque commemorating Paul Dirac was unveiled in Westminster Abbey. The memorial address was presented by Professor Stephen Hawking who was the present incumbent of the Lucasian Chair of Mathematics at Cambridge.

Professor Hawking summed up with:

"Dirac has done more than anyone this century, with the exception of Einstein, to advance physics and change our picture of the universe."

Sir Isaac Newton became Lucasian Professor of Mathematics in 1669. The Lucasian Chair was endowed by Henry Lucas MP in 1663.

The scientists and mathematicians who have contributed to our present understanding of the universe number in the tens of thousands. We cannot possibly list them here – let it suffice to briefly mention one more: Swiss born mathematician Leonhard Euler (1707–1783). Euler is considered to be the most prolific mathematical writer of all time – it scarcely seems credible that all his achievements could have been crammed into a single lifetime. The general appearance and notation of mathematics as we know it today are largely due to Euler. Students seeing for the first time the wonderful vistas opened up by complex number theory may be inclined to think 'Wow! this is really something new' but this was a stroke of genius by Euler two and a half centuries ago. His famous complex number equation linking exponential functions to trigonometric functions is a triumph of mathematical brilliance. The original definition of the complex number concept of $i = \sqrt{-1}$ comes to us from Rafaello Bombelli (1526–1572). Every engineering student will be familiar with the Euler formula for the buckling load of a slender column. Euler integrated the differential calculus of Leibniz and Newton's fluxions into mathematical analysis. He made significant contributions to pure and applied mathematics including: calculus of variations, analysis, number theory, algebra, geometry, trigonometry, analytical mechanics, acoustics, hydrodynamics and lunar theory. Euler discovered that for any polyhedron, the number of faces plus the number of vertices minus the number of edges gave the value 2. His work in this area would be taken further in Poincaré's algebraic topology. When first approached by Carl Ehler, mayor of the town of Danzig, to solve the Königsberg bridge problem, Euler thought that the problem was beneath the notice of a mathematician. The picturesque town of Königsberg in Prussia (now Kaliningrad, Russia) consisted of four separate quarters connected by seven bridges and Ehler required a path which could take groups of tourists to all four quarters but cross each bridge only once. Euler soon realized that the logic required a new branch of mathematics and solved the problem, where algebraically possible, for any number of islands and bridges making a major contribution to algebraic topology.

Euler was fortunate in obtaining his initial mathematical tuition from the famous professor Johann Bernoulli. He joined the St. Petersburg Academy in 1727 where he was in an exceptional environment surrounded by the most eminent scientists of the day. He was appointed to the senior chair of mathematics when it was vacated on the departure of Daniel Bernoulli. He remained in Russia until he accepted a position at the Berlin academy of science in 1741. Euler served as a medical lieutenant in the Russian navy from 1727 to 1730. Euler published some 900 publications in his lifetime – his collected works run to more that 70 volumes. After his death in 1783, the St. Petersburg Academy continued to publish his works for 50 years. Euler did considerable work on celestial mechanics publishing in 1772 a 775 page work on the motion of the moon as well as articles on the calculation of planetary orbits. Euler was not only a theoretician but undertook considerable practical tasks such as hydraulic engineering projects, fire engines, magnetism, ship building, cartography, science education and innumerable others. He even served as advisor to the government on state lotteries, insurance, annuities, pensions and artillery. Some of Euler's major mathematical works were:

1736   *Mechanica*

1739   *Tentamen novae thereoriae musicae*

1740   *Methodus inveniendi lineas curvas*

1748   *Introductio in analysin infinitorum*

1753   *Theorea motus lunaris*

1755   *Institutiones calculi differentialis*

1765   *Theoria motus corporum solidorum*

1768   *Lettres à une princesse d'Allemagne* (written in French, 250 letters, 800pp)

   Letters to Princess Anhalt-Dessau of Germany (3 vols.)

1770   *Institutiones calculi integralis*

1770   *Dioptrica*

1772   *Theoria motuum Lunae*

1779   *Anleitung zur Algebra*

Euler was not in the least distracted from his work by his children – he would quite happily work with a child on his lap and several others playing at his feet. In 1734 Euler married Swiss born Katherina Gsell who bore him thirteen children of whom only five survived infancy. Euler produced his phenomenal output of work despite losing the sight of his right eye in 1735 and that of his left eye in 1766. Catherine the Great referred to Euler as her Cyclops Mathematician. It is a testament to Euler's genius that the great French mathematician Pierre-Simon Laplace told his students: *"Liesez Euler, liesez Euler, c'est notre maître a tous"* (Read Euler, read Euler, he is our master in everything). Fourier remarked: *Laplace was born to perfect everything, to deepen everything, to push back all the boundaries, to solve what was thought to be insoluble.*

June 1696 saw a mobilisation of the major mathematical forces of the world when Johann Bernoulli announced a competition for the solving of the brachistochrone problem. The problem appears deceptively simple: Determine the path that a particle, acted upon only by gravity, must slide down to move in the shortest possible time from point A, to a lower point B which is not directly below. The problem was put to Newton, Jacob Bernoulli, Leibniz and de L'Hôpital. Johann Bernoulli had solved the problem before announcing the competition. Galileo had tackled the problem in 1638 and determined that the path was a circular arc, a near miss, but not correct. All of the mathematicians were able to solve the problem but not by using the same methods. Johann Bernoulli and Leibniz deliberately tempted and goaded Newton in a carefully worded letter to attempt the problem. Newton's biographer wrote:

> ... in the midst of the hurry of the great recoinage, did not come home till four (in the afternoon) from the Tower very much tired, but did not sleep till he had solved it, which was by four in the morning.

Newton was not pleased and said:

> "I do not love to be dunned and teased by foreigners about mathematical things."

The correct solution is a cycloid. A simple cycloid is the path taken by a point on the circumference of a circle when the circle is rolled along a straight line. For the present problem the circle must roll on the underside of the line. It is interesting to note that if the points have a horizontal separation of more than half a cycloid width, then the last part of the movement of the particle will be in an upwards direction. If both points are on the horizontal line then half of the movement will be downwards and the other half upwards. Two points in the same plane as a straight line, which are not on opposite sides of the line, are sufficient to define the size of a cycloid. Many other types of cycloid are also possible by having the tracing point inside or outside the rolling circle and straight or curved rolling paths. Euler also solved the brachistochrone problem but for the much more daunting case of the particle moving through a resisting medium.

As has been mentioned before, photons are without mass and cannot experience gravitational acceleration – yet they will appear to accelerate towards massive objects in exactly the same way as if they did have mass. If photons did have mass it would be impossible for them to travel at the speed of light. What is happening here? In 1916 this effect was required by Einstein's *Principle of Equivalence* which forms part of the General Theory of Relativity. With the aid of mathematician Marcel Grossmann, he found out how to write physical laws in a form that is valid for any choice of coordinates by a method involving the use of general tensor analysis. General relativity is formulated completely in the language of tensors. The word tensor was originally introduced by Hamilton but tensor calculus was developed around 1890 by Gregorio Ricci-Curbastro. Einstein entered into a correspondence with mathematician Tullio Levi-Civita for further work on tensors, writing to him in a letter:

> "I admire the elegance of your method of computation; it must be nice to ride through these fields upon the horse of true mathematics while the like of us have to make our way laboriously on foot."

A famous prediction resulting from General Relativity was that light from a star would be very slightly bent when passing by the surface of the Sun. In 1704 Isaac Newton suggested that light could be deflected by gravity. There was no possibility of performing a terrestrial experiment to prove the theory.

Sir Arthur Eddington was one of the first physicists to understand the early ideas of Einstein's relativity. On 29 May 1919 Sir Arthur performed observations of the solar eclipse on the island Principe. Unfavourable weather conditions and other problems would have wasted the opportunity had Sir Arthur not taken the precaution of arranging duplicate observations to be taken elsewhere. He managed to record the bending of starlight in the vicinity of the Sun thus proving Einstein's theory. He wrote the following parody of quatrain LXVI of the Rubaiyat of Omar Khayyam:

Oh leave the wise our measures to collate
One thing at least is certain, light has weight.
One thing is certain and the rest debate
Light rays, when near the Sun, do not go straight.

Ghiyath al-Din Abu'l-Fath Umar ibn Ibrahim al-Nisaburi al-Khayyami (c.1044–1123) poet, mathematician, astronomer and philosopher of Persia, is well known in the West for his Rubaiyat translated by Edward FitzGerald in 1859, which tends to overshadow his far greater achievements in mathematics. Far ahead of his time in mathematical methods, Omar supported his algebraic solutions by geometrical constructions and proofs. Celebrated as the astronomer poet of Persia, Omar was also the first mathematician to study and classify cubic equations and to employ conic sections in their solution. Quite remarkably, Omar was able to calculate the number of days in a year correctly to more than six decimal places. Omar's treatise on algebra has been in use as a school text in Persia for centuries. Omar's perception of the relentless irreversibility of the time vector is evidenced in quatrain LXXVI of the Rubaiyat.

The Moving Finger writes; and, having writ,
Moves on: nor all your Piety nor Wit
Shall lure it back to cancel half a line,
Nor all your Tears wash out a word of it.

In 1923 Sir Arthur wrote his 'Mathematical Theory of Relativity' which Einstein described as:

... the finest presentation of the subject in any language.

The energy source of stars remained a mystery for millennia until Sir Arthur proposed that:

Probably the simplest hypothesis ... is that there may be a slow process of annihilation of matter.

Arthur Eddington was knighted in 1930 and received academic honours and gold medals from around the world. In his later years, Eddington's reputation declined, leaving him sidelined from the mainstream of scientific study.

It is not necessary to wait for an eclipse to see the effect of gravity on light. This phenomenon is also observable in what has become known as the gravitational lens. If two stars are exactly in line with the Earth and one of the stars is about midway between, then the furthermost star will appear as a ring of light around the nearer star due to the gravitational bending of the light. This effect is also observable when observing entire galaxies instead of individual stars. This does not mean that light is actually accelerated by gravitation but that, according to Einstein, it is space-time itself that is curved by gravitation. An interesting incident occurred in 1913 when Max Planck visited Einstein who told him of the current state of his work on the General Theory of Relativity. Planck replied:

> "As an older friend I must advise you against it, for in the first place you will not succeed, and even if you succeed no one will believe you."

One of the more startling claims of the General Theory of Relativity, which deals with gravitation, is that Euclidean geometry would not precisely hold on an astronomical scale in the presence of a massive astronomical body. In particular, for a large triangle in the vicinity of a star, the three interior angles would not add up to precisely two right angles. Einstein put his theory in a nutshell with:

> "When forced to summarise the general theory of relativity in one sentence: time, space and gravitation have no separate existence from matter."

Before Einstein proposed space-time as a four dimensional continuum it was thought that time was of so fundamental a nature that it was impossible to describe time as there was nothing simpler with which to describe it. We can now create a simulated localised time simply by playing a video tape or DVD.

A nagging question still remains. Since the 19th century the Michelson Morley experiment presumably proved that the luminiferous ether did not exist and that empty space was really nothing at all. Can a space which is absolutely nothing have the properties of electrical permittivity and magnetic permeability? One could argue that these properties simply describe the actions of static and moving charges but they also perfectly describe the flight of photons, or electromagnetic energy, in the absence of any charges. One cannot but admire midshipman Albert Michelson (1852–1931) for the dedication and methodical precision with which he pursued his experiments with light until his death in 1931.   The original experiment was mounted on a large stone slab floated on a bath of mercury. Many other experiments were performed, notably the Michelson-Gale experiment in which the light beams were projected along evacuated pipes arranged in a rectangle. With this experiment it was possible to measure the rotational speed of the Earth. In a later experiment microwaves were bounced around the Earth between GPS satellites also measuring the rotational speed of the Earth. In any experiment in which it is attempted to measure deviations in the speed of light one is always confronted by a huge obstacle. The light beam must always be reflected back to the source for measurement. In an athletics event it is a simple task to measure the start and finish times with as much precision as may be required and then calculate the average speed. This is not possible with light measurements. If the start and finish points are some distance apart then it is impossible to accurately measure the time elapse as communication between the observations will also be determined by the speed of light. For this reason the light must be reflected back to the starting point which will cancel out the difference in speed between the two runs. It was shown by Fresnel and Lorentz that there should never be any detectable first order results from measuring apparatus moving through the ether but there might be second order results. This posed a tall order. The square of the velocity of light is a hundred million times greater than the square of the orbital velocity of the earth.

The Michelson Morley experiment used two beams of light at right angles to each other. The beams used multiple reflections to obtain an effective length of eleven metres. If the one beam was in the direction of the movement of the ether this would not detect any movement as the light moving in both directions would cause the upwind and downwind differences to cancel each other out. The other beam at right angles would however cause the light to travel twice along the hypotenuse of a right angled triangle in one direction only, however this second order result would be extremely difficult to measure. In the Michelson measurements the difference in speed was measured by an interferometer which could detect slight shifts of interference fringes of the light waves. Michelson was expecting the orbital speed of the Earth to cause a shift of about a third of a fringe but claimed to be able to detect a shift of one percent of a fringe. The results of the experiments were sensational: there was not the slightest difference in the speed of light in any direction putting an end to the idea of a luminiferous ether.

The Michelson interferometer has found many applications where extremely precise measurements are required. A notable application was the definition of the standard metre in a reproducible form which did not require the preservation of a prototype – a very understandable precaution in the early part of the twentieth century. The metre was defined in terms of light wavelengths of cadmium vapour. This provided a definition of the metre to an unprecedented precision of less than a micron. Since then the metre has been defined to even greater precision. The present definition of the standard metre is given in the appendix. In 1888 Armand Fizeau proposed that the sizes of stars could be measured by light wave interferometry.

In the early 1890 years Michelson used an interferometric method on the Lick Observatory's 91 cm refractor telescope to measure the sizes of the large moons of Jupiter and obtained a good result of about one arc second. In 1920 Michelson used a 6 metre interferometer on a 254 cm telescope (Hooker – Mount Wilson) to measure the diameter of Betelguese (380 million km, a gigantic star) – this was the first time that the diameter of a star had been measured. This method is in much use today on the large modern telescopes – this is described more fully in chapter six. Michelson also performed many measurements of the velocity of light using a rotating wheel with multiple flat mirrors around the periphery and a long light path of several kilometres obtaining excellent results. Einstein publicly paid tribute to Michelson's extensive contribution to science with:

> "My honoured Dr. Michelson, it was you who led the physicists into new paths, and through your marvellous experimental work paved the way for the development of relativity."

As matters stand today no one is seriously suggesting that the luminiferous ether does exist but there is a new contender - a universal substrate which forms the ultimate dimensionality of the universe. This substrate is, as are many other matters in theoretical physics, nothing more than a mathematical concept. The most obvious objection to a luminiferous ether is the question of rigidity. Sound waves in water or a metal bar are insignificantly slow compared to the speed of light. The ether would have to be millions of times more rigid than steel and yet it has not the slightest effect of slowing the planets in their orbits. The idea of light being propagated through a medium as sound waves is of no help at all in explaining photons and the photoelectric effect.

The result of the Michelson-Morley experiment was a precursor to the young Albert Einstein formulating his famous *Special Theory of Relativity* which shocked the scientific establishment and outraged the common sense of others who cared to read about it. The theory required that the speed of light would be the same in any measurement regardless of the motion of the measuring device. This theory had huge implications but there is no need to deal with it here as there is a vast body of literature dealing with the subject. This brings us back to the unanswerable question – what exactly is a photon? It is a particle if you perform an experiment to show that it is a particle; it is a wave if you perform an experiment to show that it is a wave, and how does it have the property of polarisation? The problem does not end here. Larger particles also have the particle/wave duality however the wave nature diminishes with increase in particle size. With something as large as an airliner in flight the wave nature and uncertainty practically disappear. The particle/wave duality of matter is clearly an aspect of physics which is not fully understood at the present time. The solution is probably quite near at hand. The wave/particle duality becomes more pronounced as the particle energy diminishes and uncertainty increases and is most pronounced in the case of photons. A photon in flight has an exactly defined velocity, energy and momentum so what can be uncertain? The position away from the straight and narrow path is a matter of probability. Is something as intangible as a probability zone sufficiently real to become a wave front which can be focused through a lens to expose your camera film? When the photon is captured by an electron of an atom in the film its position is exactly defined and the entire wave front of probability disappears. This is also referred to as the collapse of the wave function. Regarding a light wave as a probability function may not be as way out as it may seem – even particles which are far more substantial than photons are little more than abstract mathematical concepts with a sprinkling of energy. And what is the energy? Surely more of an idea than a 'thing'. The real/unreal nature of photons finds an echo in Jostein Gaarder's book *Sophie's World*. Ostensibly a history of philosophy since classical times, the story develops into adventurous reading, where the distinction between fictional and non-fictional characters becomes blurred and then quite uncertain.

At this point we must mention a comparatively new theory regarding the photon probability wave and the collapse of the wave function. The 'Many Worlds Theory' requires that the wave front does not merely encode all the information about the photon but has an observer independent objective existence and actually is the photon. It also requires that the probabilities actually exist and that there is no collapse of a wave function. There is a huge body of literature about this theory which will not be described here.

The configuration of electrons surrounding the atomic nucleus has come a long way since the Rutherford and Bohr models of electrons whizzing around the nucleus in orbits like a little solar system. This model persists to this day probably for no better reason than that it is an easy way to visualise an atom. Modern theory has it that the atomic nucleus is surrounded by electron orbitals, clouds of electrons as superpositions of probability density functions where the electrons can appear anywhere at any time. The electrons in turn are surrounded by clouds of 'virtual' photons which become real when the electron is ready to release energy. Can we expect anything more tangible than a mathematical probability wave for the photon? When a 'final' theory for the nature of electromagnetic waves and photons is formally and rigorously presented this will indeed be a triumph of the human intellect.

A QUANTUM MATTER OF GRAVITY

The force of gravity has apparently remained inexplicable throughout recorded history. The first major breakthrough came with Sir Isaac Newton's law of gravitation and laws of motion. These at last provided a means of calculating the elliptical orbits of planets with a high degree of accuracy sufficient for practical purposes. An additional refinement came with Einstein's theory of relativity which accounted for a slight abberation in the orbit of Mercury. Newton himself had to admit however that he could not explain why it was that objects should attract each other even over vast distances. Physicist Richard Feynman commented:

> *Newton made no comment about this; he was satisfied to find what it did without getting into the machinery of it.* No one has since given any machinery.

The force of gravity between man-made objects is vanishingly small but on a larger scale it is the most significant force in the universe. Gravity holds stars and planets in shape, keeps galaxies from flying apart and on a scale of awesome vastness attracts clusters of galaxies to each other. We can never be unaware of gravity; it is the weight that we feel on our feet and the force that makes objects fall to the floor. Gravity cannot be insulated or conducted and can never become repulsive. Gravity has traditionally been considered to be a 'field' but a unified field theory which could include gravity has never been formulated. Our quest of perception of the universe now takes us to the mysterious realm of gravitation.

Gravity is not a force, it is an acceleration resulting from superstring energy quanta passing through Kaluza-Klein dimensions to successive instants on the time vector. A force of gravity or weight only occurs when a body with mass is restrained from acceleration or an acceleration is imposed.

Let us consider two very basic principles of physics which are quite similar in some respects and diametrically opposite in others.

The first as illustrated in fig 8.1 will be instantly recognised by any student of electrical engineering. This shows a current carrying conductor in a magnetic field. When a current is passed through the conductor this will cause a force to be exerted on the conductor which is at right angles to both the magnetic field and the conductor. This arrangement is fully reversible: apply a force to move the conductor and a current will be induced in the conductor, reverse the force and the current will likewise be reversed. The various parameters here are simply related and easily calculated. The force is not actually exerted on the conductor but on the moving electric charges. This principle can also be illustrated without a conductor where the charges are allowed to flow in a vacuum such as in a magnetron. In this case the charges will be induced to move in circles. It is interesting to note that in the case of a conducting wire it is the movement of electrons relative to positive charges in the wire that is significant. If the conductor is moved lengthways so that both the positive and negative charges are moving in the opposite direction to the current this will not make the slightest difference to the force produced. This is not difficult to do.

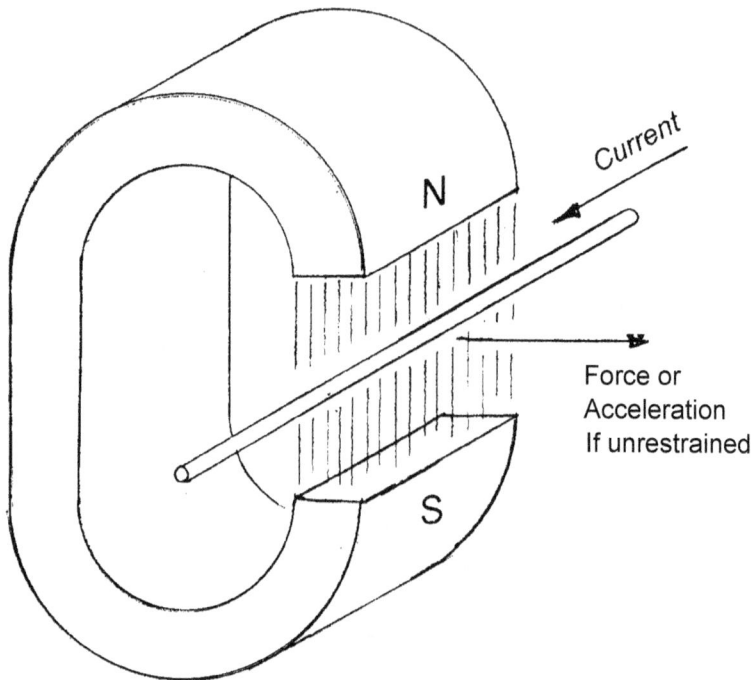

Fig 8.1 Experiment to demonstrate force on a current carrying conductor in a magnetic field.

There is a popular misconception that electric current moves at an extremely high speed along a wire. This is not the case; electrons causing an electric current move at a snail's pace, typically less than 1 mm per second – it is the pressure wave or voltage that moves at nearly the speed of light. In a wire carrying alternating current the electrons move forwards and backwards with an amplitude of less than twenty microns. In a wire carrying no current there is nevertheless considerable turbulence in the free electrons – a smooth laminar flow of electrons is difficult to obtain. There is a way of mechanically producing an electric current. A spinning disc can be given an electric charge – the surplus electrons will then constitute a current. This is not a useful way of generating electricity or magnetic fields as it is extremely difficult to put a significant charge on a disc. The electromagnetic principle described here cannot be observed in nature but only in manufactured apparatus. This is the principle that made Michael Faraday famous and gave electrical power to planet Earth.

It is interesting to consider what happens when a stream of charged particles is ejected into space. This is of course a current which is surrounded by a circular magnetic field. One might expect the particles to quickly spread out due to their mutual repulsion but the surrounding magnetic field will keep them focused. This current will then form itself into a spiral. When several streams are ejected from the Sun they can twist themselves together like a marine hawser – these are known as magnetic ropes which can connect the Sun to the Earth and reach immense size – as thick as the diameter of the Earth and carry currents of up to a million amperes. Despite this enormous current, the current density is extremely low – about ten nanoamps per square metre. When charged particles are ejected into a magnetic field they will spiral along the direction of the field. As they approach a region of higher magnetic flux density they will penetrate the region to an extent depending on their energy and then rebound spiraling back in the direction that they came from. Electric currents moving in space without voltage, resistance or even a conductor and rebounding backwards are a matter of much complexity and abstruse mathematics.

We now take a look at fig. 8.2. The sheets of graph paper are meant to represent three dimensional spaces but we will have to make do with two dimensional surfaces for practical reasons. The time axis is at right angles to any direction on the paper but in fact it is also at right angles to any direction in three dimensional space. As in the previous example, a force is exerted on objects in the space. The force is at right angles to the time vector which is in any direction in three dimensional space and

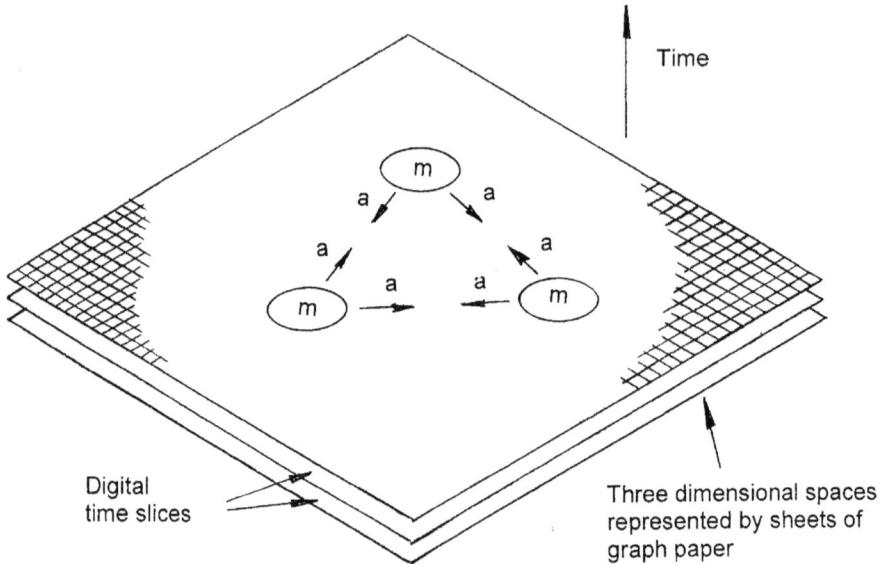

fig 8.2   Thought experiment to describe gravitational acceleration induced by superstring energy passing along the time vector.

is exerted on all energy quanta in superstring format. Superstring manifolds will be discussed in some detail in chapter 9. An analogy can be found in a large bunch of gas filled balloons held by a balloon seller. The balloons will all tend to bunch together, not because they are attracting each other, but due to the tension in the strings that are restraining them.

There is no mysterious field causing stars and planets to attract each other, they are simply being caused to accelerate together by passage along the time vector. In the balloon model all the balloons are attracted to a central point but this is not what happens with gravity. In space, the centre is everywhere and anywhere, so that small objects will appear to be attracted to larger ones. The result remains exactly the same as calculated by Newton's law of gravitation. Gravitational force and mass are not 'things', they are experiences – the experience of superstring energy passing through time. Empty space can take no part in gravity apart from being slightly curved by large gravitating bodies as predicted by Einstein's General Theory of Relativity which stated that light will be slightly deflected when passing close by the surface of a star.

In contrast to the Faraday experiment, nothing in the gravity model is reversible. The time vector cannot of course be reversed and gravity can only be attractive.

Placing any object between gravitating bodies can have no effect on gravity other than to add more gravity. The balloon analogy cannot of course be taken too far. Gravity is a complex phenomenon involving all eleven dimensions whereas the balloon bunch is a simple three dimensional model. There are several orders of magnitude between the mathematics of gravity and the simple trigonometry of balloon strings. In the foregoing discussion, for the sake of simplicity, gravity was described as a force. This is not an accurate description – gravity is in fact an acceleration – a 'force of gravity' does not actually exist. We can only experience a force due to gravity when an object is restrained from free fall. We continually feel our weight on our feet but this is because the floor is preventing us from free fall. An object in space may experience acceleration due to the presence of other massive objects but there is no force acting on it at all. In Einstein's General Theory of Relativity gravity is described as a curvature of space-time, not a force! In elementary books on applied mathematics a falling object will be described as having a 'weight' which is a gravitational force and this force will be equal to the accelerating force. This will yield perfectly good calculations but in fact there is no force at work at all! Before the object is released it has 'potential' energy which becomes 'kinetic' energy as it loses height and gains velocity. The amount of potential energy is a concept depending on how far the object is to be allowed to fall. If the object is to be allowed to fall down a hole right through the Earth it will have enormous potential energy but this does not change the object in any way. Falling objects simply do what comes naturally: accelerate.

We are ready for a DIY experiment. Take a seat at a desk and select a small unbreakable object such as an eraser. Hold the object about 20 cm above the desk. You will feel a small force pulling it down. This is its weight or more precisely the force required to prevent it from accelerating towards the earth due to matter energy passing along the time vector. Now release the object – it will come as no surprise to see that it accelerates towards planet Earth at an accurately predictable rate. If anyone wants to nit pick about air resistance, they had better try this on the moon. During its free fall the object is experiencing no forces at all – it is simply fitting in to the gravitational acceleration dictated by the matter energy contained in the Earth. The falling object will quickly come to rest on the desk. The desk will prevent any further acceleration causing the object to regain its weight and the desk will exert an upward force to match the weight of the object. Leave the desk now and take a few steps. You will notice that your feet are applying a downward force to the floor and the floor is matching this with an upward force. This is the result of the floor restraining you from natural gravitational acceleration.

When calculating the orbit of a planet the concept of mass is indispensable for calculating the centrifugal force and the gravitational force exerted by the sun on the planet. These are perfectly legitimate calculations, indeed, how else would one perform such calculations – even though these huge forces do not actually exist. It is a simple matter to explain tidal bulges on the earth in terms of gravitational and centrifugal forces – very useful concepts. Try doing this purely in terms of accelerations!

There is a common misconception that mass is a very basic property of matter and which can, in exceptional circumstances, be converted into energy. Mass is in fact merely a concept for quantifying energy. There is a very similar concept used in chemistry – the 'gram molecule' or mole. This is a convenient and indispensable concept for specifying a standard but vast number of molecules or other entities. A gram of mass is a similarly indispensable concept for specifying a vast amount of superstring energy.

When considering gravity we must of course also look at inertia. To move an object it is necessary to apply a force which will accelerate it at right angles to the time vector. The accelerating force simply provides an added component to gravity. Once accelerated, it will retain its velocity and a new set of reference coordinates until such time as another force is applied. A constantly rotating object is under continual acceleration towards the axis and centrifugal/centripetal forces are continually in effect. An imposed acceleration is simply an addition to gravity and indistinguishable. The question arises: how do we know that something is turning or moving? The simple answer is that we can see that it is moving, but how do we know exactly what is moving – is it us or the thing that we are looking at? A useful thought experiment instrument is the centrifugal governor which was commonly used on steam engines of a bygone age. The governor had weights on arms suspended from a vertical rotating shaft. As the shaft speed increased the weights would fly out further from the shaft and levers below the weights would control the supply of steam to the engine so that a controlled speed of the engine could be maintained under varying load conditions. Take our steam governor and sit on a swivel chair with the governor on your lap and get someone to set the chair spinning. The governor will appear to be stationary to the person sitting on the chair but appear rotating to the person standing next to the chair. Who must the governor try to please? The weights will fly out, showing that the person sitting on the chair has nothing to do with the experiment. The same thing can be repeated in a space ship in deep space. If the space ship is spinning about the same axis as the governor the weights will fly outwards even though the crew can see no rotation of the governor at all. The crew will of course experience centrifugal force towards the hull of the craft which is a locally induced gravitation. They will also notice the background of stars rotating about the space craft but whether the stars are there or not has nothing to do with the experiment. What then is the reference benchmark for acceleration and inertia which is quite removed from the circumstances of local observers? This brings us back to our time-slice diagram. Space dimensionality passing through successive time slices provides the benchmark reference for gravity, inertia and rotation throughout the universe! Any objects which are rotating on the sheets of graph paper can only have the axis of rotation aligned with the time vector. This applies equally to solid objects rotating in three dimensional space. The presence of other objects whether stationary or in motion has nothing to do with the space benchmark – this is purely a matter of interaction between space dimensionality (universal substrate) and the time vector. This experiment does not only apply to rotation but linear motion as well. There is an often mentioned paradox referring to Einstein's Special Theory of Relativity. An astronaut travelling at an extremely high speed away from Earth for a considerable time will find on his return that time has passed on Earth more quickly than he has experienced on the journey and that his children could even have grown older than himself. Let us repeat this situation as a thought experiment in deep space. Two spacecraft pass each other at phenomenal speed and then much later turn around and pass each other again. If any stars which could show parallax movement are obscured by dark nebulae how could one possibly know the speeds of the spacecraft or even if the one was stationary and which of the craft would have had a slower experience of time? If we park the one spacecraft within the Solar System then there would be no doubt as to which craft did the travelling. Here again the movement is relative to the space dimensionality or universal substrate. This is not to say that the Solar System is motionless. The Solar System has a motion towards the Solar Apex as well as a movement of some 900 000 km/h around the galactic centre. The Solar System is where we stay so it is quite natural to establish our definition of time units right here.

The speed at which gravity is propagated has long been disputed and remains a theoretical problem. The Kepler law for elliptic planetary orbits assumes that gravitational effects are instantaneous regardless of distance. This provides accurate calculations seemingly proving the instantaneous effect of gravitation. If gravity were propagated at the speed of light there would be a nearly 500 second gravitational delay between the Earth and the Sun resulting in an unstable orbit which is not observed and seemingly proves the instantaneous effect of gravity. General Relativity is not as simple as this and not only conveniently accommodates the 500 second delay but also requires that the Earth must radiate gravitational waves and lose energy in the process.

The velocity of electromagnetic radiation is determined by the electric permittivity and the magnetic permeability of space – must this also apply to gravitation? Is it possible to devise an experiment to prove the instantaneous effect of gravitation one way or the other? Yes and no.

We have here a similar situation to one that was tackled by Michelson and Morley in 1887. They devised one of the most interesting and significant experiments in the history of physics (The MMX). The object was to prove that light was propagated by waves moving through a medium called 'ether'. This experiment is so well known that it will suffice to summarise it as follows. Apparatus was made for the accurate measurement of differences in the speed of light in various directions. The apparatus was floated in a bath of mercury for easy change of direction without disturbing the delicate calibration. The reasoning was that as the earth is moving through the ether there would be a different speed of light measured in different directions. The experiment was performed at different times of the day and in different seasons of the year but to the astonishment of the experimenters the speed of light was the same in every case. The experimenters proved the very opposite of what they had set out to prove. We have dealt with the MMX in more detail in the chapter on photons.

How can we do something similar for gravitational waves? We can quite easily describe an experiment to do this but which is also quite impossible to carry out. All that we need is a fluctuating gravitational source and some observation and recording instrumentation. The gravity source could be a binary star but the heat would be too intense for even a thought experiment so we will select a binary orbiting planetary system. We select two medium sized planets with the plane of their rotation in line with our observation point so that for every half turn of the planets our observation point and the planets will be in a straight line. We have to move a distance away from the planets so that in the time that it takes light from the planets to reach us the planets would have moved a conveniently measurable distance. We now make and record an optical observation of the planets and also do the same with the gravitational pull.

The gravitational pull will vary considerably with each rotation and will be a maximum when one planet is close by and the other a considerable distance away. We now compare the recordings of the two observations. If they are exactly in phase then the gravitational wave will have travelled at exactly the same speed as light. If the gravitational wave is travelling at any other speed then there will be a phase shift between the two observations. If the gravitational effect is instantaneous this could easily be calculated from the phase angle between the two readings and we will have proved that gravitational waves do not exist. There is no chance of actually doing this experiment but it may be useful for clarifying one's ideas on the matter.

This experimental setup is in effect a signal transmitter and a receiving station. Does this mean that we can have instantaneous communications across vast tracts of space that would take years for light to traverse? Definitely not. The frequency and bandwidth of binary planets would be a poor choice for a TV transmission and does anyone fancy the idea of modulating a pair of planets! It is not actually impossible to perform this experiment. It could be done by using dwarf planet Pluto and its moon Charon as a fluctuating gravity source. The extreme cost of performing this experiment would however be prohibitive.

We have looked in some detail at gravitational acceleration and hopefully finally dispelled the notion of a 'gravitational field' but what of the others? The electrostatic and magnetostatic fields both have much in common with gravitation. All three of these effects have inverse square laws for their restraining forces and the 'forces' do not diminish with movement which means that all three are acceleration 'fields'. By the same token we can dispense with electric and magnetic fields in the same way as with gravitational fields. All three fields must remain as very useful concepts for calculations even though they do not physically exist! The three effects are caused in exactly the same way by passage along the time vector but differ only in three different aspects of superstring energy. The gravitation effect is caused by the 'mass' energy of the string with the addition of any nucleus binding energy. The electrostatic effect results from the rotational direction of the string causing positive and negative charges which permits both attractive and repulsive forces. The magnetic effect results from relative movement of groups of charges and also permits both attraction and repulsion. In ordinary terrestrial gravitation calculations it is quite common to regard the gravitational acceleration as a constant due to the vast distance between the points of attraction but with electrostatic attraction the distances are small so that the 'force' will change rapidly with distance. This difference gives the very different 'feel' between gravitational and electrostatic attraction even though they both function with inverse square laws. An inverse square law is a logically natural function for attraction or repulsion between point sources. This means that the force will decrease in proportion to the increase in area of a unit solid angle between the points or in more practical terms the force will decrease in the same way as the reduction in intensity of light from a point light source as the distance is increased. There is a problem with this law. If the separating distance is reduced to nearly zero as can be done with superstrings then the force will rise asymptotically towards infinity resulting in a black hole situation. Fortunately the Pauli Exclusion principle is there to save us from calamity.

The two other acceleration fields

Electrostatic Force

In fig. 8.3 the superstrings are aligned with their spirals in the direction of the time vector as must be the case for all strings as they need to 'burrow' their way in seven curled up dimensions to succeeding time frames. The strings have electric charge which is positive or negative depending on rotational direction. We do not use the word 'spin' as this has a special meaning in particle physics. The result is simply stated: like charges repel and unlike charges attract.

Electric charge is not experienced from individual superstrings as these do not normally occur freely in nature as atomic matter. Positive charge is provided by protons which contain nine strings with a nett charge of three positive string charges. Negative charge is provided by electrons, of uncertain structure, and a nett charge of three negative string charges.

The electrostatic force does not usually occur over large distances. In order to produce an electric charge it is necessary to move electrons from one point to another and the only time that this can occur with some significance is in thunderstorms and this does not continue for very long before being discharged in spectacular manner. There is no feasible way that planet sized objects of electric charge can occur. It is extremely difficult to remove electrons in any quantity from material. In the case of a capacitor electrons can be transferred from one terminal to the other but you still need the plates in close proximity separated by a dielectric.

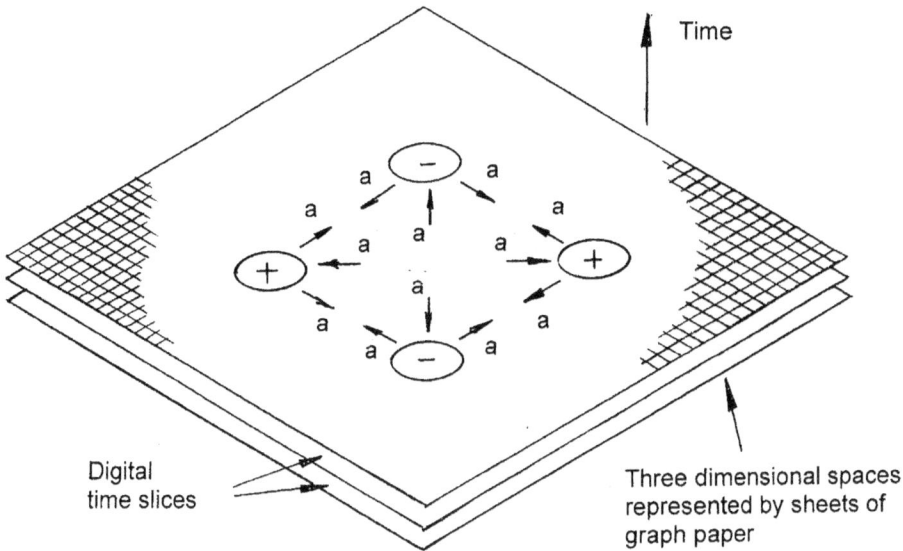

fig 8.3 Electrostatic force

Electrostatic force can be demonstrated by means of a hair comb and scraps of paper and also be rubbing one's feet on a carpet but this gives the impression that electrostatic forces are little more than a feeble curiosity – nothing could be further from the truth. Electrostatic force is one of the most powerful in the universe and of unimaginable magnitude.

Let us first take a bar of hardened tool steel and try to make an impression on it. A hacksaw or a drill will get you nowhere. About the only way you will be able to cut the bar without heat will be by means of grinding equipment (which will in any case produce much heat). It is electrical forces that are keeping the bar in shape. This is by no means an adequate demonstration of the immensity of electrical forces.

If it were possible to hold two pins slightly separated and transfer all the electrons from one pin head to the other the force between the pinheads would be of astronomical proportions! This is easily calculated but the result is unbelievable. It must be emphasized that this is a thought experiment as there is no physical possibility of removing all the electrons from the pinheads. Let us choose two ordinary dressmaker's pins with 1 mm heads and hold these separated by 10 mm. The mass of the pinheads is almost entirely due to the nucleons as the mass of electrons is insignificantly small. Each pinhead will contain about 200 coulombs of

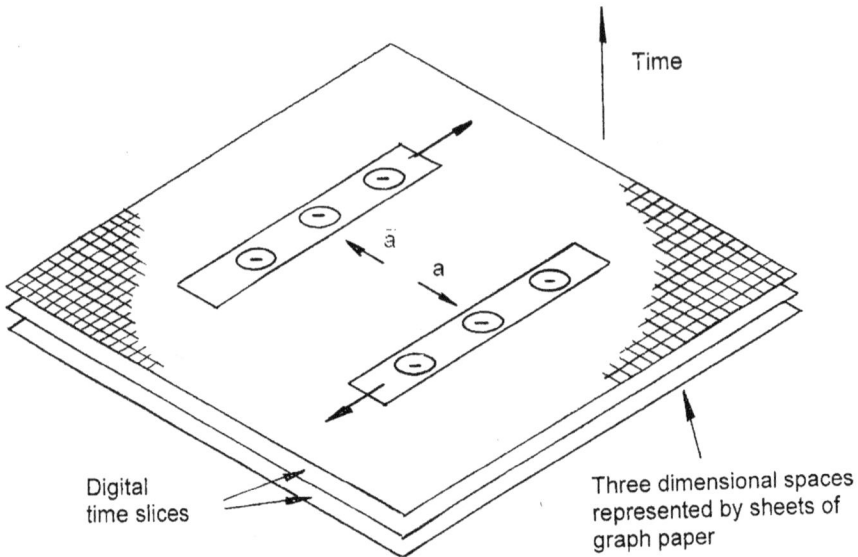

Time

Digital
time slices

Three dimensional spaces
represented by sheets of
graph paper

Fig 8.4 Magnetic force

charge - If moved by a pressure of 12 volts enough electricity to crank and start your car engine on a cold day. We now imagine that all the electrons are transferred from the one pinhead to the other. The resulting force? Enough to match the weight of several cubic kilometres of rock. This makes the forces inside even the most violent volcano pale into insignificance! The transfer of electrons will of course also create a voltage between the pins - how much? It will be millions of times greater than a lightning flash. Here again it would be physically impossible to create or contain such a voltage. Atoms do not release their electrons willingly! While we have the 200 coulombs of electrons handy let us see what will happen if we release them to the Earth. The Earth is effectively a spherical capacitor surrounded by space. To simplify the problem let us assume that the sphere has a smooth surface. This has a capacitance of 700 microfarads! If we release 200 coulombs of electrons to this capacitor the voltage will rise to 285 kV – more than a quarter of a million volts all over the Earth! A final question about these electrons – how far will they be spaced apart on this huge sphere? About 250 electrons per square centimetre! Let no one be under any illusion about how many electrons there are in a pinhead! These figures seem to be pushing the limits of absurdity - all the required constants are provided in the appendix for readers who wish to check these calculations for themselves.

Magnetic Force

In this diagram there are two sets of negatively charged particles. Movement of the charges in the same direction results in magnetic attraction and opposite directions yields repulsion. In the mid-nineteenth century, James Clerk Maxwell proposed that the magnetic field was produced by infinitesimal vortices in the ether which constituted physical matter. These vortices were thought to constitute atoms as no structure for atoms was known at the time and the idea of superstrings was a century away.

A magnetic field is only generated around a conducting wire when there is movement of electrons relative to the material of the wire. Moving electrons which are surplus to protons in the vicinity or moving through space also constitute an electric current. Ferromagnetism results from movement of electrons at atomic level. One can scarcely imagine any scientific principle more applicable to our daily lives than electromagnetism. From the small motors powering our household appliances, the various electric motors in our cars, the heavy traction motors of our trains and at the other end of the scale the huge motors driving our steel rolling mills and the gigantic alternators in our power stations. A huge bonus with electromagnetism is that it can work without a moving conductor – an alternating magnetic field will do just as well which makes it possible to have the transformer without any moving parts at all. The fact that magnetism is an acceleration field is spectacularly demonstrated by a little known device called the 'magnetic rail gun' also known as a 'Gauss gun' and 'coil gun' developed by physicist Kristian Birkeland in 1900. The word 'gun' is somewhat misleading as this device is not suited for use as a weapon. The device consists of two closely spaced conducting rails provided with powerful electromagnets along the entire length. The projectile, which must be a metallic conductor, fits between the rails. The 'gun' is fired by passing a huge electric current from the one rail and through the projectile to the other rail. The huge magnetic field produced causes the projectile to move along the rails with astonishing acceleration.

A pleasing aspect of this arrangement is that the acceleration is produced without the deafening explosion of an artillery propellant charge.

The Playing Fields of the Universe

Magnetic, Electric and Gravitational fields are aptly named as 'fields'. There is an analogy to the playing fields of a school sportsground. A field is simply a place where something can happen but is nothing in itself. A variety of sports can be played on a field, each governed by a set of rules but the players and rules are not part of the field. The rules will be available in written form but these do not influence the field. The rules governing the actual performance of the players will exist only in the form of thoughts. Take a piece of paper and place it over a magnet and then sprinkle iron filings on the paper. We will clearly see the direction of magnetic flux but without the iron filings there is nothing there at all. When the iron filings are introduced to the field they behave according to mathematical rules determined by superstring energy passing along the time vector. In this case, the rules are immutable physical laws and no referee is required. The gravitational, magnetic and electrostatic fields all imply an acceleration aspect to material coming under their influence.

The concept in Einstein's General Theory of Relativity where gravity is regarded as a curvature of space-time presents a serious challenge in terms of a readily grasped explanation. This is sometimes illustrated as a stretched elastic membrane with spheres of various masses stretching down parts of the membrane and then gravity is described in terms of the curvature of the membrane. This presents a concept of much confusion as the reader's own gravity is used to pull down the masses and the gravity being described works sideways in two dimensions. The idea of a curved space-time is extremely difficult to explain in simple terms and is best left to the mathematicians.

On 20 April 2004 a most interesting experiment was launched into space. This is the Gravity Probe B satellite orbiting 642 kilometres high with a period of 97,5 minutes in an orbit passing over the Earth's poles. This is a collaboration of Stanford University, NASA and Lockheed Martin Space Systems. The twofold mission is to determine the space-time warp caused by the gravity of the Earth and also the frame-dragging of space-time caused by the rotation of the Earth. This experiment has been many years in the making and is expected to give definite verification of Einstein's General Theory of Relativity. Frame dragging is also known as the Lense-Thirring effect. The measurements depend on minute wobbles in the spin axes of gyroscopes. The four supersensitive gyroscopes of spheres of fused quartz rotate at 4 000 r.p.m. and are the most perfectly spherical objects ever engineered. These spheres are only 38 mm in diameter and are spherical to within 40 atomic layers. The gyros are housed in a huge Dewar flask cooled with supercooled liquid helium, where they are cooled to an extremely low cryogenic temperature of 2,3 K. The spheres are coated with niobium which becomes a superconductor at this extreme temperature and this is used to electrically levitate the gyros. A stream of pure helium gas was used to spin-up the gyros and the gas was then evacuated to a few molecules. The gyros are separated from their housings by only 32 microns! The spin-down time of the gyros is estimated at 15000 years.

It is expected that the space warp should cause spin axis change of 6,6 arcseconds per year and the drift for frame dragging should be 42 milliarcseconds in the direction of the Earth's rotation. Measurements of such extreme precision are almost unbelievable. In an ordinary survey theodolite the circles give direct readings of degrees and minutes but seconds can only be estimated to the nearest five. The star 1M Pegasi is locked on by the satellite's telescope for use as a reference point for measurements. The pointing accuracy required is only 0,1 milliarcsecond; to achieve this the target star's motion is mapped relative to quasars so that the gyroscopes measurements can be related to the distant universe.

The first public announcement made in April 2007 confirmed the geodetic effect to be verified to within 1%. The frame dragging measurements are ongoing and are expected to continue through 2010. A mission update on 12 November 2009 declared that the frame dragging was clearly visible. Previous attempts at frame dragging measurements were performed by means of the LAGEOS satellites. The GRACE satellite was used to obtain an accurate gravitational model of the Earth as mountain ranges and ocean trenches have a measurable effect on the orbits of satellites. The LAGEOS satellites (Laser Geodynamics Satellites) are scientific research satellites designed to provide high precision orbiting laser ranging benchmarks for geodynamical studies. These are 60 cm diameter brass spheres coated with aluminium and have a mass of 411 kg. They are festooned with reflectors giving them the appearance of golf balls. These satellites orbit at an altitude of 5900 km. Ground stations are based in many countries including the US, Mexico, France, Germany, Poland, Australia, Egypt, China, Peru, Italy and Japan. The stability of the orbits permit extremely accurate measurements of points on Earth. The satellites are expected to remain in orbit for 8,4 million years.

According to Einstein's General Theory of Relativity, massive accelerating objects in space should emit gravitational waves and which should also drain energy from the source object. A spinning spherical object will not radiate gravitational waves but a non-spherical one, and more particularly, orbiting objects will. Gravitational waves must not be confused with 'gravity waves' which is a term used in hydrodynamics. To put matters in perspective let us consider gravitational waves emitted by the Earth orbiting the Sun. Using Einstein's quadrupole equation, the gravitional power emitted by the Earth would be about 300 watts, hardly enough to power the microwave oven in your kitchen! The frequency period of this wave would of course be one year. When passing through physical objects, the gravitational wave would cause alternate lengthening in one direction and shortening in another, between pairs of widely spaced objects, at right angles to each other, in a plane at right angles to the direction of the wave. It is this dimensional oscillation that gravitational astronomers intend using to detect gravitational waves. In the case of the Earth-Sun system, at a distance of one light year, the stretch-shrink strain factor of a ring of particles would be 10^{-26} , an impossible dream as far as practical measurement is concerned. There remains however a possibility that gravitational waves can be detected from extreme sources such as neutron stars or black holes orbiting as close coupled binary systems. The loss of energy would cause a slight inspiralling of the objects which could possibly be detected by a slight change of frequency. In 1974 Russel Alan Hulse and professor Joseph Hooton Taylor, both of Princeton, using the 305 metre Arecibo radio telescope in Puerto Rico discovered a binary pulsar thus providing the first step in the new science of Gravitational Astronomy. Hulse and Taylor were awarded the 1993 Nobel Prize in Physics for:

> The discovery of a new type of pulsar, a discovery that has opened up new possibilities for the study of gravitation.

Professor Carl Nordling of the Royal Swedish Academy of Sciences concluded his presentation address with:

> Dr. Hulse, Professor Taylor,
> You have been awarded the 1993 Nobel Prize in Physics for your discovery of the first binary pulsar, PSR 1913+16, a discovery which has had a great impact on gravitational physics. It is my privilege to convey to you the heartiest congratulations of the Royal Swedish Academy of Sciences, and I now ask you to receive the Prize from the hands of His Majesty the King.

The Nobel Banquet address of professor Joseph Taylor given on 10, December, 1993 is of particular interest:

Your Majesties, your Royal Highnesses, Ladies and Gentlemen,

We have heard earlier today that scientific discoveries come at unpredictable times. Just as a person cannot say "I shall write poetry", another cannot say "I shall make a scientific discovery."

Russel Hulse and I did not set out in 1973 to detect gravitational waves, or even to conduct experiments into the fundamental nature of gravity. Instead, we set out to chart the celestial globe with a new type of star – aware only that we were sailing a route that none had explored before, and that wondrous new lands might be revealed beyond the next horizon. We were young, well-prepared, and receptive, but not yet wise. We were playing a detective game, gathering clues and solving logical puzzles as they presented themselves. One special new island, at first only faintly visible in our telescopes, later showed its bounty in full relativistic glory. When its treasures were gathered and brought home, some after many years of labour, they provided keys to long-locked gates and added new notes to the symphony of natural law. In discovering this new island and gathering its exotic fruits, Russel Hulse and I, and other colleagues in later years, were enjoying the privilege of doing what we liked best: satisfying our own curiosities, by asking and answering questions. We sought no other reward than the pleasure of an exciting journey. To be honoured by being here tonight is beyond our wildest youthful dreams of nineteen years ago, and brings us joy that mere words cannot express.

The details of the Hulse-Taylor binary neutron stars are extreme beyond anything we can experience on Earth. Let us look at some figures posted by Robert Johnston in 2004.

Distance from Earth	21 000 light years
Mass of pulsar	1,441 times Sun mass
Mass of companion	1,387 times Sun mass
Pulsar rotational period	59,029 997 929 milliseconds (about 1016 r.p.m.)
Diameters of neutron stars	20 km
Orbital period (year length)	7,751 939 106 hours
Periastron separation	746,600 km

Apastron separation	3 153,6 km
Rate of decrease of orbital period	76,5 microseconds per year
Rate of decrease of semimajor axis	3,5 metres per year
Lifetime to final inspiral	300 million years

This is a most convenient source of gravitational waves but the development of instruments to detect the waves is challenging to the point of absurdity. To detect gravitational waves it is necessary to use two evacuated tubes at right angles to each other with proof masses at the ends. The passage of a gravitational wave will cause alternate increasing and decreasing in the distances between the masses. The obvious and only instrument that can be used for this purpose is a Michelson interferometer but this has its limitations. The light used for interferometry must be in the visible or infrared spectrum, of about 400 to 1000 nanometre wavelength. Using a shorter wavelength will not only present the inconvenience of invisible fringes but the X-Rays will pass straight through the instrument's mirrors making interferometry impossible. The usual interferometer configuration cannot detect movements below the nanometre range. The gravitational instruments require measurements in the zeptometre range – millionths of millionths of the size of an atom. It is hoped that this will be achieved by means of interferometry using nano-scored diffraction gratings and sophisticated interferometric computer software such as FINESSE (Frequency domain INterfErometry Simulation SoftwarE). The data will be further improved by recycling the laser beams. The task of insulating the proof masses from even the slightest seismic vibration must be an engineering nightmare. The proof masses and mirrors must be absolutely unresponsive to magnetic fields and also be electrically non conducting as eddy currents could make them sensitive to fluctuating magnetic fields. An additional source of unwanted noise could come from thermal vibration of molecules within the mirror material and also 'shot noise' from photon quanta impacting the mirrors. The masses are suspended by thin fibre glass filaments. The measurements are so sensitive that waves breaking on a beach hundreds of kilometres away could overwhelm the instruments with unwanted noise. A Richter magnitude six earthquake anywhere on the Earth is sufficient to temporarily disable a gravitational observatory. To get an idea of the extreme precision being attempted we can consider the result of a sizeable gravitational wave passing through the solar system. The wave would cause the distance between the Sun and the Earth to increase and decrease by the size of a single atom for a few hundredths of a second. It is astonishing that detecting gravitational waves is even being contemplated. Analysing the vast quantities of data from the instruments is very similar to the processing of data from a SETI project. The data is mostly noise with the slight possibility that there may be useful information embedded somewhere. The data of at least two widely spaced instruments must be analysed simultaneously lest a cat pouncing on a mouse outside the building be interpreted as significant data. In order to spread the burden of the vast computer processing required, data can be downloaded, as with SETI, to members of the public who are willing to run an analysing program as a screensaver thus utilising computer time which would otherwise be wasted. A team of the Smithsonian Astrophysical Observatory discovered, in April 2011, an eclipsing pair of binary white dwarf stars some 3000 light years distant. This permits the gravitational effects to be measured with unparalleled accuracy in optical wavelengths. The stars orbit in thirteen minutes at about a third of the Earth-Moon distance. The gravitational effects have, since discovery, resulted in the eclipses occurring 20 seconds earlier. Team member Warren Brown commented: "This result marks one of the cleanest and strongest detections of the effect of gravitational waves". The stars will eventually merge in about two million years.

A single gravitational wave instrument will not be able to give any indication of the direction of the wave. The wave could even come from below right through the Earth as if there were nothing in its path. In order to detect the direction of a gravitational wave it is necessary to have at least three, but preferably more, widely separated instruments.

There is much speculation that the December 2004 earthquake and tsunami which devastated Indonesia and South East Asia could have been triggered by the passage of gravitational waves, as well as a conjunction of planets including Jupiter. The gravitational waves could have come from a massive gamma ray burst which was detected 44,6 hours after the 9,3 Richter earthquake. The gamma burst was 100 times more powerful than any yet recorded equaling the brightness of the full Moon. It declined over a period of 5 minutes with a pulsation of 7,57 seconds. It was determined that the burst originated from a soft gamma repeater from 20 000 to 32 000 light years away in our own galaxy. It is not considered improbable that the gravitational wave could arrive before the associated gamma burst. The gamma rays could be slightly slowed by galactic dust and cosmic ray particles and the gravitational waves could even have had superluminal speed at the time of the burst. A difference of 44,6 hours after travelling for more than 20 000 years is hardly significant. No gravitational waves were detected at the time of the earthquake as no gravitational observatories were yet operational. A gravitational wave cannot actually cause an earthquake, there is only a slight possibility that it could trigger one which was about to occur anyway.

The LIGO (Laser Interferometer Gravitational wave Observatory) is a highly sensitive set of three devices using pairs of evacuated light storage tubes which are from two to four kilometres in length. One is situated in Livingstone, Louisiana and the other two (using the same vacuum tubes) in Hanford, Washington. Even with using such long light paths it is estimated that a gravitational wave will cause a lengthening and shortening of the path by a hundredth of a femtometre, much less than a millionth of the size of an atom. It is expected that sensitivity in the zeptometre range will eventually be achieved.

The VIRGO 3 km gravitational interferometric antennae form part of the EGO (European Gravitational Observatory) which will optimistically detect gravitational; waves from the Virgo star cluster. Multiple reflections are used to increase the interferometric path length to 120 km. The observatory is situated near Pisa in Italy and is funded by a consortium of CNRS (France) and INFN (Italy). The observatory involves 13 laboratories in France and Italy and more than 150 scientists. The observatory will detect gravitational waves with a frequency range of from 10 Hz to 6 kHz. Insulation of the instruments from seismic noise is achieved by means of a 10 m high elaborate system of compound pendulums. The observatory was completed in 2003.

GEO600 (Gravitational European Observatory) is a joint British-German gravitational wave observatory situated near Hannover in Lower Saxony. The observatory utilises two six hundred metre evacuated tubes of sixty centimetres diameter. The GEO600 is funded by: The federal state of Lower Saxony, Volkswagen Foundation, Max Planck Society, German Federal Ministry of Education and Research and the UK Particle Physics and Astronomy Research Council (PPARC).

An early form of gravitational wave detector, the Weber Bar, has been in use since 1960. This consists of a large aluminium cylinder kept at cryogenic temperature in a vacuum. When a gravitational wave passes through the bar this causes it to ring at its resonant frequency. Two bars separated by a great distance must be used together so that terrestrial effects can be cancelled out.

A very exciting gravity space mission, the LISA (Laser Interferometer Space

Antennae) is to be launched by ESA/NASA as an ESA Cornerstone mission in 2015. This is intended to detect gravitational waves from massive black holes and galactic binaries. The frequencies of these waves will be extremely low with periods from a few seconds to a few hours. The three spacecraft will act as a Michelson interferometer in a triangle with sides of 5 million kilometres. The triangle of spacecraft will face the sun and be tilted 60 degrees to the plane of the earth's orbit and follow 20 degrees behind the earth. This distance was chosen to minimise the gravitational influence of the earth but still allowing good communications between the Earth and the spacecraft. The three spacecraft will each have two free floating solid blocks known as proof masses which will be made of a special platinum-gold alloy with vanishingly small magnetic susceptibility. A gravitation wave passing through the spacecraft will cause the blocks to vibrate and these slight movements will be detected by laser interferometry. The spacecraft will be kept centred on their test masses by using drag-free technology and field-emission electric propulsion. A precursor mission to be launched in 2014, the LISA Pathfinder, will test new and highly innovative technology required by LISA. The LTP (LISA Test Package) will have a payload of two test masses of 46mm cubes of gold-platinum. The motion of the masses relative to the vehicle will be controlled to within one nanometre – the motion between the masses will be measured to within one picometre, about 1% of the size of a hydrogen atom.

Before we take leave of gravity, let us take a quick look at the realm where science fact and fiction intermingle and also take a look into the life and times of Nikola Tesla.

To take a payload from Earth to the Moon and then back again to the same spot means that the payload is in the same energy state as it started from and that the vast quantities of chemical fuel used on the trip have been dissipated as heat. We need to travel with less dissipation. Rocket motors are fuel wise extremely inefficient especially at low velocities and most of the fuel is used to transport more rocket fuel. We need to find a better way. The energy required to hoist the payload by crane for an equivalent distance would be only a small fraction of the rocket energy required.

We now need to look at the TIG-effect (Tesla Induced Gravitation). Gravitational acceleration occurs when superstring energy passes along the time vector causing superstrings to accelerate towards one another until opposed when a weight force is produced. We need to apply intense magnetic and electric fields to a sizeable piece of material in such a way that the intensity and direction of the acceleration can be controlled. Travel would now be a matter of freely falling in a gravitational acceleration without any force or fuel apart from the electrical excitation of the craft. Speed will not be a problem as the craft will be able handle phenomenal accelerations without any discomfort to the occupants.

How does the TIG-effect work? We can take a clue from the traditional UFO design. The flat circular saucer shape indicates that the windings (Tesla coil) for the intense magnetic field are housed in the periphery of the craft and the hump in the middle contains a spherical cabin for pressurised travel in space and protection from radiation. The underside of the craft contains the three heavy blocks of ballast and circuitry for inducing the intense but controlled gravitational acceleration.

The control of gravitation and unlimited energy coming out of nowhere? Is this a foolish fantasy? – no, it is going to happen but hopefully the international political situation will be prepared for such development. These developments in the hands of an irresponsible military would be a catastrophe the likes of which the Earth has never seen.

We must now mention one of the most remarkable scientist-inventors that the world has ever seen. Like Leonardo da Vinci before him, Nikola Tesla (Никола Тесла 1856–1943) was an alarming creature seemingly in possession of universal knowledge. Tesla, an ethnic Serb born in Smiljan, Croatian Krajina, was born at midnight to the accompaniment of the lightning flashes of a thunderstorm. The midwife commented that he would be 'the child of the storm' to which his mother responded: 'No, of light'. Prophetic words indeed for the child who would become the electrical scientist known as 'the man who switched on the twentieth century'.

Regarding energy coming out from nowhere, Tesla made in 1891, the following comments in a speech before the American Institute of Electrical Engineers:

> *Ere many generations pass, our machinery will be driven by a power obtainable at any point in the universe. This idea is not novel... We find it in the delightful myth of Antheus, who derives power from the earth; we find it among the subtle speculations of one of your splendid mathematicians... Throughout space there is energy. Is this energy static or kinetic? If static our hopes are in vain; if kinetic – and this we know it is, for certain – then it is a mere question of time when men will succeed in attaching their machinery to the very wheelwork of nature.*

These words were spoken more than a hundred years ago but remain as relevant today as they were then. Modern quantum theory predicts the existence of a universal sea of zero-point energy at every point in the universe! This theory is based on Heisenberg's uncertainty principle with the end result that it is impossible for any point in space to be completely devoid of electromagnetic radiation. This energy is essentially inaccessible. What could Tesla have had in mind? At the time the idea of zero-point energy was still decades away. The background microwave energy of the universe is not much help either as this is too feeble to be of practical use. Dutch physicist Hendrik Casimir proposed, in 1948, an experiment in which it would be possible to measure the force between two plates resulting from zero-point waves. This has become known as the Casimir Effect. It has also been suggested that electromagnetic zero-point energy could be the underlying cause of inertia and gravitation. The Casimir force between two closely spaced plates is caused by interference of waves resulting in a calming of waves between the plates. A similar effect can be found when two ships draw alongside each other – the calming of the swell between the ships can cause the dangerous situation of the ships being drawn together. The force produced by the Casimir Effect is inversely proportional to the fourth power of the distance so that a marked increase occurs as the separation approaches atomic dimensions. The Casimir experiment is an extremely delicate affair: gold coated quartz plates, one mounted on a piezo-electric tripod and the other on a torsion pendulum of which the movement can be measured by means of laser interferometry to within ten nanometres! These extremely delicate measurements give no hint of the enormity of the matter in hand. The zero-point waves would find themselves at the very edge of the granularity of space and time with a wavelength equal to the Planck length and a frequency period equal to the Planck time. These are mind blowing figures – an atom sized point in space would have as much energy as the entire physical universe! Let us leave the matter here – this is far more than most of us can believe! There is clearly a problem here somewhat reminiscent of the 'ultraviolet catastrophe' of a hundred years ago and which was laid to rest by Max Planck's quantum theory. The slight force and movement detected in the Casimir experiment implies that a slight amount of energy has indeed been extracted from the zero-point field – perhaps there is yet a possibility for Tesla's idea of attaching our machinery to the wheelwork of nature!

Let us look at a quote from Paul Severing, 1910 (from *Marvels of Modern Science*):

Mighty, sublime, wonderful, as have been the achievements of past science, as yet we are but on the verge of the continents of discovery. Where is the wizard who can tell what lies in the womb of time? Just as our conceptions of many things have been revolutionized in the past, those which we hold to-day of the cosmic processes may have to be remodeled in the future. The men of fifty years hence may laugh at the circumscribed knowledge of the present and shake their wise heads in contemplation of what they will term our crudities, and which we now call progress. Science is ever on the march and what is new to-day will be old to-morrow.

Fifty years have passed twice over since this quotation was made and it is still quite relevant.

Tesla filed more than seven hundred patents and changed the way of life for the people of planet Earth forever. Every student of electrical engineering will immediately recognise the word tesla as the unit of magnetic flux density. One cannot flip a switch or turn on a motor without a debt of gratitude to Tesla. Of his most significant inventions were alternating current polyphase power distribution, the rotating field induction motor and alternator, radio transmitting and receiving, and fluorescent tube lighting. In 1898, Tesla demonstrated a robot controlled by radio waves for which he was granted a patent. Eleven years later Marconi was awarded a Nobel prize for the discovery of radio transmitting and receiving. Tesla also had a vision of a world wide network of radio communications which could function as a single brain. This idea had no hope without modern computers and communication satellites. At the end of the nineteenth century Tesla made some astonishing claims – these included the means to electrically control gravity to levitate vehicles and to draw unlimited power from empty space and later a death ray which could zap aircraft out of the sky. He even claimed to be able to split the earth like an apple. Fortunately the details of these inventions were all kept secret lest they bring about the destruction of mankind. With so many wonderful inventions to his credit, one would expect a life of international fame and fabulous wealth to be assured. This was not to be.

Tesla could not bring himself to accept Einstein's relativity and also insisted on the existence of a luminiferous ether. Relativity is in and the luminiferous ether is out! Einstein's Special Theory of relativity states that the energy in a piece of matter is a function of its mass and the velocity of light. The theory also states that the velocity of light is constant for all observers regardless of their relative movement. Tesla was adamant that matter contained no energy other than that which was imposed on it and that the luminiferous ether was necessary for the propagation of light. These two concepts of the universe seem poles apart and totally irreconcilable. Really? No! Let us first look at the energy contained in matter. No one will be in any doubt about the vast amount of energy in matter after considering the result of transforming a small quantity of uranium into energy. The formula for mass energy is of similar form to the formula for kinetic energy of a moving object. If we consider the mass energy to be the result of superstrings moving along the time vector at the speed of light then Einstein and Tesla's opinions are reconcilable. What about the luminiferous ether? This has been discussed earlier and it should be plain for all concerned that light propagated through a medium like sound waves is a hopeless case. Photons travelling at the speed of light with mathematical probability waves fit the bill nicely. The mathematical universal substrate is necessary for Einstein's relativity and should also suit Tesla's requirement for an ether.

Tesla took a keen interest in the Vedas, ancient writings in Sanskrit written 5000 years ago. In addition to hymns, prayers and myths the Vedas also deal with science, nature, matter, antimatter and atomic structure. Tesla's introduction to the Vedas probably came from Swami Vivekananda who went, in 1893, on a three year tour of Europe and the US. During his tour the Swami met with many well-known scientists including Lord Kelvin and Nikola Tesla. Tesla probably first met the Swami at a party given by actress Sarah Bernhardt. Swami Nikhilananda describes Tesla's contact with Vivekananda as follows:

Nikola Tesla, the great scientist who specialised in the field of electricity, was much impressed to hear from the Swami his explanation of the Samkhaya cosmogeny and the theory of cycles given by the Hindus. He was particularly struck by the resemblance between the Samkhaya theory of matter and energy and that of modern physics. The Swami also met in New York Sir William Thompson, afterwards Lord Kelvin, and Professor Helmholtz, two leading representatives of western science. Sarah Bernhardt, the famous French actress had an interview with the Swami and greatly admired his teachings.

The concept of matter and energy being equivalent comes to us from ancient Sanskrit writings. In February 1896, Swami Vivekananda noted the following:

... Mr Tesla was charmed to hear about the Vedantic Prana and Akasha and the Kalpas, which according to him are the only theories modern science can entertain

... Mr Tesla thinks he can demonstrate mathematically that force and matter are reducible to potential energy. I am to go see him next week to get this mathematical demonstration.

The Swami was hopeful that Tesla would be able to reconcile the teachings of the Vedas with modern science by demonstrating the equivalence between matter and energy. Tesla was unable to do this so it fell to Albert Einstein several years later, with his Special Theory of Relativity, to formulate his three symbol equation which is probably the most famous equation in all of physics and which has finally equated Prana and Akasha to Energy and Matter.

Tesla probably shot himself in the foot when he railed against the work of Maxwell, Lorentz and Hertz. Tesla also took a great interest in 'life forces' and vortices which could enhance the health and state of well-being of people and also improve the feng shui of buildings. Such ideas did not accord well with the scientific and engineering communities and Tesla's reputation plummeted leaving him with a 'mad scientist' image. He died in New York in January 1943 aged 86. On a few occasions, Tesla had been cheated out of large sums of money owed to him. Other notable events of 1943 were:

- The end of the battles of Stalingrad and Kursk which turned the tide of battle and settled the ultimate outcome of WWII. This ghastly military blunder resulting in slaughter, horror and misery on a vast scale, costing the winners far more dearly than the losers, was almost a repeat performance of Napoleon's disastrous invasion of Russia. Napoleon's foolhardy adventure has been thunderously commemorated in Tchaikovsky's 1812 Overture.

- The 'Dam Buster' raids on the Ruhr. This thrilling episode of desperation, innovation, courage, disaster and destruction has provided much material for books, documentaries and movies.

- The Manhattan Project rushing planet Earth headlong into the age of nuclear weapons and the horrific ultimate consummation of Einstein's chilling three symbol equation.

- The death of composer and concert pianist Sergei Rachmaninov who, in the opinion of the author, was the last of the truly great masters of musical composition.

- The death of Beatrix Potter. Nearly all the people that I know had their first introduction to the wonderful world of Bookland through the delightfully illustrated animal stories and many of us had a first porridge bowl illustrated with pictures of Peter Rabbit. Bookland has its own country code in the ISBN barcode.

At this time the world was embroiled in the most horrific military conflict in history. Whether genius or crackpot, the government of the day was in no mood to take any chances with Nikola Tesla. Shortly after his death, all of his working research material was seized and classified 'most secret'. Curiously, some of Tesla's subtle energy life force gadgets are available for sale on the Internet and there are before and after kirlian photos to show the efficacy. Tesla plates are made from aluminium sheetmetal anodised on one side and metallic Tesla spirals look eerily similar to a single spiral loop of the ten loop Leadbeater superstring.

There is a more down-to-earth idea of getting energy out of nowhere. Nearly a century and a half ago James Clerk Maxwell came up with the idea of Maxwell's Demon. A clever little fellow but not quite clever enough. If you take two containers of fluid at different temperatures you have available heat which can be used to drive a heat engine. If the containers are first brought together the temperatures will equalise and the energy will be unavailable for use even though none of the energy has been lost. This is the problem of entropy or availability of energy as formulated by the second law of thermodynamics. The lower the entropy the higher the availability of energy. Maxwell's little demon is a challenge to the laws of thermodynamics. All quantities used in thermodynamics are statistical values of heat and fluid properties without regard to the molecular nature of matter. Maxwell's demon has the job of controlling a door between two containers of gas. The gas molecules in each container do not all have the same velocity. The velocity of a molecule is a measure of its temperature. The demon's task is to control the door in such a way as to let cold molecules pass through the one way and hot ones the other thereby creating a temperature difference between the gases and voila, energy for nothing! There are many problems with the demon idea, notably quantum constraints and evaluating the speed of the molecules. The demon idea was finally laid to rest by physicist Leo Szilard in 1929 when he showed that units of information are also subject to the law of entropy. There is another way. We can make something similar to a transistor but instead of electrons tunneling through a semiconductor we can use ionised molecules passing through a charged permeable membrane. The fast molecules will be able to overcome the repulsive charge on the membrane and pass through and the slow ones will be repelled. There may be many problems getting this to work as providing the charges may overshadow the heat energy gain and the hot molecules that have passed through the membrane will have to be removed to prevent them from bouncing back to the other side where they came from, but who knows, it might work. To power an automobile in this way would be fabulous. All the energy would come from the air and all energy expended in driving and applying the brakes would simply return to the air where it came from.

A sobering thought from Sir Arthur Eddington:

> If your theory is found to be against the second law of thermodynamics, I give you no hope; there is nothing for it but to collapse in deepest humiliation.

If the TIG-effect can be rediscovered it will quickly render the entire Earth's fuel industries redundant, dramatically reduce the need for wheeled vehicles, roads, railways, aircraft and shipping, reduce pollution and ease pressure on the natural ecosystems. It would also be quite easy to transport icebergs to relieve drought stricken communities. What of tourism? It would be possible to take parties of martini sipping tourists to most parts of the Solar System. Piloting a TIG craft would be simplicity itself. It would only be necessary to climb aboard, type in the destination on the keypad and then relax for a few minutes until the craft has arrived and parked. For fully automated navigation it will of course be necessary to have a world-wide network of land or space based radio beacons but this is not a big deal – there is already something similar in place in the form of GPS satellites and mobile phone antennae. Collisions between TIG craft should be impossible with computer controlled navigation centres. It seems logical that all privately-owned TIG craft will be licensed and programmed to take off and land at a defined list of parking spots. This will considerably ease the problem of illegal cross-border traffic and also simplify tracking of criminal movements. Fly-anywhere status would only be available to professional pilots with appropriate security clearance. TIG craft will bring a whole new dimension to mobile home dwelling: one could park one's home in the most inhospitable place on Earth and any city on Earth would be no more than a few minutes away. TIG craft will not replace all forms of transport. There will still be a need for compact vehicles for short distance travelling. Small vehicles using hydrogen fuelled engines would be very economical as the cost of producing hydrogen from TIG electricity would be almost negligible. Hydrogen burning cars are pollution free – simply burning hydrogen to water where it came from in the first place. Hydrogen burning cars are a reality; the present models use fuel cells and electric motors rather than internal combustion engines. Before hydrogen fuelled cars can become ubiquitous there remains a problem to be overcome – the storage of sufficient hydrogen to be able to travel a reasonable distance before refuelling. Hydrogen can be stored in liquefied form or under extreme pressure but not everyone would feel comfortable driving a car fitted with serious pressure vessels. The ideal solution would be to use compact TIG motors but it remains to be seen whether these can be developed.

The need to find new renewable energy sources free of pollution and carbon dioxide emission has become urgent to the point of desperation. Carbon dioxide emission has placed life on planet Earth on a course to catastrophe and the point of no return no more than one or two decades away. The main culprits are the fossil fuel burners, coal fired power stations, aircraft, road traffic and shipping. The wanton destruction of rain forests not only causes carbon emission but also destroys the means of recovery. The destruction of rain forests is a crime against life on planet Earth which should be opposed by military force. Nuclear power stations do not cause carbon emissions but are beset with other, possibly worse, problems. There are beautifully clean sources of electric power such as hydroelectric plants, geothermal power stations and wind turbine farms but these depend on specific site and climatic conditions. Direct solar power is very effective for water heating but not of much value for large scale power generation. Photovoltaic panels provide the ideal means of powering space satellite instruments but until recently have been quite useless for generating power on an industrial scale. This is set to change – Concentrated Solar Power (CSP) holds great promise in areas with infrequent cloud cover. An exciting announcement by a MIT research team was made in July 2008. They had discovered a dye which could be applied to ordinary window glass. The dye would cause solar energy to deflect through the glass, much like a wave guide, which would then be captured by photovoltaic cells around the edge of the glass. This Luminescent Solar Concentrator (LSC) promises a huge improvement in efficiency and cost reduction of solar power. Surprisingly, burning wood need not necessarily cause a carbon emission problem. If trees are growing faster than they are being harvested for firewood then this will not cause a carbon buildup. Great advances have been made in the efficiency and cost reduction of wind turbine farms so that the onshore and offshore deployment of these in windy locations holds great promise. Wind turbines come in a large assortment of sizes from a domestic unit of less than a kilowatt to gigantic units of several megawatts. Power generation from sea waves has been somewhat neglected but now appears to hold tremendous potential and can provide perfectly clean and inexhaustible electricity. This is effectively an indirect way of utilising wind power which is in turn indirect solar power. Several designs of sea power machines have been developed: some utilising the vertical movement of floating buoys, others funnelling waves to higher level reservoirs for turbine use and others using the ebb and flow of water to produce rotary power. A most remarkable and successful machine is the pelamis. This is aptly named after a surface swimming sea serpent. At a distance it looks like a four coach tube train which has left its burrow and lost its wheels. Four cylindrical tubes hinged together float on the sea and the flexing of the hinges forces rams to provide high pressure fluid to drive hydraulic motors which in turn drive alternators. The power then passes to a transformer and by cable to a land based installation. The lengths of the tubes are made to approximate half the distance between wave crests for maximum flexing. A single pelamis can typically provide 750 kW of power so that a farm of 40 units will give a substantial 30 MW. It cannot be expected that tumultuous seas and tempestuous winds will usually occur when power demand is at its highest so pelamis and wind turbine farms can conveniently be used together to provide a more constant power supply. It will also be necessary to use a pumped storage hydroelectric plant to accommodate the highs and lows of power demand. All of this will be more expensive than a coal fired power station but with the future of human civilisation in the balance it will certainly be a bargain. The wind turbines have brought wind power full circle from the windmills introduced to Europe and the Baltic states from Saracen countries centuries ago. The windmills were widely used for grain milling, pumping, lumber sawing and other applications. The windmills became obsolete with the introduction of steam engines, internal combustion engines and finally electric motors. Many old windmills can still be seen on display in Holland, Greece, France and Spain. The modern wind turbines do not have the romantic associations of Don Quixote de la Mancha, Daudet's Windmill, the Moulin Rouge, de Kinderdijk Molen or the Moonspinners of Mykonos, but they do bring a clean and easily distributed source of power. The windmills of Holland were built in a variety of styles, some even being used as a dwelling for the miller and his family. The Dutch windmills also served as a means of communication: the sails would always rotate in an anticlockwise direction when viewed from the sail side and when not in use would be stopped with one sail pointing vertically. Stopping the sail at angles before or after the vertical position would signal various messages. There is much interest in the study and preservation of windmills – the International Molinology Society being founded for this purpose. Water power was also in widespread use for driving mills; in this case a river would be diverted to a waterwheel and gravity would do the work. Rustic watermills are no match for the vast power of a large hydroelectric power station, but these cannot match the tranquil beauty of the Mill on the Floss.

Maggie loved to linger in the great spaces of the mill, and often came out with her black hair powdered to a soft whiteness that made her dark eyes flash out with new fire. The resolute din ... the meal ever pouring, pouring ... helped to make Maggie feel that the mill was a little world apart from her outside every day life.

George Eliot

A working water mill can be found in the town Lyme Regis on the Dorset coast. A mill in this town is mentioned as far back as the Domesday Book. The present mill had become derelict but was fortunately saved from demolition and has been painstakingly restored to full working order. Visitors can now, in addition to viewing this relic from the past, also buy stone-milled meal and freshly baked bread.

The modern green and clean machines for producing power are not without carbon emission. In any large engineering enterprise huge quantities of steel and concrete are required. In concrete construction much carbon emission is caused in the manufacture of cement and also in the quarrying and transport of aggregates. Steelmaking is an intensely polluting industry. Smelting iron in a blast furnace requires huge quantities of iron ore, coke and limestone. The coke is produced from coal in coke ovens which also produce coal gas, which is used for heating throughout the steelworks. The blast furnace produces huge quantities of blast furnace gas which is rich in carbon monoxide. This gas is used to heat the hot blast stoves which provide air at high temperature to the inferno within the furnace. A blast furnace produces pig-iron which is high in carbon content requiring burn-off in the steel making process. The coal gas is also used in the steel mill soaking pits for heating the steel to rolling temperature. Coal gas yields several valuable by-products such as tar, pitch, benzole, naphthalene, ammonia etc.

With the successful development of TIG levitation one of the first familiar sights to disappear would be heavy road haulage – unlimited loads could be swiftly transported from anywhere to anywhere in driverless craft. A brief moment of megalomania: How about a TIG space observatory? We could have a trio of telescopes with 100 metre mirrors with full spectrum observation from infra red to ultraviolet and kilometre long interferometric baselines. This could be augmented with microwave radio telescopes as well as gamma ray and X-ray astronomy. The crew could be housed in huge residential quarters with Earth strength gravity, keep fit gymnasium and canteens to suit every taste. It would not be necessary for the observatory to be in orbit – it could be stationed at any convenient spot in the

Solar System well away from the paths of comets and meteoroids. The observatory could also be an interesting stopping point for solar system tourists and school groups. Another very exciting possibility would be to have a matrix of gravitational observatories in space with separations of several astronomical units. With such a configuration it would be possible to realise the impossible dream of gravitational imaging.

How far are we away from TIG? It might be right under our noses! From the scant information available it would seem that the windings of the craft might be part of a high frequency oscillator required to produce the high voltage electrostatic oscillation on the underside of the craft. It may be necessary to use superconducting material for the windings in order to obtain a high 'q' value for the resonant circuit. A steady high voltage would be applied to the topside to produce the gravitational acceleration. Evidence of these high voltages can sometimes be seen as corona discharges around UFOs in flight. A word of caution should you decide to build your own TIG craft. No information is available regarding electromagnetic radiation from TIG propulsion. The visible corona light is perfectly harmless but microwave radiation is another matter! The craft should not be flown low over populated areas or even over farmlands. Radiation from the underside of the craft could cause plant stalks to partially cook resulting in crops collapsing in circles, and hovering over fields of popcorn could cause the maize kernels to pop. The radiation could also impair the fertility of the topsoil. In recent years, there has been much confusion surrounding the appearance of crop circles. Most crop circles are actually fakes made by rope toting, plank stomping individuals perpetrating hoaxes or by agents providing dis-information to draw attention away from genuine circles and clues about TIG propulsion. The fake patterns sometimes have the most bizarre geometrical shapes. Circles produced by carelessly flown TIG craft are easily recognisable. These make simple circular patterns of partly cooked plant stalks, and any seeds present will have the appearance of being cooked in a microwave oven. Further clues will be found in the topsoil. Worms, crickets and other small inhabitants of the soil within the circle will have met a tragic end. There is an unpleasant matter concerning TIG craft that we must also mention here – cattle mutilations. TIG craft being test flown over farmlands have on numerous occasions caused the deaths of farm animals, mostly female neat and also some horses. The farm cows are the main casualties as they are the most numerous of the animals. The TIG crews are desperate to cover their traces and the craft being tested are not equipped for transporting heavy animal carcasses – so what to do? The deaths of the animals are caused by radiation damage to the brain, and soft tissue thinly covering bone and some extremities will also show visible damage. The problem is dealt with by removing parts showing obvious damage and then grotesquely mutilating the carcass in the hope that the owner will not think of having the animal examined for internal microwave trauma. The mutilations are obviously done after the deaths of the animals as there are no signs of blood loss even when major blood vessels have been severed. The owners of the livestock are left feeling intimidated by unknown forces of evil and with no prospect of being compensated for their loss. Who knows what becomes of any farm workers who are zapped by microwaves? There is an ominous side to this matter. The number of livestock mutilations has soared into many thousands, indicating that TIG craft are being secretly manufactured and tested on a very large scale. We are left with the chilling question: by whom and to what purpose?

We have here a dichotomy. On the one hand we need to research TIG right from first principles and take it through all the development model stages up to industrialisation, and on the other hand we have TIG craft secretly flying about and desperately trying to cover their traces. Are we to assume that there is a link between the seizure of Tesla's research material and the flurry of UFO sightings over the US and UK less than two years later? What is going on? The answer may be found in the 'Vibes of the Universe' section of the appendix in the discussion on EMP. Let us say no more.

If anyone is planning to experiment with TIG we can suggest some starting parameters: The corona discharges around UFOs in flight suggest a voltage of about 100 kV for the topside electrostatic charge as well as the underside oscillation. The microwave oven effect on growing crops indicates an oscillation of 2 or 3 gigahertz. TIG experiments should only be conducted in a controlled laboratory environment as extremely dangerous voltages and microwave radiation will be generated. How does one obtain a high voltage at high frequency? Simple – use a Tesla coil, but with great care – energising a Tesla coil could be a life threatening exercise. If the craft is intended for underwater use as well as flight then an additional waterproof housing will be required. Underwater travel is possible as seen from the Shag Harbour UFO incident. In this incident, the UFO was quite likely of local origin. According to Tesla, the maximum speed for TIG propulsion, also known as electropulsion, would allow a trip to the Moon and back in less than an hour. What about travel outside the Solar System? There is little hope of this happening. Even travelling to the nearest stars would require a journey of many years. Would anyone be foolish enough to spend a large part of their lives travelling to a neighbouring star and its planets without knowing what to expect there and quite likely find a star looking like most others with an odd assortment of barren planets. Even communication with Earth would eventually become hopeless as the time for message transmission and reply would stretch into years. The only way to travel to other solar systems would be by taking a short cut through time, but travelling this way is still a matter of science fiction – or is it?

There is a remarkable story regarding teleportation and time travel which has more than 1,5 million sites on the Web. This is the 'Philadelphia Experiment' or 'Project Rainbow' as it was originally called. The objective was to create a field around a naval vessel causing light and radar waves to bend around the ship rendering it invisible to enemy observation. It is claimed that the method was based on Einstein's Unifield Field Theory of Gravitation and Electromagnetism despite the fact that the completion of such a theory has never been generally acknowledged. It is claimed that Nikola Tesla headed the project but withdrew before the experiment took place because of horrific effects it might have on the crew. Tesla died, aged 86, in a New York hotel room several months before the fateful experiment. The first experiment was conducted in June 1943 in Philadelphia harbour and also at sea. The equipment, installed on USS *Eldridge*, consisted of large generators, some huge Tesla coils, 3000 power amplifier tubes and many tons of heavy electrical cables. In the early hours of 13 August 1943 the *Eldrige*, accompanied by a tender vessel, moved down the Delaware and out to sea to its destiny of unspeakable horror and with the crew blissfully unaware of the nature of the secret mission. The experimenters got more than they bargained for. The ship became enveloped in a green cloud, became invisible and then instantly disappeared only to reappear hundreds of kilometres away. Later it reappeared in its original position. The remote place where the ship appeared, was where it had actually been months before, suggesting teleportation through time into the past. Worse was to come. The deck was a lazar field of human remains. Of the crew of 181 men only 21 survived. Forty died from radiation exposure, extreme burns or electrocution, many went missing without a trace, some went insane and others were found embedded in the steel structure of the ship. Some of the embedded sailors were still alive but were mercifully dispatched by pistol shot. This sounds like a very tall story but it does fit in with the ability of UFOs to instantly disappear and who knows: does this give them the ability to teleport without the restriction of the velocity of light? The *Eldridge* later came to an ignominious end in a Greek scrapyard, or did it? Some claim that the scrapped warship was a look-alike decoy and that the original has been carefully preserved. It has been claimed that Einstein and John von Neumann were also deeply involved in the experiment. A report of Einstein's involvement states that a crew-cut crewman, on seeing Einstein come aboard the *Eldridge*, remarked, "That guy needs a haircut!" It seems somewhat unlikely that Tesla, a frail and elderly man at the time, could have headed a major military research project. An official announcement stated that the ship's hull was merely being fitted with de-gaussing cables as a protection from magnetic mines and that the *Eldridge* had not at any time put in to Philadelphia harbour. No crew member of the *Eldridge* has ever confirmed that the experiment actually took place. This story, as with Roswell, is so loaded with cover-up theories, hype and snake oil that we will probably never get to know what actually happened. The story must surely be a hoax but it is a sensational idea that will not go away. Another story closely linked to the Philadelphia Experiment is the Montauk Project. This is an ultra secret underground facility experimenting with psychotronics (interfacing mind and machine), particle beams, electromagnetic mind control, teleportation and inter dimensional time travel. This too seems a very tall story and too horrible to describe here. One bit of information conspicuously absent from the many accounts of the Philadelphia experiment is noise. The loud report of a pistol shot is caused by the grains of propellant in the cartridge rapidly expanding to a few litres of gas. A kilogram of explosive is sufficient to blow an automobile to smithereens – this blast is caused by the rapid expansion of the charge to a few cubic metres of hot gas. The difference between high explosives and less fearsome combustibles is in the speed in which the material becomes gaseous. There is little difference in the energy per kilogram between petrol and dynamite. If something as large as a ship were to instantly disappear the vacuum left by the displacement of thousands of cubic metres of air would cause an implosion which would be heard a great distance away. Is there any way in which a large object could instantly disappear quietly?

Tesla and Gauss have both been honoured by having units of magnetic flux density named after them. In 1956, on the centenary of Tesla's birth, the International Electrotechnical Commision issued the following statement:

> *The Committee of Action of the I.E.C desires the President of the I.E.C. as its personal representative at the Tesla centenary Celebration to convey to all who are there assembled, the warm greetings of the I.E.C. on this occasion in commemoration of the great Tesla.*

> *It is always with a sense of profound respect and admiration that the name Tesla is remembered throughout the electrical world and the I.E.C. is very mindful that*

its work today for international agreement in the electrical field is dependent in a very large measure on the fundamental scientific work of Nikola Tesla.

The I.E.C. is very happy that this fact has been marked this year by the agreement they have reached for the world unit of magnetic flux density in the Giorgi system to be called the 'tesla'.

The Giorgi system was introduced to overcome problems with the Centimetre Gram Second system of units and which later was known as the Metre Kilogram Second system and later to become the SI system – see appendix for more detail.

John von Neumann (1903–1957) was a most astonishing phenomenon. Born in Budapest as Janos von Neumann he showed exceptional abilities from an early age. Aged six he could mentally divide eight digit numbers and exchange jokes with his father in classical Greek. The family would astonish visitors with little Jancsi's memory. He would be shown a column in the phone book and would then be able to remember the number of any name from the column that was mentioned. By the age of eight he had mastered calculus! Not only did von Neumann have a brilliant career in mathematics, he also made a huge contribution to quantum mechanics. Von Neumann's name has become synonymous with the sequential processing of computer instructions. He was also the first to think of storing data in computer memory by means of binary digits and of storing instructions and data in the same memory. During the war years he made a contribution to the implosion of nuclear fuel to cause a fission explosion and also to the development of the hydrogen bomb. At the time of the Philadelphia experiment, he was a member of the Navy Bureau of Ordnance. The list of von Neumann's awards and academic honours is almost unbelievable.

Time for a little piece of science trivia: Thomas Edison once claimed that he could accurately determine Tesla's weight, without using a scale, and after feeling Tesla all over came up with the correct figure. How did he do it? Edison once worked at a slaughterhouse where he had to weigh hogs by the thousand and in the process became expert at estimating the weights of carcasses.

Thomas Edison, one of the most famous inventors of modern times, was a 'Direct Current' man with associations with General Electric. Nikola Tesla was an 'Alternating Current' and high voltage man with associations with Westinghouse. Edison improved the incandescent light bulb and Tesla invented the fluorescent tube light and enjoyed demonstrating that he could turn the tube on while holding it in his hand without any connection to a power source. (I have myself seen this astonishing feat done at a university open-day.) Sir Humphrey Davy invented the original carbon filament lamp as early as 1800 but it was only after Edison's improvements and the availability of electric power that it came into widespread use. It was only in 1910 that William Coolidge introduced the tungsten filament which outlasted all other filaments in use. Alternating current soon became the obvious way to transmit high voltage electrical power over long distances. Edison, piqued by the success of Tesla's AC, launched a road show going from town to town electrocuting animals in public to demonstrate the danger of Tesla's alternating current. Even an elephant became victim to this mindless cruelty. Edison omitted to mention to the horrified onlookers that direct current was equally if not more dangerous. Edison would later admit that he realised all along that polyphase AC was the obvious way to distribute electrical power. Tesla was actually employed by Edison for a short time to perfect the design of the DC generator for an agreed sum of $50 000, but when Tesla asked why he wasn't being paid Edison replied, "You don't understand American humour." Using high voltage step up and step down transformers meant great savings in the cost of conductors and heat loss of power, and using three phase power brought further great saving by reducing the need for a return conductor. The ubiquitous three phase induction motor also meant huge cost and maintenance advantages over DC motors. The idea of a rotating field induction motor came to Tesla in 1890 while walking with a friend in a park in Budapest and observing the sunset.

Tesla chose a frequency of 60 Hz for alternating current because he liked the number. Sixty is the product of 3, 4 and 5 which can also be the sides of a right-angled triangle. For an induction motor with a two-pole wound stator this gives a synchronous speed of 3,600 r.p.m. A four-pole stator would give half this speed. The numbers 3, 4 and 5 are the numbers of sides of the only regular polygons which can form regular polyhedra in a three dimensional space giving the five Platonic solids. The numbers 3, 4 and 5 also suggest the theorem of Pythagoras which later became the 47th proposition of Euclid. The 48th proposition is the inverse. The 47th proposition is easily demonstrated geometrically and is valid for any right-angled triangle. There are also a large number of other proofs for the proposition. This also leads on to the last theorem of Fermat, a fascinating topic, but let us not digress too far.

There is yet another right-angled triangle aspect to alternating current. There are three basic electrical components to an alternating current circuit: resistance, inductance and capacitance. The inductance of an inductor is determined by the magnetic permeability of the core material and its physical configuration. The capacitance of a capacitor is determined by the electrical permittivity of the dielectric material and its physical configuration. The magnetic permeability and electrical permittivity of space are fundamental properties of the universe which together determine the velocity of light. This Maxwellian matter is discussed further in the appendix. In the rotating vector diagram of an alternating current, the inductive component lags the resistive by a right angle and the capacitive leads the resistive by a right angle. The vector sum of these determines the reactive component of the circuit. When the inductive and capacitive current components are equal the circuit becomes resonant. The reactive, resistive and resultant values of the alternating current are related as the sides of a right angled triangle.

For readers with an interest in the ancient Egyptian Mysteries the 3, 4 and 5 unit triangle has further profound significance. According to Plutarch, the ancient Egyptians took the 4 unit side as a base and assigned this to Isis; the vertical 3 unit side was assigned to Osiris and the hypotenuse assigned to Horus. The three units of Osiris have been attributed to the alchemical principles of Salt, Sulphur and Mercury (Hindu concepts – Tamas, Rajas and Sattva). The Four units of Isis were attributed to the alchemical elements of earth, water, air and fire (Hindu concepts – prithivi, vayu, apas and tejas). The five unit hypotenuse of Horus was attributed to the kingdoms of mineral, plant, animal, human and the Fifth Kingdom. There is a curious connection between the fundamental electrical components of capacitor, inductor and resistor with a group of three ancient secret hand signs used by a well-known fraternal society. The signs illustrate in turn the capacitor, inductor and resistor and when the resistor is cut its value is raised from the finite to the infinite. Possibly few of the worshipful brethren will have noticed this electrical connection, but they will certainly have noticed that the numbers 3, 4 and 5 when multiplied by three give the sizes of 'proper steps' in inches. There is a 3, 4 and 5 nicety about the Egyptian pyramids: the triangular sides have three vertices, the square base four and the solid shape five vertices.

In addition to Tesla's 60 Hz frequency, an AC frequency of 50 Hz is also in common use. Power frequencies above 60 Hz or below 50 Hz are not considered practical for several design and efficiency considerations, however high frequency power is used in aircraft and in some specialised industrial and military applications. Direct current has remained in use for powering trains and other traction applications where the motor is required to operate at full load over a wide range of speeds. The gigantic DC motors of steel rolling mills are an awesome sight. I can remember, as a teenager, climbing up a ladder to seat new carbon brushes on the enormous commutator of the generator of an Ilgner set supplying the mill motors. Alternating current motors are best suited to single speed applications. Improvements in control have brought about widespread inroads of AC into traction applications. Some trains are actually equipped to operate using AC or DC supply at more than one voltage along the same track but still using DC traction motors. Other widespread uses of DC are in automobiles, metal refining and electroplating.

A brief comment of Tesla at a time of hubris:

I do not think there is any thrill that can go through the human heart like that felt by the inventor as he sees some creation of the brain unfolding to success.

Thomas Edison considered his improvement of the incandescent light bulb as his greatest achievement. This has been the primary source of domestic lighting in the twentieth century. Public buildings and offices were nearly all illuminated by Tesla's fluorescent tube lights due to their greater efficiency and better light colour. Incandescent bulbs are now rapidly being superseded by energy saving compact tube lights which use only a fifth of the power for the same light output and have a considerably longer working life. Incandescent bulbs give a light output of only 5% of the energy input. Edison's bulbs came mainly with three sizes of caps – small Edison screw, Edison screw and giant Edison screw. In years to come all that will remain of Edison's light bulb will be the screw cap. Tesla's greatly improved compact tube lights are available with either bayonet or Edison screw caps uniting the inventors in a one-sided showdown – there is poetry to be read in this situation. The CFLs (compact fluorescent lights) do however contain a small quantity of mercury which can pose a waste disposal problem. Incandescent bulbs contain no mercury but if they receive power from a coal fired power station they can actually cause the release of more mercury to the atmosphere than is contained in the corresponding CFL and its power station emission.

When TIG transport and power generation are fully developed this will herald a new golden age of prosperity such as has never been seen on this planet. We need however to proceed with caution. The fine human qualities of integrity, strength of character and intellectual stature usually thrive better in times of hardship than luxury. An out-of-control situation of runaway prosperity and luxury could well be a recipe for disaster! We can take a lesson from the gold and diamond rushes of the nineteenth century. These did bring fabulous wealth to some but most of the people involved found little more than degradation and misery. In the Fairmont

Hotel conference of 1995, the concepts of a 20/80 society and Zbigniew Brzezinski's 'tittytainment' emerged. This described an industrialised society of the future where twenty percent of the population would lead educated, meaningful and productive lives and the remaining eighty percent would have a shallow soap-opera type of existence and be pacified by 'tittytainment' (a contraction of tits and entertainment). This is something akin to the 'bread and circuses' of ancient Rome intended to appease the masses and distract them from ideas of revolt. This gloomy forecast was made even without regard to the phenomenal changes that we can expect from TIG power generation and propulsion. An unfortunate sign of artistic decay is the general debasement of nearly all TV channels, news and serious documentaries alike, with the widespread use of a primitive drumbeat throughout the programs. Do the producers of these programs really think that viewers who are not offended by the primitive thumping are likely to be interested in news and serious documentary programs? Persons unable to tolerate the offensive noise are left with a choice between silent TV or none at all. Another appalling sign of artistic decay is the reworking of timeless musical masterworks into short life trash – what kind of musical bankruptcy can think that this is a good idea? Where will we find a modern Robert Schumann to lead the Davidsbündler against the Philistines? Are we really to believe a forecast of a population with most people living a zombie-like existence? No, surely not. All people have an innate aspiration to Cosmic Consciousness which must eventually emerge and cannot be suppressed.

A very useful possibility for TIG transport would be to jettison our nasty repositories of nuclear waste into the Sun well out of harm's way. Electrical power would be only a pleasure – as much as you need wherever you need it and without billing. No more power stations and unsightly power lines. Power generation from nuclear fission reactions is a no-no – the sooner this is removed from our planet the better!

We have looked into some huge terrestrial and space projects launched into fundamental gravitational research, as well as some sensational ideas on electromagnetically induced gravitation, yet we still have no final response to Richard Feynman's lament:

"We have no machinery."

SUPERSTRINGS AND CONSCIOUSNESS

Why do we need to mention consciousness in a discussion on particle physics? For the simple reason that the two go together as much as the wave and particle properties of a photon go together. It may come as a surprise to many that research into particle physics right down to superstrings is not new – in fact this has been studied in minute detail more than a hundred years ago. As early as November 1895 a detailed and documented study of superstrings was published in London. At this time the term superstring was not in use – the strings were called by their ancient Sanskrit name 'aahnoo'. How? The study was made using a special Yoga technique learned from an Indian guru. This permitted the study of particles from an elevated level of consciousness outside the space-time continuum. This may seem unorthodox, heterodox or whateverdox but the results of the study are irrefutable and will not go away – we cannot dismiss them and therefore must deal with them here. Our quest of perception of the universe now takes us to the realm of pure consciousness.

It is perfectly natural for professional people to be reluctant to discuss their field of expertise with the unqualified. This is usually not a problem but there have been cases where this has resulted in major discoveries remaining unrecognised for many years. Philosopher Schopenhauer had this to say on the matter:

All truth passes through three stages:
First, it is ridiculed.
Second, it is violently opposed.
Third, it is accepted as being self evident.

The following are a few of such cases of professional or authoritarian rejection:

- Anaxagoras on Solar System more than 2000 years ago.

- Galileo's model of planets orbiting the sun.

- Van Leeuwenhoek's microscopic observation of microorganisms in 1676.

- Darwin's evolution of species by natural selection.
- Gregor Mendel's discovery of dominant and recessive traits.
- Tesla's reception of radio signals from the cosmos a century ago.
- Birkeland's Auroral Currents proposed in 1908.
- Wegener's theory of continental drift in 1915.
- Bretz's Missoula ice dam burst proposed in 1920.
- Dart's fossil discoveries in 1924.
- Lemâitre's expanding universe proposed in 1927.
- Leadbeater's quarks and superstrings.

It is the last item of this list that we will deal with here. It is almost beyond belief that this most astonishing and fundamental scientific breakthrough has remained unrecognised by the scientific community for more than a century.

We now take a quantum leap from 'what the butler saw' in chapter one to 'what the occultist saw'.

Studying anything from outside the space-time continuum has two tremendous advantages: firstly, the observer is not bound to the progress of the time vector so that the exceptionally rapid motion of particles is not a problem; secondly, the size of the object of study does not matter at all. The observer can study galaxies and superstrings in much the same way as these can be visualised in the imagination. The superstrings were studied right down to the structure of the string filaments.

When reading material dating from the end of the Victorian era it is necessary to have some feeling for the *Zeitgeist* of the time. This was a period when communications were by handwritten letters collected and delivered a few times a day; when transport was powered mostly by horse or steam and when little girls and old ladies could see fairies at the bottom of the garden. This was also the time before the 'Titanic' and WWI.

Most of the source material for this research was produced by the early Theosophical writers H. P. Blavatsky, C.W. Leadbeater and A. Besant. Renowned or notorious depending on one's point of view; relativity does not only apply to physics. The most convenient reference is the book *Occult Chemistry* co-authored by C.W. Leadbeater and A. Besant. A large book written over some forty years of intermittent part-time writing. Only a small portion of the book is of particular relevance to our present concern. A large part of the book is devoted to the study of chemical compounds. Unfortunately for the authors, the Heisenberg Uncertainty principle was two decades away and the authors had no way of knowing that arresting the molecular motion and fixing the position of particles was an impossible state of matter. This caused the nuclei of compounds to collapse on each other rendering this part of the study scientifically worthless. We can come to no other conclusion – the theory of atomic electrons and energy levels is too well established for us to consider chemical reactions taking place directly between the nuclei.

The study of individual atomic nuclei is another matter altogether. The drawing of the hydrogen nucleus is awesome. It shows a proton consisting of two u-quarks and a d-quark tightly bound in a triangle and the quarks each containing three superstrings. The investigators were also able to identify positive and negative electric charge and labelled all the strings accordingly. It gets even more remarkable. Also included in the drawing is an anti-proton slightly overlapping the proton. An exact mirror image with all the string charges reversed – clearly the quantum-uncertainty result of slowing the proton and fixing its position. The investigators did not assign any name to quarks but merely showed them as groupings of three strings.

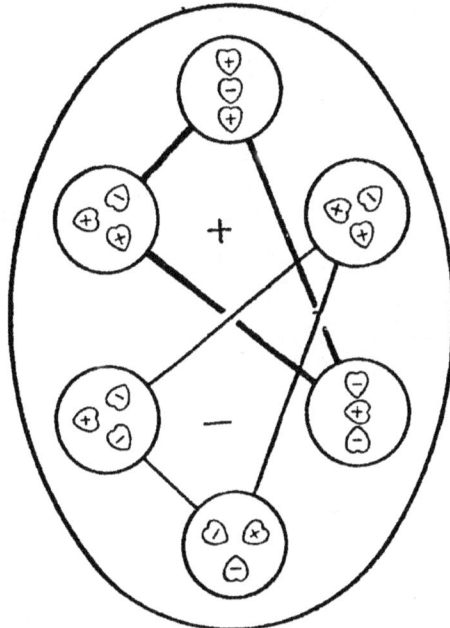

Fig. 7.1 Diagram of hydrogen nucleus taken from "Occult Chemistry."

This is the end of familiar territory. Nuclei heavier than hydrogen take strange shapes and configurations of superstrings with no hint of the time honoured proton-neutron structure. The only remaining conformity to be found is the ratio of number of strings to atomic weight. At this point the temptation to include a verse from a nonsense poem by Louis MacNeice is irresistible.

> It's no go the yogi man,
> it's no go Blavatsky,
> All we want is a bank balance
> and a bit of skirt in a taxi.

Big deal – we need no ghost from a hundred years ago to tell us that the proton consists of three quarks. This is only the beginning. The investigators went further to study the superstring in detail right down to point zero.

The following drawing of positive and negative superstrings is taken from *Occult Chemistry.*

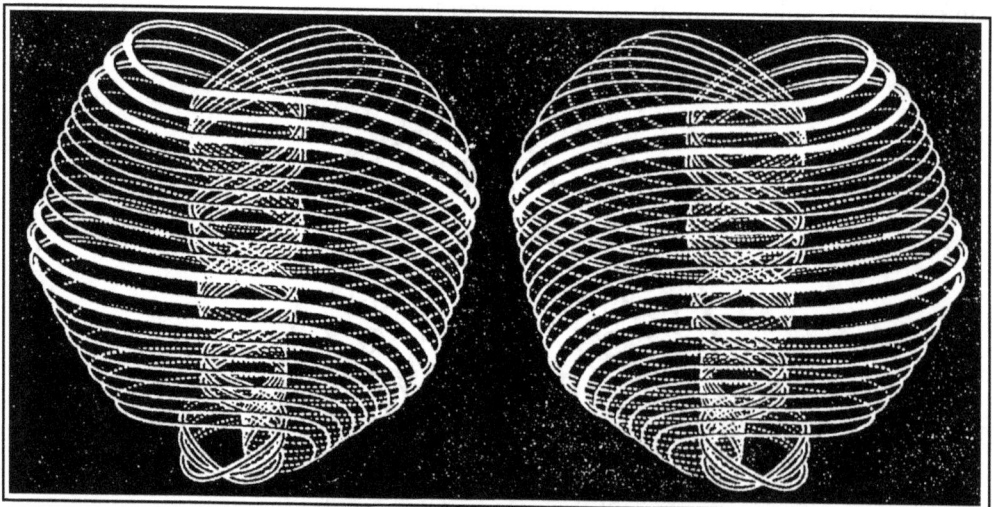

Negative   Positive

Fig. 7.2 Drawings of superstrings or "aahnoo" as described in "Occult Chemistry."

It should not be supposed that this is a drawing of a physical object. It is a representation of an eleven dimensional mathematical manifold of which one of the dimensions is time. This manifold is beyond human visualization.

The aahnoo consists of ten closed loops – three of slightly thicker filaments corresponding to the three spatial dimensions and the seven thinner filaments corresponding to the Kaluza-Klein dimensions. Closer investigation showed that the filaments are multiple helical in structure and even closer investigation showed that the ultimate structure of the filaments is, wait for it, a string of dimensionless points.

> There you have it - the ultimate nature of matter is nothing but energy.

It would be most interesting to see if anyone can express the aahnoo mathematically as an eleven dimensional manifold. A mathematical manifold is essentially a thought. With the aahnoo we come to the ultimate eleven dimensional interface between the Universal Consciousness and the objective universe in one time dimension, three spatial dimensions and seven Kaluza-Klein dimensions. It may be instructive to take another look at Blake's painting 'The Ancient of Days'. The universe is being created by means of a pair of compasses, or rather a pair of dividers, with two sharp points. The dividers are making circles of dimensionless points, which when drawn along the time vector, become spirals of points and hence the creation of all matter. There is no reason to suppose that Blake had this in mind when he produced this famous painting but it is quite possible that Blake, like many other artists of genius, was able to work with more wisdom than he was aware of.

It is interesting to note that in 1867, Scottish physicist Sir William Thompson wrote at length on 'vortex atoms' based on a paper by Helmholtz on *Wirlbewegung* written in 1858. In 1892, Thompson was honoured with the title Baron Kelvin of Largs. Kelvin is the name of a river passing through the grounds of Glasgow University.

Since classical times the circle was considered to be the simplest perfect shape. Greek philosopher Proclus (412–485) declared:

The circle is the first, the simplest and most perfect form.

This is however not true for the simplest and smallest shape in a digital space-time. A perfectly smooth small sphere is simply not possible in a digital universe. You cannot even make a sphere of equally distributed points. Take a circle of equally spaced points, add another at right angles and again another at right angles to the other two. You now have an equator with two longitude lines and eight curved triangular spaces with curved edges. It is impossible to fill these shapes with equally distributed points due to π being a transcendental number. Circles of digitally spaced points spiralling along the time vector provide a beautiful basis for matter.

Mr Leadbeater went to great pains to record as much detail of the aahnoo as possible. Each of the ten loops is twisted through two and a half turns. Each string has in its initial helix 1490 loops and a total of millions of dimensionless points. He spent much effort in estimating the structure of the strings. In all, the filaments were helixed seven times over; by counting the points in the finest level of helix the total number of points for the entire aahnoo were estimated at 7^{12} – nearly fourteen thousand million. Another astonishing observation was that the string filaments were seen to be glowing in brilliant colour. This is not as strange as it may seem. Colour is not a property of light waves; different wavelengths signal colour information to the visual cortex but the actual perception of colour takes place in the consciousness. It is quite easy to recall to the memory a painting in a distant location and consider the various colours chosen by the artist. One can 'see' these colours in the imagination without the involvement of light waves. Isaac Newton in a letter to Harry Oldenburg wrote:

> "... to determine by what modes or actions light produceth in our minds the phantasm of colour is not so easie."

What of the quarks? The proton of the hydrogen nucleus was shown with two u-quarks and a d-quark each with three superstrings. This is not the end of the story – there is a whole farmyard of quarks containing from two to seven strings. The charges of the strings do not give a direct indication of the electrical charge of the quark; this is also determined by the configuration and connectivity within the quark. Quarks with the strings pointing outward are generally positive and those pointing inwards are generally negative. Quarks comprising three strings can either have their strings arranged in a triangle or in-line and also have several connectivity arrangements.

Why this huge collection of quarks – what are they for and who needs them? If you are a scientist searching for 'dark matter' there is plenty of it here. Dark matter is not merely a theoretical curiosity – it constitutes the major quantity of matter in the universe. The stars in galaxies are so sparsely spread out that without dark matter there would be insufficient material for gravity to prevent the galaxies from flying apart due to their rotation. Many of the quarks are free ranging without any need to be trapped in a hadron. Each one has its use but this is spooky territory so let us leave them be and move on.

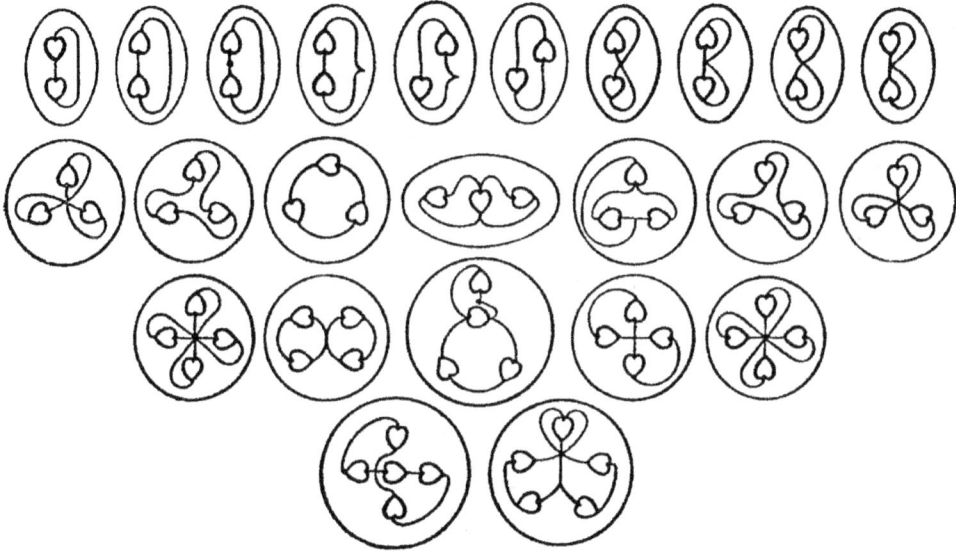

Fig 7.3 Part of the farmyard of quarks as described in Occult Chemistry. *In terms of mathematical topology some of the quarks appear identical due to subtleties and nodes which are not described here.*

There is also documentation of 'large quarks' associated with the nuclei of heavier elements in contrast to the generally accepted proton-neutron structure. Perhaps this is the result of quantum-uncertainty associated with the method of observation – who knows? We cannot dismiss these too lightly. These large quarks are made up from the innards of several nucleons – a better name would be 'superhadrons'. Is it too outlandish for collections of nucleons to reconfigure themselves into larger groupings? Some lateral thinking is called for here. We have all been conditioned from school science class days to the nuclei of heavy elements being depicted as a spherical agglomeration of smooth balls, sometimes coloured or shaded to show that half are protons and the other half neutrons. Has the nucleus ever been depicted in any other way? This is the schoolboy's bag of marbles model. It is generally accepted that this extremely unstable grouping is restrained by the 'strong force' but how does the strong force work? Does it overpower the strong electrostatic repulsion of the protons and make them mutually attractive? If this is the case then the protons would clump together and shuffle the neutrons to the outside where the thermal vibration of the atom would shake them off much like a wet dog shaking the water from its coat. Clearly the strong force must apply equally to both protons and neutrons and are we really sure that neutrons are not protons with an entrapped electron. This is a strange world where particles can change identity in an instant.

Protons cannot survive in a nucleus without neutrons – with the exception of the very light elements a nucleus always has more neutrons than protons. Except for the lightest elements there is a range of neutron surplus which will provide stable isotopes – outside this range the nucleus becomes radioactive. What of clumps of neutrons – this never happens. In a neutron star, matter receives the ultimate compression forcing electrons and protons to form neutrons resulting in material of unimaginable density. Would a small quantity of this material be unstable? – there is no unbalanced electrostatic repulsion to force it apart and the strong force should still be there. If small quantities of such matter exist they would fall to the centre of the nearest star or planet and oscillate as if there was nothing in the way. An individual neutron is unstable and after only a few minutes will eject an electron and become a proton.

A brief moment of wishful thinking. The electrostatic force and gravitation force both vary in the same way with distance so that gravitation cannot overtake electrostatic repulsion at close quarters. What if gravitation has a trick up its sleeve and has a different function for attraction in the sub-femtometre range – this would then take care of the strong force and simplify our lives considerably. This can easily be done – simply add an inverse cube portion to the gravitation force formula. This won't make any difference to the usual gravity calculations and will provide the strong force to keep the nuclei contained.

An interesting aspect to gravitation is the 'cosmological constant'. Einstein's original gravitational theory required that the universe be constantly expanding. Astronomical observations at the time could not confirm this so Einstein introduced the 'cosmological constant' to accommodate a non-expanding universe. Einstein considered this the greatest blunder of his career when the expanding universe was subsequently confirmed by astronomical observation. More recently, observations indicate the universe expanding at an increasing rate, which suggests a repulsive aspect to gravity. To accommodate this, the concept of 'Dark Energy' has been introduced. Newton's simple gravitational law may well expand into a series of exponential terms. A summation of terms would however seem a very inelegant formula for something as fundamental as gravitation. Shall we roll in a new 'cosmological constant?' No, let us leave this to the experts rather than put together a cockamamie gravity equation.

There is another problem with the electrical and gravitational attraction equations. Inverse square equations are exponential. In practical life one simply cannot get near enough for this to be a problem but inside a nucleon things are quite different. As separating distances approach zero the forces are poised to rise to infinity. Divide the smallest number you can find by 10^{-70} (planck length squared) and you will get an immense value. If superstrings were able to start clumping together in the region of the Planck length this could result in a runaway black-hole situation. This fortunately cannot happen – there is the Pauli exclusion principle which applies to a wide class of particles called fermions. Fermions are particles which include quarks, leptons and composites such as protons. The principle requires that two particles within uncertainty range cannot occupy the same state causing the particles to keep at a safe distance. In practical terms this means that only one electron can occupy a particular orbital. The Pauli exclusion principle also has astronomical implications. When a neutron star forms the immense pressure cannot force the atomic electrons into shared orbitals so the protons of the stellar material accept the electrons to form neutrons. This results in a material of immense density called neutronium which is in effect a continuous atomic nucleus.

One cannot of course apply mechanical engineering methods to quanta – this is a world of extremes. Extreme energy densities, extremely fast vibrations and rotational velocities and bewildering uncertainties. You cannot simply say that this thing attracts that thing and pushes the other one out of the way. There is another curious property of particles that needs mentioning – they can be mildly attractive at large distances and highly repulsive at close quarters. At this point we should make mention of an astonishing claim in *Occult Chemistry* which states that all matter energy exists as bubbles in a space of infinite energy density called 'koilon'. Strangely this is also mentioned in modern papers but it would be fruitless to pursue the matter further here.

Let us consider what happens when fission takes place in a heavy radioactive nucleus. If the nucleus were a random agglomeration of nucleons then the fission products would be a random assortment of light element nuclei. This is not what happens. The remnants of fission are few and specific indicating that the heavy nuclei have definite fracture zones and that all nuclei of the same element have the same ordered structure. Mr Leadbeater's fanciful geometric nuclei constructed from strange looking superhadrons don't look so bad after all. Now that we have lost the marbles there is a gap requiring some serious theoretical work. This is not to suggest that the atomic nucleus has been neglected – a vast amount of research has been done in this field – it is the elusive structure of the nucleus that has yet to be sorted out. When a neutron moving at the correct velocity strikes a U-235 nucleus it can be absorbed so that the nucleus becomes an extremely unstable U-236 nucleus which will then split into two or sometimes three smaller nuclei and also release some neutrons and radiation. At the time when nuclear fission was discovered it was thought that the nucleus was something like a drop of liquid which, if destabilised by absorbing a slow moving neutron, would pinch itself together in the middle and then split into two nuclei. The fission product nuclei are unfortunately not naturally occurring isotopes so that these can be highly radioactive posing a severe health hazard.

It is also interesting to note what Mr Leadbeater was unable to do. He had been asked to record details of the electron but was unable to locate an atomic electron – or so it seemed. At the time he was expected to see small particles orbiting the nucleus much like a little solar system and of course there was no such thing. The electrons were there all along for everyone to see – the nuclei were shown with envelopes which neatly describe the modern idea of the electron probability orbital. Orbiting particulate electrons have long ago faded into history except for icons and logos which are intended to denote atomic matters. Whizzing particles are graphically much easier to handle than mathematical concepts. To make matters even worse for the investigator, the electrons do not move about in the orbital – they have a probability of appearing in it anywhere at any time. It is to his great credit that Mr. Leadbeater nevertheless showed his atoms, complete with their envelopes even though at the time no explanation could be offered for this puzzling phenomenon. At a later stage another clairvoyant, Geoffrey Hodson, was asked to observe electrons flowing in a cathode ray tube. He claimed to succeed in this by 'playing tricks with time'. His observation was that the electron was similar to the aahnoo but smaller and that it proceeded along the tube in a spiral motion.

At first glance the Leadbeater atomic nuclei look very strange indeed. Spindly articulated structures with almost insect like appearances. There is actually no need for nuclei to be of ruggedised military appearance. Spacecraft which are not required to return to Earth unscathed are of gangly design even though they may travel more than twenty times as fast as a jet aircraft. Atomic nuclei are in little danger of being damaged in collisions as they are well cushioned by their electron orbitals. The nucleus material is configured into several groups each enclosed in its own electron orbital and many other orbitals are attached much like the petals of a flower. This arrangement may well provide valuable clues as to the working of nuclear fission and fusion reactions. What remains unclear is which of the orbitals are responsible for photon ejection. The spread out structure of the Leadbeater nuclei gives a hint that the Pauli exclusion principle may well be fundamental to the structure of the nucleus.

We are left with a shocking question – what happened to the nucleons? The only atom shown with the accepted proton/quark configuration is hydrogen. In all the heavier elements the nucleons reconfigure themselves into larger structures. Are we to believe that protons exist only in hydrogen or on their own? Another fundamental shock is what are we to do with the 'strong nuclear force' – do we need it at all? We are getting into deep water here – let us leave this matter for another time. Before we take leave of Mr Leadbeater's atom there is another most interesting matter to consider. Not only did he show an electron orbital for the nucleus, the quarks were shown as three superstrings also enclosed in envelopes. This suggests very strongly that quarks are enclosed in lepton probability domains. All atomic nuclei have a positive charge and are surrounded by a cloud consisting of a probability density function of negatively charged leptons. What of the quarks? In a nucleon there are positive and negative quarks. This would imply that a negative quark should be surrounded by a positive lepton. If someone can write a paper verifying that the positron is an integral part of nucleons this would be a remarkable scientific breakthrough indeed. There is another slight problem with the lepton envelopes. The leptons surrounding atomic nuclei are electrons with a charge of three negative strings. If the quarks are also surrounded by leptons these would have to be of one or two positive or negative string charges. This would introduce a major complication to the lepton family. One is instinctively inclined to think that the electron, one of the most fundamental and common particles of all matter, should have a single indivisible unit of electric charge. Much work remains to be done here. The electrical charge arithmetic of nucleons is quite simple if there are no leptons surrounding the quarks. It all works out quite nicely if the superstrings have single unit charges and the electrons three. If we are to consider lepton orbitals for the quarks things get much more complex. The quarks are smaller than protons by several orders of magnitude requiring high energy leptons with wavelengths in the gamma ray spectrum. Does anyone fancy the idea of introducing new animals to the particle farmyard? If the existence of a gamma-lepton surrounding quarks is proven it would be most appropriate to name it in honour and memory of Charles Webster Leadbeater who 'saw' it more than a hundred years before its discovery. It could be called a 'Webster'. 'Charlie' would be too flippant and lead has connotations of dull heaviness quite inappropriate to something as sprightly as a lepton. There is actually a strong case for a quark lepton orbital. The size of quarks is currently estimated to be in the zeptometre range making them smaller than protons by a factor of thousands. This makes quarks thousands of millions of times larger than their three superstring components. What, other than a lepton probability cloud, could give quarks such a vast size? There is another small matter to consider – the neutron. By all accounts the neutron is actually a proton with an electron tucked away. It would be most useful if a parking spot could be found for this electron. Protons are phenomenally stable particles. If they do indeed contain quarks with electron and positron orbitals in their innards then they have potential for violent disintegration. When an electron and a positron are able to attract each other they annihilate themselves and release their energy as gamma ray photons. In a neutron star atoms are so compressed that protons and electrons become neutrons and the star material becomes effectively a continuous atomic nucleus. For a proton to become a neutron one of its u-quarks must become a d-quark. Does this not mean that the u-quark acquires an electron in its lepton orbital? The quark, a composite of three charged superstrings and a gamma-lepton probability orbital and observed more than a hundred years ago – this will take some beating.

Study of the aahnoo showed that it had several degrees of freedom. The rotational direction about the axis (chirality) resulted in positive or negative electric charge. The string as a whole had a volumetric pulsating motion as well as a precession motion. The filaments were also seen to be moving through a fourth dimension. An interesting additional observation was that by breaking a filament the aahnoo would instantly disappear and by providing a suitable 'space' a new one would spontaneously appear out of nowhere. Breaking a string by thought power may be quite a tall order – it has been estimated that this may require immense force. There may be a link here to the 'virtual particle' concept in particle physics. The lifetime of virtual particles is determined by the Heisenberg uncertainty principle. They can appear out of nowhere but must disappear again in less time than they can be noticed. This is rather like having a ghost who only makes himself visible when no one is looking. One may well ask, 'who needs a ghost like that?' but virtual particles are a serious matter in particle physics.

The superstrings have in fact degrees of freedom in all eleven dimensions of space-time and herein lies the key to the ultimate nature of gravity, electromagnetism and the passage of time. What happens when a positive and a negative aahnoo find themselves on a collision course – do they annihilate each other in a matter and anti-matter death wish? No, they start spinning around each other and voila – we have a two-string quark. Not only Mr. Leadbeater but other of his co-workers claimed to be able to see the particles. They did not seem to think that these powers were exceptional and thought that these abilities would become commonplace as the consciousness development of the public became more advanced. The clairvoyant observation of particles apparently did not require much effort on the part of the investigators. Mr. Leadbeater would simply make himself comfortable in a chair and dictate his observations to an assistant.

At this point, if I may be excused the impertinence, I would like to suggest that the spacing of the dimensionless points in the strings corresponds to the Planck length which makes the string considerably larger than in papers where the size of the entire string is equated to the Planck length.

I do not suggest that everyone should rush out to get a copy of *Occult Chemistry*. This book may be very difficult to find and even more difficult to read, however, it has quite a good showing on the Web. There is considerable comment about *Occult Chemistry* on the Web ranging through the whole spectrum of comment from adulation to outright hostility. To modern readers, much of the book may seem gobbledegook, and some parts may make even the most enthusiastic readers squirm with embarrassment. Most of the observations of the book were made by Mr. Leadbeater himself; his co-author of *Occult Chemistry*, Dr Annie Besant, had little patience for counting dots on wide open spaces and preferred to occupy herself with social matters such as introducing family planning to India. Her attempts to promote family planning in England got her into serious trouble with the law. Dr Annie Wood Besant was a most astonishing phenomenon of the early twentieth century and her biography makes fascinating reading. The interaction between George Bernard Shaw and his 'amazing Annie' is of particular interest. Shaw described her as the best public speaker in London and possibly all of Europe. Shaw and Besant would on occasion play piano duets together. Their close relationship came to an abrupt end when Shaw was frightened off by a marriage proposal from Besant. All the dictation for the project was taken by Mr. C. Jinarajadasa MA (Cantab) who also wrote the lengthy introduction to the third edition. Curiously, Mr Jinarajadasa never acquired the ability to observe the atomic particles himself. In *Occult Chemistry* Mr. Leadbeater renders the Sanskrit word 'aahnoo' as 'anu'. I prefer to avoid this spelling as it becomes repeatedly necessary to mention that the word does not take an 's' in the plural. Why was this huge project ever undertaken? Presumably because it 'had to be done' and nobody else was able or willing to do it. This effort brought Leadbeater little but embarrassment to his supporters and ridicule from his opponents. At the time, particles were thought to be indestructible balls of immense density, and the idea of atoms consisting of complex structures within structures and containing no 'solid' matter at all seemed utterly ludicrous. I do not suggest that anyone should rush off to India for a crash course in Planck length observation skills. This might require a premature awakening of the *Kundalini fire;* a more frighteningly dangerous course of action can scarcely be imagined.

The Sanskrit 'anu' is mentioned in the Bhagavad Gita (chapter 8, verse 9):

One meditates on the omniscient, primordial, the controller, smaller than the atom, yet the maintainer of everything; whose form is inconceivable, resplendent like the sun and totally transcendent to material nature.

Occult Chemistry includes many pages of notes made during recording sessions which make fascinating reading. The last recording session was an attempt to track down and record details of an electron. After reading this, one is left with a feeling of loss. We can only wonder what might have been achieved if the work could have been carried out with the assistance of a modern scientist.

While we are considering Mr Leadbeater's remarkable psychic powers it is interesting to consider what else he claimed to be able to do. An obvious subject of great interest is the Akashic Record or simply the 'past'. This he claimed to be able to observe in much the same way that we can use a VCR. He had all the controls of search, pause, play, fast forward and rewind. This did not only apply to his own previous experience but to any point in history. Let me throw a bone to the skeptics and admit that I do not know of a single living person who is able to do this, but that he was able to observe subatomic particles there can be no doubt. A point worth considering is that there are many reports of persons in a near death situation who claim to have seen a fast-forward replay of their entire lives. Sadly, Mr

Leadbeater has left no legacy of his claimed clairvoyance into the Akashic Record which makes his claim somewhat suspect. Even a small portion of the time spent on *Occult Chemistry,* if focused on the Akashic Record, could have yielded treasures of inestimable value to historians and archaeologists and which would have been susceptible to positive verification. There are many web sites referring to *Occult Chemistry* and some of these go to great lengths to denounce the book as the work of charlatans. There is no need to comment on chemical reactions taking place directly between atomic nuclei – it would be very brave or foolish to champion this cause at the present time. We need only look in awed silence at the descriptions of the superstring, quark, proton and electron orbital and marvel at this astonishing demonstration of psychic powers.

Pierre Teilhard de Chardin (1881–1955) palaeontologist, biologist, geologist and Jesuit priest and author of several books sought to integrate scientific research with a religious vocation. His work *Le Phenomene Humain* (The Phenomenon of Man) was published posthumously. His perception of the interaction between matter and consciousness is passionately expressed in the following quotation:

> *All around us, to right and left, in front and behind, above and below, we have only to go a little beyond the frontier of sensible appearances in order to see the divine welling up and showing through. But it is not only close to us, in front of us, that the divine presence has revealed itself. It has sprung up universally, and we find ourselves so surrounded and transfixed by it, that there is no room left to fall down and adore it, even within ourselves.*

Teilhard outraged both the scientific and religious establishments with his views which were even more iconoclastic than Galileo would ever have dared to utter. He will long be remembered as the priest who sought to teach the church how to 'see'. Teilhard saw an evolutionary process starting from subatomic particles going right up to the ultimate fulfilment of human consciousness. The starting and end points were named Alpha and Omega. This idea did certainly not please everyone but would seem quite natural to the Theosophists.

A similar idea to Teilhard's Omega point comes to us from Giordano Bruno (1548-1600) who had a pantheistic view of a divine consciousness pervading and being part of every aspect of the universe. Bruno was a man of towering genius, but sadly, his advanced ideas resulted in his martyrdom at the stake.

Teilhard has nevertheless, in his several books, made a considerable contribution to the philosophy of science and consciousness. This may well receive greater acknowledgement in time to come. Teilhard spent many years of his life on geological work in China. Let us take a look at a brief quotation written in Peking in 1934:

> *The day will come when, after harnessing space, the winds, the tides, gravitation, we shall harness for God the energies of love, and, on that day, for the second time in the history of the world, man will have discovered fire.*

Despite the horror with which the ecclesiastical authorities regarded Teilhard's heretical views, he nevertheless remained fervently devout as can be seen from the description of the Eucharist at the moment of Consecration of the Host. The writing is, as before, passionate and unrestrained:

> *It is done. Once again the Fire has penetrated the earth. Not with sudden crash of thunderbolt, riving the mountaintops: does the Master break down doors to enter his own home? Without earthquake or thunderclap: the flame has lit up the whole world from within. All things individually and collectively are penetrated and flooded by it, from the inmost core of the tiniest atom to the mighty sweep of the most universal laws of being: so naturally has it flooded every element, every energy, every connecting-link in the unity of our cosmos; that one might suppose the cosmos to have burst spontaneously into flame.*

Leadbeater was also a celebrant at mass, and wrote the following in his book *The Science of the Sacraments* describing the same moment of the Consecration of the Host. The writing styles of the two priests could scarcely be more different:

> *At the moment of consecration the host glowed with the most dazzling brightness; it became in fact a veritable sun to the eyes of the clairvoyant, and as the Priest lifted it above the heads of the people I noticed that two distinct varieties of spiritual force poured forth from it, which might perhaps be taken as roughly corresponding to the light of the sun and the streamers of his corona.*

Teilhard was enthralled with the concept of a coalescence of consciousness which would lead to a state of peace and planetary unity. This global consciousness he named the 'Noosphere' (from the Greek *nous* meaning "mind") as a logical extension to 'Geosphere' and 'Biosphere'. This was a decade before Dr. James Lovelock proposed his Gaia Hypothesis which suggests that the Earth is actually a living creature. Teilhard's noosphere suggests that the Earth has its own autonomous personality and being the prime centre and director of our future.

> We have reached a crossroads in human evolution where the only road which leads forward is towards a common passion... To continue to place our hopes in a social order achieved by external violence would simply amount to our giving up all hope of carrying the Spirit of the Earth to its limits.
>
> The Age of Nations is past. The task before us now, if we would not perish, is to build the Earth.

Physicist-philosopher Peter Russell in his book *The Global Brain Awakens* also takes the idea of Teilhard's Omega Point:

> Inner evolution is not an aside to the overall process of evolution. Conscious inner evolution is the particular phase of evolution that we, in our corner of the universe, are currently passing through. From this perspective, the movement toward a social superorganism and the mystical urge to know an inner unity are complementary aspects of the same single process, the thrust of evolution toward higher degrees of wholeness.

Interestingly, Leadbeater also wrote much about world consciousness which showed concentrations in major geographical features such as mountain ranges, islands, forests etc. These living manifestations he claimed would attract colourful auras about themselves which were visible to the clairvoyant observer. These have also been written about and lavishly illustrated by Theosophist Geoffrey Hodson. The concept of a noosphere finds a close similarity in Theosophical writing in the idea of a Planetary Logos. The noosphere or global consciousness must not be confused with the idea of a universal consciousness. The planet earth is but an infinitesimal speck in our galaxy which is in turn an infinitesimal speck in the universe.

Teilhard's idea of a noosphere received a remarkable boost with the launching of the Global Consciousness Project (GCP) in 1998 directed by psychology professor Dr. Roger D. Nelson of Princeton. This is also known as the EGG Project from ElectroGaiaGram. The method employed is to use electronic random event generators (REGs) distributed around the globe. The REGs use an input of 'white noise' which is generated electronically and then converted to binary digits giving a stream of 'random walk' data. 'White noise' refers to a signal containing a continuous range of frequencies in much the same way as white light from the sun. 'Random walk' data is a stream of binary digits representing steps either forwards or backwards. If the data is purely random then the steps will concentrate about a central point. The data is transferred through the Internet to a central facility in Princeton where it is subjected to a series of mathematical statistics processing. The statistical data is then examined for anomalies which are unlikely to arise from random noise. The information obtained from the EGG project is then correlated with major world wide events which could disturb the tranquillity of the Global Consciousness and also presumably influence the randomness of the white noise. The results have been sensational. We list here a few of the successful correlations:

An early result from the EGG project came in Sept. 1997 with the funeral of Diana, Princess of Wales, which was a time of world wide grief and compassion and presumably greatly moved the Global Consciousness. The death of Mother Teresa a week later did not however give any significant EGG result, presumably because of the dignified, natural and peaceful circumstances of this event.

- The August 1998 bombing of American embassies in Africa.
- The March 1999 NATO bombing in Yugoslavia.
- The August 1999 India train crash.
- The October 1999 Billion Person Meditation.
- The Papal visit to the Middle East in March 2000 – the data continued for six days.
- The August 2000 Kursk submarine disaster.
- The October 2000 Funeral of Pierre Trudeau.
- The March 2003 Candlelight Vigil
- The March 2004 Mali terrorist attacks
- The September 2004 Russian school hostage incident.
- The December 2004 undersea earthquake which resulted in the tsunami which devastated South East Asia and Indonesia was detected some time before the earthquake actually took place.

- The April 2005 Funeral of Pope John Paul.

- The April 2005 Oraworld Resonance

- The August 2005 Baghdad stampede

- The October 2005 Bali bombing.

- The May 2006 Indonesian earthquake

- The February 2009 Valentine meditations

- The January 2010 Haiti earthquake

- The August 2010 Chilean miners' rescue

- The January 2011 Brazilian mudslides

- The September 2011 International Peace day

- The April 2012 Mogadishu bombing

This list is obviously not a full account of major worldwide events over the given time period but it does give an indication that there may well be merit in the experimental method employed.

One of the most remarkable correlations occurred in 2001 with the 9/11 terrorist attacks. This was a time of world-wide shock, horror and outrage. The graphs show clearly the times of each tower being struck, the times of the tower collapses and also the third plane crash into the Pentagon. The graphs also showed the times of world wide memorial services and coordinated prayer. A strange feature of the graphic data was that it became apparent some four hours before the first plane crash. At present there is no explanation of why there should be a global consciousness or how it could be capable of influencing electronic devices and change the entropy of white noise, but the results speak for themselves. Clearly, we can expect a great deal more from the EGG project, other similar endeavours in the future, and ultimately the fulfilment of Teilhard's dream. It must be mentioned that there are detractors of the EGG project who claim that the statistical data that is captured is being shoehorned to fit global events.

A fascinating newcomer to the consciousness scene is the work of Dr. Masaru Emoto dealing with a strange interaction between consciousness and water. The results are produced by freezing samples of water and then examining the patterns of ice crystals that are formed. Water samples which were polluted or chemically contaminated were found to be incapable of producing ice crystals. The interesting part of the experiments comes when the water samples are subjected to music, blessings, curses and even written words affixed to the sample containers. A large number of beautiful patterns have been photographed which generally have the hexagonal structure of snowflakes. An unfortunate aspect of the work is that the ice crystals are prepared with the experimenter being aware of the sample source and also with expectations of the outcome. In work dealing with such subtle influences it is essential that the experiments be conducted using a scientific 'double blind' approach; hopefully this problem will be attended to. Another potential problem is the temperature of the water and the speed of ice formation. The formation of snowflakes is greatly influenced by the ambient temperature. This research must not be neglected and can fortunately be done with minimal expense.

A line of research very similar to the work of Dr. Emoto has been based on the Kirlian photography of water drops by professor Konstantin Korotkov. Kirlian photography is done without the use of illumination or lenses; the object is simply placed on a screen over a photographic film and a high voltage applied to the object causes a corona discharge which is recorded by the film. The pioneering work on electrical corona discharges was done by Nikola Tesla in the nineteenth century. It has been claimed by some that the 'etheric' auras of living creatures can also be recorded by means of Kirlian photography. It is claimed that water drops which have been subjected to blessings or meditation give Kirlian results which are spectacularly different to those of water drops which have been ignored. We have here a possible practical reason for the custom in several cultures since ancient times, to give thanks for food before it is consumed.

If the consciousness interaction with water can be rigorously proven, this will open a sensational new avenue of research and may also link up with the sacramental use of 'holy water'. Another possibility is that the water content of brain tissue could provide a clue to the experience of elevated consciousness and psychic phenomena. The idea that water might provide the link between physical and psychic phenomena is both revolutionary and exciting. It is interesting to note that in many cases of an awareness of psychic phenomena there is no requirement for a transfer of energy but only a change of entropy. Entropy is a fundamental irreversible process in nature; entropy can only increase with time – never decrease. This is also known as the second law of thermodynamics. To lower the entropy of a closed system without the input of energy implies a movement through time into the past.

An interesting line of research claims that the experience of consciousness and retention of memory is not contained within the brain, but is actually imprinted on the zero-point energy field of the universe, making the transitive consciousness of the present the leading edge of the akashic record and the rest becomes the past. It must be emphasized that the zero-point energy field of the universe is part of the objective space-time continuum of the universe and has nothing to do with hyperdimensionality. Zero-point forces can actually be detected by means of the Casimir experiment. (See chapter 8) The zero-point energy field of the universe is a sea of chaotic electromagnetic vibration of unimaginable energy density – it hardly seems possible that this could be imprinted with any kind of ordered psychic information. The zero-point energy field is a requirement of the Heisenberg uncertainty principle which requires that no point in space can be completely devoid of energy. By the same token this also implies that this energy cannot be removed, dashing the hopes of those seeking a way to get unlimited energy for nothing. We should not meddle too much with the zero-point energy field. This is at the outer limits of the physical universe, possibly beyond the purview of science and at the borderline of creative consciousness. The idea that zero-point energy can be equated to the akashic record does not appear to be viable. It has been suggested that a change of entropy to the zero-point field could provide a repulsive aspect to the Casimir effect giving a possible explanation of poltergeist activity.

There is a fascinating similarity between the water crystal experiments of Dr. Emoto and homeopathy. Homeopathy comes to us from Samuel Hahnemann (1755–1843) who proposed that ailments could be cured by extremely diluted preparations of substances which, in concentration, would cause the same symptoms. Homeopathy has developed into a huge world-wide industry and the remedies are used by patients in their millions despite the denunciations of people who claim it to be ridiculous quackery. From a chemical point of view the preparations are indeed bizarre. The remedies are diluted way beyond Avogadro's number to oblivion so that it is unlikely that a single atom of the substance being diluted remains in the liquid – usually water. The water will of course contain legions of other atoms as it is impossible to purify water to this degree. We are clearly not dealing with chemistry here. The remedies are not strictly designated to a particular use but will also depend on the 'type' of patient as determined by the practitioner. It would appear that the remedies are imprinted at a psychic level for their intended use. An interesting phenomenon was discovered when it was found that certain people, when coming into proximity of homeopathic remedies, would unknowingly erase the efficacy of the remedies. Another bizarre type of homeopathic remedy consists simply of a piece of paper with the name of the remedy written upon it, and it is then kept in a pocket of the patient. This links up neatly with Dr. Emoto's ice crystal experiments and also the use of Tibetan prayer wheels.

It is a sad fact of life that there are hoaxers who are, for the most unworthy of reasons, prepared to perpetrate the most elaborate frauds to bring disrepute to serious research and topple prestigious reputations. A noteworthy example is the infamous and despicable Piltdown skull hoax which caused huge embarrassment and wastage of time and energy of serious researchers. Even after nearly a century it is still not certain which of the suspects are definitely responsible. Teilhard de Chardin was intimately involved in the matter but it is extremely unlikely that he was aware of the conspiracy. Another prominent person involved was Sir Arthur Conan Doyle, author of the *Sherlock Holmes* stories and who was well qualified and placed to perpetrate the hoax. He actually wrote a book with cryptic clues regarding the hoax, confirming his culpability. Two other suspects are Charles Dawson, amateur fossil hunter and Martin A. C. Hinton, keeper of the geology collection at the Natural History Museum. In all, some 25 suspects have been suggested.

Doyle had a keen interest in spiritual matters and had a highly suspicious involvement in the Cottingly Fairies hoax. The hoax started off as a harmless prank by two schoolgirls who took photographs of each other posing with fairies which they had cut out from a magazine and propped up with hatpins. The photographs received much publicity and the matter spun out of control sensationalised by a gullible and war-weary public. The girls were left with little option but to play along with the hoax. Sir Arthur Conan Doyle lost no opportunity to bask in this newly generated publicity and claim that the girls had the psychic power to materialise the fairies. Doyle's friend Harry Houdini was not fooled by this nonsense. An ironic twist to the incident was that Doyle himself contributed to the magazine from which the fairy cutouts were taken. It seems inconceivable that he could not have known the source of the fairies. Arthur Conan Doyle was knighted in 1902 for writing a booklet in favour of the Boer War at a time when this infamous conflict was coming under international condemnation. Doyle caused another mischief when he anonymously published a short story in the *Cornhill* magazine. This fanciful fiction of piracy and murder on the high seas involved a ship called the *Marie Celeste*. This caused much confusion and became publicly identified with the brigantine *Mary Celeste* which was found abandoned off the Azores. The Theosophical Society took a foolhardy leap onto the fairy bandwagon endorsing the fairy photographs as proof of the girls' psychic powers and also as proof of the existence of invisible 'little folk'. Things were to get worse. The Leadbeater pederasty scandal dealt a blow to the society from which it has never fully recovered. There was yet another huge embarrassment to be endured by the Theosophical Society. The emblem of the society included an ancient symbol of good connotations used by various cultures for three thousand years – the swastika. This symbol had to be discreetly removed when it was usurped by the Third Reich and became publicly identified with brutal racism and genocide.

Another fraud that we can mention here is the astonishing mathematician and forger case. A prestigious mathematician, Michel Chasles (1793–1880) became professor at the Sorbonne in 1846 and was elected as full member to the *Académie des Sciences* in 1851. His keen interest in historical matters led him to become a victim of Denis Vrain-Lucas, a respectable law clerk who had become a forger of historical documents on a vast scale. Over a period of sixteen years Vrain-Lucas sold 27 000 autographed forgeries to a gullible professor Chasles. Vrain-Lucas would work long hours every day in the libraries of Paris writing his most astonishing letters. These included: 175 from Pascal to Newton, 139 from Pascal to Galileo, others from Descartes, Newton, Boyle, Rabelais, Joan of Arc, and Louis XIV. There were even letters from ancient history: Cleopatra to Julius Caesar, Alexander the Great to Aristotle, Lazarus to Mary the Magdalen (his sister?), Lazarus to St. Peter, also letters by Pontius Pilate, Attila the Hun, Herod, Cicero, Pompey, Socrates and many more. In order to make a reasonable attempt at such forgery the perpetrator would need to have excellent writing skills in ancient Greek, Latin, Hebrew, Aramaic and whatever. This the forger did not have and simply wrote the letters in contemporary French. Even a child would have laughed at this absurdity. We can only suppose that the august professor did not bother to look at much of his wonderful and expensive collection of precious documents. The forger did however take the trouble to give his creations an old look by means of the dexterous application of a candle flame. The fraud fell apart when the patriotic professor presented the *Académie des Sciences* with proof that Pascal had discovered the law of universal gravitation years before Newton. The letters had absurd anachronisms and were not in Pascal's handwriting. In 1870, Vrain-Lucas was sentenced to two years in jail, a 500 franc fine and payment of all costs. Not all hoaxes are bad. A matter that has recently become very prominent is the 'secret' document planted in the Paris *Bibliothèque Nationale*, and 'discovered' in 1975, regarding a fictitious order, 'The Priory of Sion'. This document includes a list of all Grandmasters of the Priory from 1150 which includes familiar names such as Leonardo, Nostradamus, Robert Boyle, Isaac Newton and Claude Debussy. This hoax has provided a huge bonanza to the publishing, printing and bookselling industries and spawned wide scale public interest in medieval and ancient historical fiction and also in the 'Order of the Knights Templar'. Newton does not look out of place in the list – he spent many years of his life immersed in Biblical studies and also dabbling in alchemy. Leonardo does not fit so well. He had a very mischievous streak and a profound love-hate relationship with the ecclesiastical authorities. A man of towering genius who surely felt himself above authoritarian strictures. Much of Leonardo's artistic output was of religious subjects but he also drew much of anatomical studies and explicitly detailed drawings of human genitalia. His painting of Virgin and Child with St Anne, when turned sideways, shows a picture of a vulture. His painting of a grinning St John the Baptist pointing with his index finger to heaven, possibly his last painting, seems to have an "up yours, mate" look about it. Leonardo treasured this painting and kept it, together with the Mona Lisa, in his possession until his death. Leonardo's naughtiest drawing must surely be the study of a man's head composed entirely of phalluses, leaving no doubt as to what Leonardo supposed that the man had on his mind. A strong case has been made for a hoax perpetrated by Leonardo – the Turin Shroud. This is a long piece of linen showing a faint impression of a man, front and back. The impression is further embellished showing injuries corresponding to those of the Crucifixion. A puzzling feature of the impression is that it is not made with pigments but scorch marks. A way of doing this is to make bas-relief models of the front and back of a body and then heat the models sufficiently to scorch the linen. An alternative suggestion is that the image was actually made by means of a camera obscura – in effect a photograph. Light sensitive chemicals suited to this purpose were available at the time and Leonardo took a keen interest in optics. There is a slight discrepancy in the height of the front and back impressions – this error would be more likely to occur with the photographic projection than the sculpted reliefs. There is also evidence that the image of the face was formed separately. The face of the figure could well be a self portrait by Leonardo. If Leonardo did indeed manufacture the shroud then he has mocked the ecclesiastical authorities on a grand scale and bamboozled the devout in their millions – something which would quite likely have given him great satisfaction. Radio carbon tests on shroud material indicate that the shroud dates from the time of Leonardo. One of Leonardo's most famous paintings 'The Last Supper', has unfortunately suffered much damage and was actually considered lost in Leonardo's lifetime. It has since seen much further damage at the hands of restorers and vandals. There is much current speculation that the figure meant to represent the young disciple John is actually a woman but no reason can be given for John being absent from the table. Another source of endless speculation and discussion is that there is no sacramental chalice on the table. It has also been suggested that Leonardo included a portrait of himself in the painting. The painting has remained a masterpiece of composition and together with the poses and groupings of the figures provides a wealth of information for detailed study.

It seems appropriate to give here a brief summary of 'auras' and levels of consciousness which have been written about in great detail by Mr. Leadbeater. This is given *voetstoots* without any claim of verification and the use of unfamiliar Sanskrit words will be avoided.

The first aura which is seen as an extension of the physical body is called the etheric or health aura. It has been claimed this can actually be recorded by means of Kirlian photography and can also be observed on plants and even inanimate objects. This aura is most useful for observing a person's state of health as it will show medical conditions long before the appearance of any symptoms. At the first stage of the development of elevated consciousness this aura is the first thing that will be noticed. When looking at a person's entire body seven vortices called *chakras* will be noticed. With further development of consciousness the emotional or *astral* aura will become visible. This is said to appear in glowing colours making a person's emotional state appear readable as an open book. The next level is the 'mental' aura which will reveal the person's state of intellectual development and mental activity. This is the last of the auras associated with the physical body. At the next level we have the consciousness which survives from one lifetime to the next and it is presumably from here that one can observe the universe from outside the space-time continuum. This is not the highest level of consciousness. A detailed account of auras and the hierarchy of levels of consciousness is given by Leadbeater in his book *Man Visible and Invisible*. Leadbeater has also written in detail about thoughts in his book *Thought Forms* in which he maintains that thoughts can manifest themselves as independent entities and attract colourful auras about themselves which are visible to the clairvoyant observer. Interestingly, thought forms are frequently used in comic strips where vocabulary is of secondary importance. The forms are used to illustrate thoughts and passions such as surprise, anger, gloom, desire etc. We need not however pursue this matter in detail here.

Mr Leadbeater had an astonishingly varied life and engaged in far more activities than we can mention here, and the study of atomic particles was little more than a spare time interest. Two matters in which he took great interest and occupied much of his energy were reincarnation and karma. Karma could perhaps do with a brief explanation. This is something like a current account of offences against fellow beings and which is usually heavily in overdraft. The balance fluctuations equate to the setbacks and successes of daily experience. This process, when taken with reincarnation will continue from one lifetime to the next until there is no further need for physical experience. What then of the akashic record – does this mean that we are to be embarrassed by our past to all eternity? No. This matter is dealt with in several ways by the various ideologies but let us give this a miss lest we stir up a hornet's nest. The psalmist must have had an intuitive feel for the nature of time when he wrote, as paraphrased by Isaac Watts, in his most famous hymn:

A thousand ages in Thy sight
Are like an evening gone;
Short as the watch that ends the night
Before the rising sun.

Emily Brontë, supreme poetess of the nineteenth century and master of the 'inner life' must have had experience of higher consciousness when in her beautiful and mystical poem 'Oh, thy bright eyes must answer now' she wrote:

... ever present phantom thing –
My slave, my comrade, and my king.
A slave, because I rule thee still;
Incline thee to my changeful will,
And make thy influence good or ill;
A comrade, for by day and night
Thou art my intimate delight,

Before we take leave of the Leadbeater aahnoo perhaps we should reflect on what exactly did the occultist see. If the observations were made from outside the space-time continuum then he would have been able to observe all the physical dimensions of space and also the progress along the time vector. He clearly was able to slow-play the time vector in order to see movement. Observing movement in Planck length steps along the time vector at the speed of light is a prospect terrifying beyond belief. If the observations were actually made within a three dimensional space then we will need to add the three to the Kaluza Klein dimensions that we already have. It seems an almost obvious possibility that the ten loops of the superstring correspond to the dimensionality of space. What of the three dimensions that define the space with which we are familiar? Are they 'straight' or also curled up into spirals? In the final analysis the only difference between a straight line and a spiral is a choice of mathematical co-ordinates. In a domain consisting of only one dimension there is no difference at all between a straight line and a spiral or even a line tangled and knotted. The spiral nature of the fundamental dimensions is due to the dimensions being circular in nature but simultaneously drawn along the time vector to produce spirals. Are we really to consider the possibility of the universe being curled up into almost nothing? This idea may seem offensive in its absurdity yet we cannot discard it outright.

It should not be supposed that it is possible to make an exact drawing or model of the aahnoo. The Leadbeater model gives us valuable information about the number of dimensions, the multiple helical structure of the filaments and the ultimate digital nature of space and time but the ultimate mathematical manifold in eleven dimensions is quite beyond human visualization. The mathematics we are dealing with here is in the upper stratosphere of intellectual endeavour and we must look to the mathematicians to show us the way.

An obvious question to ask is what other accounts are there of clairvoyant investigations into atoms. There is unfortunately very little other material to work with. Edwin D. Babbitt (1828–1905) also investigated superstrings producing drawings very similar to the Leadbeater aahnoo except that these have some extraneous clutter and have little to add to the subject. Throughout the ages there have been many instances of art and architectural decoration containing images looking very much like superstrings and quarks but nothing nearly as explicit as in *Occult Chemistry*. Interesting examples of quark art in the middle ages were produced by Abbess Hildegard of Bingen (1098–1179) also known as the Sybil of the Rhine. An astonishing person, visionary, philosopher, healer, herbalist, artist, songwriter and musician. Many of Hildegard's angelic songs are available in CD shops. What present day musician would dare to hope that his work would still be on sale in nine hundred year's time? Astonishingly, Hildegard even wrote in detail about copulation and female orgasm – a matter of which she could surely not have had any experience and remarkably daring writing for the middle ages. It is thought by some that Hildegard's visions were migraine related but if this is the case then she certainly turned her affliction to sublime divine use.

There are of course many people today who claim to have clairvoyant powers but how many of them if calibrated in SI units would score better than the pico or nano range? There is a remarkable similarity between the luminiferous ether and micro spirals of Nikola Tesla and the koilon and aahnoo of Leadbeater. It is not known if there has ever been any communication between Tesla and Leadbeater. Tesla took an interest in Vedic writings and used Sanskrit words in cases where there were no western equivalents. The following quotation from Tesla, probably written in 1907, is of particular interest:

> There manifests itself in the fully developed being, Man, a desire mysterious, inscrutable and irresistible: to imitate nature, to create, to work himself the wonders he perceives. Long ago he recognised that all perceptible matter comes from a primary substance, of tenuity beyond conception, filling all space, the Akasha or luminiferous ether, which is acted upon by the life giving Prana or creative force, calling into existence, in never ending cycles all things and phenomena. The primary substance, thrown into infinitesimal whirls of prodigious velocity, becomes gross matter; the force subsiding, the motion ceases and matter disappears, reverting to the primary substance.

We have dealt with Tesla's work in some detail in chapter eight.

Anaxagoras (500–428 BC) neglected and lost his possessions to study science and philosophy. What could he have had in mind, when discussing atoms, he wrote more than two thousand years ago:

> So the things rotate and are separated by force and swiftness. And swiftness produces force, and the swiftness is in no way like the swiftness of the things now existing among men, but it is certainly many times as swift.

Did not Isaac Newton have some uncanny foresight of Relativity and the photoelectric effect when he wrote:

> Are not gross bodies and light convertible into one another, and may not bodies receive much of their activity from particles of light which enter into their composition? The changing of bodies into light, and light into bodies is very comfortable to the course of Nature, which seems delighted with transmutations.

After looking in some detail at the structure of the atom are we now in a position to make a scale drawing of the hydrogen atom? No – not at all. The obstacles are insurmountable.

For starters, let us first look at scale. A drawing of any part of an atom would make all other parts too large or too small by a factor of thousands. We can look at the smallest part – the superstring but we can hardly do it artistic justice. An infinitesimal entity throbbing and spinning fast beyond human comprehension in eleven dimensions – is anyone ambitious enough to draw in more that three dimensions? And what is there to draw? The superstring is little more than a mathematical function of exceptional complexity without any 'solid' substance at all. The next stage is the quark – three superstrings tightly bound together by invisible forces. Perhaps we could have a stab at drawing the forces with an air brush. The quark is enveloped in a lepton probability zone – another mathematical function which hardly lends itself to portrait painting. We now come to the proton after scaling up by thousands. Again we have invisible forces linking three quarks and on a vastly larger scale a mathematical probability function for the electron orbital. And what of these electrons which can appear anywhere at any time? Yet another probability function envelope, this time of virtual photons on the borderline of being real or unreal. A picture of an atom is best left in the imagination.

The problem of visualisation and objectification of matter at quantum level was described by physicist Werner Heisenberg (1901–1976) in his brilliant Nobel laureate award lecture in 1932:

> However the development proceeds in detail, the path so far traced by the quantum theory indicates that an understanding of these still unclassified features of atomic physics can only be acquired by foregoing visualisation and objectification to an extent greater than that customary hitherto. We have probably no reason to regret this, because the thought of the great epistemological difficulties with which the visual atom concept of earlier physics had to contend gives us the hope that the abstracter atomic physics developing at present will one day fit more harmoniously into the great edifice of Science.

Tagore mentioned this matter many years before. Albert Einstein and his friend Rabindranath Tagore are regarded by many as two of the greatest minds of the twentieth century. Tagore, philosopher, mystic, poet and playwright is probably best known for his masterpiece the *Gitanjali* but let us see what he has to say in a quotation from his book *Personality – Lectures delivered in America* first published in 1917.

> ... *But with science it is different. For she tries to do away altogether with that central personality, in relation to which the world is a world. Science sets up an impersonal and unalterable standard of space and time which is not the standard of creation. Therefore at its fatal touch the reality of the world is so hopelessly disturbed that it vanishes in an abstraction where things become nothing at all. For the world is not atoms and molecules or radio-activity or other forces, the diamond is not carbon, and light is not vibrations of ether. You can never come to the reality of creation by contemplating it from the point of view of destruction.*

Let us look at a quotation from Albert Einstein on consciousness:

> *A human being is a part of this whole, called by us 'Universe', a part limited in time and space. He experiences himself, his thoughts and feelings as something separated from the rest – a kind of optical delusion of consciousness. This delusion is a kind of prison for us, restricting us to our personal desires and to affection for a few persons nearest to us. Our task is to free ourselves from this prison by widening our circle of compassion to embrace all living creatures and the whole of nature in its beauty.*

These wonderful words are reflected in Beethoven's glorious musical setting of Schiller's:

Seid umschlungen, Millionen!	*O ye millions, I embrace ye!*
Diesen Kuß der ganzen Welt!	*Here's a kiss to all the world!*

A quote by Albert Einstein in Mein Weltbild, 1934.

The scientist's religious feeling takes the form of a rapturous amazement at the harmony of natural law, which reveals an intelligence of such superiority that, compared with it, all the systematic thinking and acting of human beings is an utterly insignificant reflection. This feeling is the guiding principle of his life and work... It is beyond question closely akin to that which has possessed the religious geniuses of all ages.

The great French philosopher, mathematician and scientist René Descartes (1596–1650) is famous for the statement on consciousness, *'Cogito, ergo sum'* which is usually rendered in English as, '*I think, therefore I am*'. Descartes gives this in French as '*je pense, donc je suis*'. Descartes constructed a huge edifice of modern philosophy using mathematical methods of deduction and proof. Let us take a look at his fundamental reasoning as to whether he exists or not:

> But I have convinced myself that there is absolutely nothing in the world, no sky, no earth, no minds, no bodies. Does it now follow that I too do not exist? No: if I convinced myself of something [or thought anything at all] then I certainly existed. But there is a deceiver of supreme power and cunning who is deliberately and constantly deceiving me. In that case I too undoubtedly exist, if he is deceiving me; and let him deceive me as much as he can, he will never bring it about that I am nothing so long as I think that I am something. So, after considering everything very thoroughly, I must finally conclude that the proposition, I am, I exist, is necessarily true whenever it is put forward by me or conceived in my mind.

A dispute between giants arose when Descartes and his contemporary, Blaise Pascal (1623–1662), could not agree on whether it was possible for a vacuum to exist. Suggesting that a vacuum was possible at that time was a serious and dangerous religious matter. Descartes maintained that the entire universe was filled with a tenuous substance and huge solar system sized vortices in this substance were responsible for the force of gravity. Pascal insisted that a vacuum was simply a space containing no atoms of matter. He claimed that the space at the top of a Torricelli tube (mercury column barometer) was such a vacuum. He showed that the height of the column was due to atmospheric pressure and was not influenced by the size of the space at the top. He conducted a famous experiment where two tubes had the same height of column when brought together but when one tube was taken up a mountain, to his great delight, the column would gradually become shorter with increase of altitude. Pascal further deduced that with increase of altitude, the pressure would eventually become zero, and from there the earth would be surrounded by a vacuum. Pascal was a devoutly religious man but he did not regard the existence of a vacuum as a threat to his religion. Of these two giants, who had the better idea? Both were right and both were wrong. Descartes's huge vortices may now seem a quaint and foolish idea but there is an uncanny resemblance to the infinitesimal vortices of superstrings which are fundamental to matter, mass, energy and gravity. From an engineering point of view the space at the top of a mercury column is indeed a vacuum, maugre a few atoms released from the surface of the mercury or the sides of the tube. The average density of the universe is something like one atom of matter per cubic metre of space. Considering the vast amounts of matter contained in the galaxies one can safely assume that most of the universe is a perfect vacuum with no atoms of matter at all. The universe is of course flooded with background microwave radiation from the Big Bang but it is possible to exclude this from a suitably designed container. The universe is also pervaded by the zero-point energy field which cannot be excluded anywhere. This may seem a fuss about nothing, which indeed it is, but this is not an end to the matter. Descartes has the last laugh – the nothingness of space is taken very seriously by modern theoretical physicists. The initial state of space is seen as a seething mass of virtual particles which can borrow extremely brief existences from antiparticles. This is only the beginning – it is even speculated that there are several more states of nothingness. We need not delve deeper into this matter – let us take our hats and run. We have taken a look at the Casimir Effect and zero-point energy in chapter eight.

S. P. Sirag has taken a mathematical approach to modelling consciousness. Relativity unites space and time in a four dimensional continuum. Sirag has developed a hyperspace model of consciousness utilising seven Kaluza Klein dimensions linking consciousness at a deep level with physical reality. In work on unified field theory it is generally believed that space-time is hyperdimensional, with all but four of the dimensions being invisible. Besides the space-time dimensions there are also other internal dimensions called 'gauge dimensions'. In Sirag's view both the extra space-time dimensions and the gauge dimensions are real, providing scope for considering ordinary reality to be a substructure within hyperdimensional reality. This is a modern version of Plato's cave dweller parable where the cave dwellers can see the outside world only as two dimensional shadows on a screen stretched across the cave entrance. A further innovation in Sirag's approach is that his version of unified field theory embeds both spacetime and gauge space in an algebra whose basis is a finite group. This group which directly models certain symmetries of particle physics, is a symmetry group of one of the Platonic solids – the octahedron. Thus it is a mathematical entity contained in the reflection space hierarchy. In fact the reflection space corresponding to the octahedron is seven dimensional and is also a superstring type reflection space, so that a link with the most popular version of unified field theory is provided.

In C. W. Leadbeaters *Occult Chemistry* all five Platonic solids are cited as being fundamental to the structure of atomic nuclei. In a three dimensional geometric space only three regular polygons can construct regular polyhedra and the only regular polyhedra possible are the Platonic solids. The five Platonic solids are as follows:

Tetrahedron	Four triangles
Hexahedron (cube)	Six squares
Octahedron	Eight triangles
Dodecahedron	Twelve pentagons
Icosahedron	Twenty triangles

The central postulate of Sirag's work is that this seven dimensional reflection space is a universal consciousness, and that individual consciousnesses tap into this universal consciousness. This implies that the high level of consciousness enjoyed by humans is due to the complex network of connections to the underlying reflection space afforded by a highly developed brain. The hierarchy of reflection spaces suggests a hierarchy of realms (or states) of consciousness. Each realm would correspond to a different unified field theory with different sets of forces. Strong stuff, it almost seems unusual to be able to read about these ideas without having to puzzle over the meanings of Sanskrit words. Modern physics has no vocabulary for discussions on a hierarchy of levels of consciousness – it is here that Sanskrit has a rich contribution to make. The dependence of some quantum mechanics processes on observation for their existence is extended by some physicists to the macro state suggesting that physical reality does not objectively exist independently of the participating observers.

Einstein's response to this absurdity is a classic:

When I look at the moon, I like to think that
it will still be there when I am not looking.

The lack of an observer for the existence of physical reality is possibly neatly dealt with by Sirag's idea of universal consciousness. The reflection space of the universal mathematical substrate providing universal consciousness should be more than enough observation to ensure the sustained existence of the entire objective universe. The idea of an observer in quantum mechanics does not mean that a live human is actually looking, or even capable of looking at the particles – observation will normally mean a collision with another particle or a photon from which results can be deduced. The deeply philosophical nature of modern physics bears little resemblance to Newtonian applied mathematics. Saul Paul Sirag's idea of a reflection space universal consciousness into which individual consciousness can tap is also mirrored in David's Psalm 139:

> Thou knowest my downsitting and mine uprising, thou understandest my thought afar off.

> Whither shall I go from thy spirit? or whither shall I flee from thy presence? If I ascend up into heaven, thou are there:
> if I make my bed in hell, behold, thou art there.

> How precious also are thy thoughts unto me, o God! how great is the sum of them

Samuel Avery, in his book *Transcendence of the Western Mind: Physics, Metaphysis, and Life on Earth* writes: "The dimensional structure of consciousness is based on a new definition of reality that is based on all of experience – subjective and objective, perceptual and conceptual; and on the fundamental assumption that each realm of perceptual consciousness corresponds to a space or time dimension. It explains new experience that human beings have had at extremes of space, time and mass. It has made itself necessary because of these experiences. It will create new myth that changes who we are in relation to other living things and to ourselves. With this there is hope."

Physicist Evan Harris Walker has put forth an observational theory that equates the conscious mind with the 'hidden variables' of quantum theory.

> *The measurement problem in Quantum Mechanics has existed virtually from the inception of quantum theory. It has engendered a thousand scientific papers in fruitless efforts to resolve the problem. One of the central features of the controversy has been the argument that characteristics of Quantum Mechanics imply that an observer's thoughts can affect an objective apparatus directly, which in turn implies the reality not only of consciousness but of psi phenomena. I have written several papers saying that such a feature of Quantum Mechanics is not a fault, but rather represents a solution to the problems that go beyond the usual purview of physics. Thus, I have developed a theory of consciousness and psi phenomena that arise directly from these bizarre findings to Quantum Mechanics, findings*

now supported by specific tests of the principles of objective reality and/or Einstein locality.

We have taken a look into Einstein locality/non-locality, hidden variables and the EPR Paradox in chapter seven.

Gottfried Wilhelm Leibniz (1646–1716) was one of the truly great polymaths of the seventeenth century making huge contributions to philosophy and mathematics. His discovery of calculus, independently of Newton, is of inestimable value to modern physics and engineering. His notation, an elongated 's' for integration (L. *summa*) and 'd' for differentiation (L. *differentia*) remains in world wide use. Calculus also saw huge advances made by later mathematicians. His discovery of the binary numbering system remains the foundation of modern computer architecture. Leibniz preferred to do most of his writing in Latin and French. He taught himself Latin by the age of twelve. He was with René Descartes and Baruch Spinoza one of the three greatest seventeenth century rationalists. He made major contributions to physics, technology, biology, medicine, geology, probability theory, psychology, linguistics and information science. He also wrote on politics, law, ethics, theology, history and philology.

Leibniz proposed a theory of monads. These are the ultimate elements of the universe, substantial forms of being which are eternal, indecomposable, individual and subject to their own laws each reflecting the entire universe in a pre-established harmony. Monads are centres of force; substance is force, while space, matter, and motion are merely phenomenal. Leibniz's theory of monadology did initially not receive much recognition but this idea of universal consciousness pervading all of space, inanimate matter and living creatures would be fully endorsed by the Theosophical Society two centuries later. The term 'monad' must be used with care as it has different meanings in philosophy, science, mathematics and music.

Consciousness is at present in an indescribable state.

What is it?

At the most basic level one is in no doubt that a person able to speak is conscious and one who is concussed or under general anaesthetic is not. Likewise a cat prowling about the refrigerator is more conscious than one sleeping on the mat. These simplistic examples of intransitive consciousness are not what it is all about. If you enter 'consciousness' as a parameter for an Internet search engine you will get more than forty million hits. This is not a subject without interest. The physical dimensions of the universal mathematical substrate provide a basis for the existence of the physical universe but the dimensionality of the universe is nothing in itself; likewise the reflection spaces of the universe may provide a basis for a universal consciousness but bring us no closer to answering our fundamental question. In the sixteenth century a model for the structure of the universe was at a critical stage. Copernicus had made a seven-point statement regarding the universe which Galileo enthusiastically supported after his remarkable discoveries with his home made telescope. The statements of Copernicus are given in chapter six under the heading 'Galileo'. Let us aim for some clarification and structure by making a statement list regarding consciousness – the following points are simply a list cropping up in the flow of writing and do not represent a manifesto or creed of any particular ideology:

1. All living creatures have a basic consciousness which is primarily a brain function. It is not necessary for us to consider plant life or creatures bordering on plant and animal life at this time.

2. There is a hierarchy of levels of consciousness. This has been part of oriental cultures since antiquity. This does not mean that we have conscious-nesses like a troop of monkeys in a tree – individuals each have only one consciousness which is stamped on their passports to eternity.

3. Consciousness at a certain level can repeatedly return to physical incarnation.

   This idea is expressed by Rabindranath Tagore in the opening lines of his *Gitanjali*:

   Thou hast made me endless, such is thy pleasure. This frail vessel thou emptiest again and again, and fillest it ever with fresh life.

4. Consciousness at a high level may be free from the need to return to physical incarnation.

5. Acquiring consciousness at a higher level may require a mystical initiation process. Consciousness active at a high level has easy top-down access to any lower level. Bottom-up access is in discrete stages and working your way up can take millennia. Most humans and animals have a rudimentary

capability for extra-sensory perception implying a slight access to a higher level of consciousness. Some individuals have extraordinary clairvoyant powers for communication with the consciousnesses of persons either in or out of physical incarnation and free of the constraints of the space-time continuum.

6. Individuals leaving physical incarnation for the last time may experience initiation and martyrdom. In this case martyrdom would be the final stage of wiping clean the slate of karmic debt.

7. There is a universal consciousness to which individuals can gain access at some levels. There is evidence of this in the extraordinary creative powers of some musicians, artists, writers, mathematicians, philosophers etc. The highest levels of the universal consciousness are totally inaccessible to humans in physical incarnation, so that of these, physicists and mystics can know nothing. We are left with the conclusion that at the highest level of consciousness there is no individuation and that all individuals are actually the same consciousness. There is no adjective to describe the enormity of this prospect. We have here an incentive to be compassionate to the guy next door. There is no need to be silly about this. For individuals in physical incarnation there remains a need to respond to crime with a karmic hand in punishment and rehabilitation.

The non-individuation of consciousness at cosmic level is vividly illustrated in Matt.25:37

... when saw we thee an hungered, and fed thee? or thirsty, and gave thee drink? When saw we thee a stranger, and took thee in? or naked, and clothed thee?

Or when saw we thee sick, or in prison, and came unto thee?

A quote from physicist Erwin Schrödinger's *Meine Weltansicht* dealing with the individuation of consciousness:

In all the world, there is no kind of framework within which we can find consciousness in the plural; this is simply something we construct because of the spatio-temporal plurality of individuals, but it is a false construction. Because of it, all philosophy succumbs again and again in the hopeless conflict between the theoretically unavoidable acceptance of Berkeleian idealism and its complete uselessness for understanding the real world. The only solution to this conflict, in so far as any is available to us at all, lies in the ancient wisdom of the Upanishads.

When seeing someone in a state of extreme misfortune it is quite natural to take comfort in the words of English Protestant Reformer and martyr John Bradford (1510–1555), when seeing a prisoner being led to his execution:

"There but for the grace of God go I."

These words however suggest a plurality and individualisation of consciousness. It would be more to the point to simply say: "There go I." John Bradford was arrested and imprisoned in the Tower on a charge which was both false and trivial. He went to his martyrdom at the stake with sublime serenity and courage. Before the fire was lit he forgave all those who had wronged him and asked forgiveness from any whom he might have wronged. He said to his companion chained with him to the stake: "Be of good comfort brother; for we shall have a merry supper with the Lord this night!"

Another famous quotation dealing with the individuation of consciousness comes from Jacobean poet and preacher John Donne (1572–1631):

" ... send not to know for whom the bell tolls; it tolls for thee."

A reference to this quotation became the title of the famous novel by Ernest Hemingway set in the Spanish civil war. Several hundred of John Donne's sermons remain available in print to this day. John Donne's memorial is the only sculpture to survive the destruction of the old St. Paul's in the Great Fire and can now be seen in Wren's masterpiece, the new St. Paul's. London's St. Paul's remains one of the greatest architectural masterpieces on Earth. What could be more glorious than the sound of the great diapasons reverberating through the vast space.

As part of the consciousness package, animals are equipped with a basic instinct for mate acquisition and reproduction. In humans this can range from the brutal Heathcliff-Catherine passion to the sublime soul-mate relationship approaching Nirvanic consciousness re-unification. The full spectrum of these passions is vividly portrayed in Emily Brontë's powerful masterpiece *Wuthering Heights*. Another reference to sublime consciousness re-unification can be found in Beethoven's famous letter to his 'immortal beloved': *"My angel, my all, my very self ... "*

There are a great many accounts of near death experiences – let us take a look at an extract of that of psychic healer Denise Linn, vividly portrayed in her book *Sacred Space*. It is of particular interest that she mentions the non-individuation of consciousness and the absence of the passage of time. Denise had been viciously attacked by an armed assailant and left for dead.

> *Life is so very precious. At the hospital everything seemed amplified. The lights appeared glaring and bright. Searing pain. Shrill harsh voices. Slowly, the lights began to dim. Pain subsided. Voices faded into stillness. I found myself in a soft womb-like darkness. I felt as if I were being drawn deep down within a velvet black cocoon.*
>
> *Instantly the black bubble seemed to burst. The most brilliant luminous golden light enveloped me. It was so vibrant that the brightest sun would pale in comparison. Everywhere around me, into infinity, was light. Infused in the light with crystalline delicacy was pure sweet music. This liquid light symphony was ebbing and flowing throughout the universe in perfect harmony. The fluid harmonics pervaded my being until I merged with the light and sound. Light and sound were not separate from each other. I was light-sound. And this surrounding, all pervasive universe of warmth and light and music seemed completely natural and completely familiar. Everything seemed more real than anything else I had ever experienced! It was as if my teenage life up to that time had only been a dream. Just as when you awake in the morning and your dream begins to fade in the 'reality' of the day, my entire life up to that time seemed to dissolve into a fine mist as I stepped into this new 'hyper-real' reality. My previous life seemed nothing more than an illusion to me.*
>
> *All time seemed to flow in a continuous, everlasting 'Now'. There was no past and no future. Everything was contained within an infinite present. I remember trying to think of the past and I couldn't, because it was inconceivable. It literally didn't exist. It was as impossible for me to imagine linear reality when I was 'there' as it is for me to fully understand non-linear reality when I am 'here'.*
>
> *Completely infused within this world of light/ sound/ infinite now, was 'Love'. This was so very different from the way we usually think of love. Our culture's conception of love involves loving someone or something as an entity separate from ourselves. The love I experienced was infinite and limitless. There wasn't anything that wasn't Love. The love I experienced was not separate from anyone*

or anything. It was as natural as breathing. Everything simply was Love, a part of it, without any separation. It was a love beyond form, without boundaries. And I wasn't alone. You were there with me. Everyone was there. There wasn't anyone who wasn't there. We were all One. We weren't separate. There was no beginning, no end, just infinite eternal light. No longer confined to my body, I experienced being one with all things and all beings. I was everyone that I had ever loved and everyone that I had ever hurt. I was everyone that I had known and I was everyone that I would never know. I was the hungry beggar on a side-street in Delhi. I was the thief in New York City. I was the baby held in her mother's arms in Kenya. I was the spiritual adept in a mountain temple in Japan. I was everyone and everyone was me.

Let us take a look at a literary gem – the 'Book of Job'. This is a masterpiece in its own right, rich in symbolism, susceptible to much interpretation, and certainly the story of the elevation of consciousness to a sublime level. Remarkably, this story has been illustrated in 1825 by William Blake in twenty two engravings – a masterpiece and his last major work done at the age of sixty five. Blake's engravings have been commented on by Carl Jung providing valuable additional insight into the masterpieces. According to Jung, as with most great works of art, Blake expressed far more than he knew. In these pictures the objective psyche speaks directly to us. We cannot deal with the whole story and engravings in detail here. Readers who wish to follow this further will be richly rewarded. There is an excellent commentary on Jung's interpretation by Edward F. Edinger in his booklet *Encounter with the Self.* Let us take brief looks at the first and last pictures:

In the first picture, Job and his wife are seated in front of a tree with open books on their laps and surrounded by their kneeling children. The animals in the foreground are asleep and musical instruments are hanging in the tree. The sun is setting and the moon is in its last phase. In the background are large flocks and barns showing Job's wealth. Job is a good, orthodox, prosperous and charitable man. There are subtle clues in the picture portentous of unspeakable calamities to come.

In the final picture Job, his wife and new family are gathered around the tree of life once again. Everything is different. The books are gone, everyone is standing, the animals are awake, the musical instruments are being played. The sun is now rising on the right and the moon is waxing on the left.

The lesson of the Job story for modern man is described by Jung in his letter of 30 June 1956 to Elined Kotschnig, who had asked for an answer to "the problem of an unconscious, ignorant creator-god." Jung replies:

> We have become participants of the divine life and have to assume a new responsibility, viz. the continuation of the divine self-realisation, which expresses itself in the task of our individuation. Individuation does not only mean that man has become truly human as distinct from animal, but that he is to become partially divine as well. This means practically that he becomes adult, responsible for his existence, knowing that he does not only depend on God but that God also depends on man. Man's relation to God probably has to undergo a certain important change: Instead of the propitiating praise to an unpredictable king of the child's prayer to a loving father, the responsible living and fulfilling of the divine will in us will be our form of worship of and commerce with God. His goodness means grace and light and His dark side the terrible temptation of power. Although the divine incarnation is a cosmic and absolute event, it only manifests empirically in those relatively few individuals capable of enough consciousness to make ethical decisions, i.e., to decide for the Good. Therefore God can be called good only inasmuch as He is able to manifest His goodness in individuals. His moral quality depends on individuals. That is why He incarnates. Individuation and individual existence are indispensable for the transformation of God the Creator.

Another literary gem dealing with the elevation of consciousness is John Bunyan's *Pilgrim's Progress*. A highly allegorical story with characters having explicit names. The story starts with:

> O my dear Wife, said he, and you the Children of my bowels, I your dear friend, am in myself undone by reason of a Burden that lieth hard upon me; moreover, I am for certain informed that this our City will be burned with fire from Heaven; in which fearful overthrow, both myself, with thee my Wife, and you my sweet Babes, shall miserably come to ruine, except (the which yet I see not) some way of escape can be found, whereby we may be delivered.

He sets off on his own; this is definitely a do-it-yourself mission. His family thinks that he is crazy and will not go with him. His friends follow him and try to persuade him to go home but without success. Pilgrim carries on with his mission through trials and tribulation until he reaches his ultimate goal. At the end of the story Bunyan declares that it was but a dream. Bunyan wrote his masterpiece while languishing in jail – imprisoned for his non-conformist views; a similar persecution to that experienced by Galileo. He remained in jail from 1660 until the non-conformist laws were suspended by Charles II in 1672. He wrote nearly sixty books and died in 1688.

There is another story dealing with mystical initiation and elevation of consciousness that we can mention here: *Snow White and the Seven Dwarfs*. The Disney movie of this story is a classic. The original story comes from the Grimm brothers compilation after much revision and sanitization over the years to suit the mores of the day. Jacob Ludwig Carl Grimm (1785–1863) and Wilhelm Carl Grimm (1786–1859) made major contributions to European literature but they are probably best known for their compilations of hundreds of legends, fables and fairy tales. The fairy tales are set in enchanted forests and castles and peopled with princes, princesses, magical witches and fairies and also huntsmen, wolves, foxes, frogs and also unpleasant stepmothers. The appeal of this romantic escapism has never waned and has taken a remarkable resurgence in the phenomenal success of the Harry Potter stories. In early fairy tales Snow White is called 'Snowdrop'. The name change is unfortunate as a flower name for a girl seems far preferable to a cadaverous pallor. The Disney movie of *Snow White and the seven Dwarfs* has much in common with Mozart's *The Magic Flute*. Both of these are musicals which fall into the category of comic opera yet both deal with mystical initiation and elevation of human consciousness and are loaded with Masonic symbolism. The only significant difference is in the intellectual quality of the music and librettos. Snow White embarks on her mission into a dark forest surrounded by dangers and terrors that she cannot understand. She is discovered by seven dwarfs, suggesting with their comical adjectival names, the seven human temperaments of the Theosophists. This is linked to the seven candle menorah and also the church altar candles and cross. The dwarfs' 'work' is to uncover polished gemstones in the earth. The original idea of 'seven dwarfs' comes to us from ancient Norse mythology. After Snow White's martyrdom she is raised by the Prince and taken to dwell in a castle in the sky. The fate of the evil witch is strikingly similar to the downcasting of the devil into hell as illustrated by Blake taking some liberty with the Book of Job. One may well ask why this material is dressed for the amusement of children but this is a psychological issue which would digress from the focus of this book.

The Disney animal stories have become a remarkable if not strange phenomenon of the twentieth century. None of the animal characters are married and all the children have no obvious parents but live with uncles and aunts avoiding social biology altogether. Another curious phenomenon is that all the characters wear white gloves with four fingers but this is probably nothing more than a convenient way of giving the animals hands instead of paws. There are also several other fairy tales (Sleeping Beauty, Cinderella etc.) which also have cryptic meaning and symbolism.

Sir James Hopwood Jeans (1877–1946) was one of the greatest mathematician scientists of the early twentieth century. Let us take a look at a quote from his *The Mysterious Universe* (1930):

> *The terrestrial pure mathematician does not concern himself with material substance but with pure thought. His creations are not only created by thought but are pure thought… And the concepts which now seem to be fundamental to our understanding of nature… four dimensional space, a space which expands forever; a sequence of events which follows the laws of probability instead of the laws of causation; all these concepts seem to my mind to be structures of pure thought. To my mind the laws which nature obeys are less suggestive of those which a machine obeys in its motion than those which a musician obeys in writing a fugue, or a poet in composing a sonnet… If all this is so, then the universe can best be pictured, although still very imperfectly and inadequately, as consisting of pure thought, the thought of what, for want of a wider word, we must describe as a mathematical thinker.*

A universal consciousness holding the whole objective universe together – a new idea? No, not at all. Irish empirical philosopher Bishop George Berkeley (1685–1753) claimed that no objective physical reality exists at all; the world being a sort of collective dream, and God as the over-consciousness holding the whole thing together.

Let us look at a few pithy quotations of Sir Arthur Eddington:

Not only is the universe stranger than we imagine, it is stranger than we can imagine.

We are bits of stellar matter that got cold by accident, bits of a star gone wrong.

Science is one thing, wisdom is another. Science is an edged tool, with which men play like children, and cut their own fingers.

Something unknown is doing we don't know what.

It is impossible to trap modern physics into predicting anything with perfect determinism because it deals with probabilities from the outset.

Every body continues in its state of rest or uniform motion in a straight line, except insofar as it doesn't.

We have found a strange footprint on the shores of the unknown. We have devised profound theories, one after another, to account for its origins. At last, we have succeeded in reconstructing the creature that made the footprint. And lo! It is our own.

I ask you to look both ways. For the road to a knowledge of the stars leads through the atom; and important knowledge of the atom has been reached through the stars.

We used to think that if we knew one, we knew two, because one and one are two. We are finding that we must learn a great deal more about 'and'.

Proof is the idol before whom the pure mathematician tortures himself

The mathematics is not there till we put it there.

A rainbow described in the symbolism of physics is a band of ethereal vibrations arranged in systemic order to wave-lengths from about 0,00004 centimeters to 0,000072 centimeters. From one point of view, we are paltering with the truth whenever we admire the gorgeous bow of colour, and should strive to reduce our minds to such a state that we receive the same impression from the rainbow as from a table of wave-lengths. But although that is how the rainbow impresses itself on an impersonal spectroscope, we are not giving the whole truth and significance of experience if we suppress the factors wherein we ourselves differ from the spectroscope. We cannot say that the rainbow, as part of the world, was meant to convey the vivid effects of colour; but we can perhaps say that the human mind, as part of the world, was meant to perceive it that way.

A little relief from physicist Niels Bohr:

There are some things so serious that you have to laugh at them.

Physicist, Sir James Jeans writes:

So long as we adhere to the conventional notions of mind and matter, we are condemned to a view of perception which is miraculous. We suppose that a physical process starts from a visible object, travels to the eye, there changes into another physical process, causes yet another physical process in the optic nerve, and finally produces some effect in the brain, simultaneously with which we see the object from which the process started, the seeing being something "mental," totally different in character from the physical processes which precede and accompany it. This view is so queer that metaphysicians have invented all sorts of theories designed to substitute something less incredible ...

Everything that we can directly observe from the physical world happens inside our heads, and consists of mental events which form part of the physical world The development of this point of view will lead us to the conclusion that the distinction between mind and matter is illusory. The stuff of the world may be called physical or mental or both or neither as we please; in fact the words serve no purpose.

Even if the two entities which we have hitherto described, as mind and matter are of the same general nature, there remains the question as to which is the more fundamental of the two. Is mind only a by-product of matter, as the materialists claimed? Or is it, as Berkeley claimed, the creator and controller of matter?

Before the latter alternative can be seriously considered, some answer must be found to the problem of how objects can continue to exist when they are not being perceived in any human mind. There must, as Berkeley says, be 'some other mind in which they exist.' Some still wish to describe this, with Berkeley, as the mind of God; others with Hegel as a universal or absolute mind in which all our individual minds are comprised. The new quantum mechanics may perhaps give a hint, although nothing more than a hint, as to how this can be.

It seems, at least, conceivable, that what is true of perceived objects may also be true of perceiving minds; just as there are wave-pictures for light and electricity, so there may be a corresponding picture for consciousness. When we view ourselves in space and time, our consciousness is obviously the separate individuals of a particle-picture, but when we pass beyond space and time, they may perhaps form ingredients of a single continuous stream of life. As it is with light and electricity, so it may be with life; the phenomena may be individuals carrying on separate existences in space and time, while in the deeper reality beyond space and time we may all be members of one body. In brief, modern physics is not altogether antagonistic to an objective idealism like that of Hegel.

To believe that the universe consists of only those elements and forces that are perceptible to our senses or detected by our instruments is to belie the latest assessments of science. The very size and the extent of the Universe, the new formations discovered in the sky and the problems created by them, the marvels of the ultra-microscopic world and the possibility of even superior types of life in other parts of the Cosmos provide more than sufficient material to make it clear that the creation round us is too complex, too vast and too full of unsolved riddles to make us complacent about the fact that what our senses perceive or minds apprehend is all that exists in it. Such an attitude of mind at this stage of our knowledge can only emanate from one not in touch with the progress of today.

The first impact of genuine mystical experience on the mind of the experiencer is something like this; that the world he was perceiving and his own individuality, as he was conscious of it so far, were not true realities but only the figures of, say, a relatively speaking, dream state from which he has just awakened to the full bloom of another sun shining on a splendrous world, entirely unlike the one which his senses were revealing to him before. It should be remembered that for this state of cognition, it is not necessary that the percipient should be insensible to the sensory world Not at all. What makes mystical ecstasy an increasing wonder is the incredible fact that both the sensory and supersensory worlds can be perceived simultaneously. But how? Like the radiant sky showing a mirage on it, both visible side by side.

The most significant book dealing with mysticism and elevation of consciousness must surely be the *Bhagavad Gita*. Originally written over five thousand years ago in Sanskrit, an English translation was written by Charles Wilkins in 1785. Einstein remarked that when reading the *Bhagavad Gita* he thought about how God created the universe and then everything else seemed so superfluous.

Canadian psychiatrist Dr. Richard Maurice Bucke (1837–1902) published his treatise *Cosmic Consciousness* in 1901. Bucke envisioned only three stages of consciousness in the course of evolutionary development.

1. Simple instinctive consciousness.

2. Self consciousness, that self-awareness of a human to realise himself as a distinct entity.

3.   Cosmic consciousness at the pinnacle of human evolution.

Bucke apparently had an experience of Cosmic Consciousness and he drew up a rather short list of people throughout history whom he considered might have had the same experience. This experience is described as follows:

> *Like a flash there is presented to his consciousness a clear conception in outline of the meaning and drift of the universe. He sees and knows that the cosmos is in fact, in very truth a living presence. He sees that instead of men being, as it were, patches of life scattered through an infinite sea of non-living substance, they are in reality specks of relative death in an infinite ocean of life. He sees that the life which is in man is as immortal as God is; that the universe is so built and ordered that without any peradventure all things work together for good of each and all; that the foundation principle of the world is what we call love, and that the happiness of every individual is in the long run absolutely certain. The person who passes through this experience will learn in the few minutes, or even moments, of its continuance more than in months or years of study, and he will learn much that no study ever taught or can teach. Especially does he obtain such a conception of 'the whole'. Along with moral elevation and intellectual illumination comes what must be called, for want of a better term, a sense of immortality.*

Physicist Werner Heisenberg, in his *Ordnung der Wirklichkeit*, writes of access to higher consciousness through music:

> *... This is also, particularly now in our own time, relevant to many people who do not belong to any religious community, and for whom there is an encounter with the other world for the first time in the sounds of a Bach fugue, perhaps, or in a flash of scientific illumination.*

We must mention here a significant writer of the early twentieth century, Evelyn Underhill (1875–1941) who wrote prolifically on mysticism, consciousness and spirituality while remaining within the orthodoxy of the time. Underhill wrote 39 books and 350 articles. Her first important book *Mysticism* appeared in 1911. She was a skilled bookbinder and enjoyed going on yachting holidays with her husband. She also had a keen appreciation of fine arts going for many years on springtime holidays to Europe to experience the art treasures of France and Italy. She became a pupil under the spiritual guidance of Baron Friedrich von Hügel. She was a dedicated worker of spiritual upliftment spending some time of every week of her adult life in the slums of London. Let us take a look at a quotation from her *Mysticism*.

> *The mystical side of ecstasy represents the greatest possible extension of the spiritual consciousness in the direction of Pure Being: the blind intent stretching here receives its reward in a profound experience of Eternal Life. In this experience the departmental activities of thought and feeling, the consciousness of I-hood, of space and time, all that belongs to the world of becoming and our own place therein, are suspended. The vitality which we are accustomed to split amongst these various things, is gathered up to form a state of pure apprehension, a vivid intuition of the Transcendent. This is that perfect unity of consciousness, that utter concentration on an experience of love, which excludes all conceptual and analytic acts. Hence, when the mystic says that his faculties were suspended, that he 'knew all and knew nought' he really means that we are so concentrated on the Absolute that he ceased to consider his separate existence, so merged in it that he could not perceive it as an object of thought, as the bird cannot see the air which supports it, nor the fish the ocean in which it swims. He really 'knows all but thinks nought, perceives all, but conceives nought.'*

Underhill's letter to Archbishop Cosmo Lang of Canterbury is as relevant today as it was then. Let us take a look at an excerpt from this letter:

> *I desire very humbly to suggest with bishops assembled at Lambeth that the greatest and most necessary work they could do at the present time for the spiritual renewal of the Anglican Church would be to call the clergy as a whole, solemnly and insistently to a greater interiority and cultivation of the personal life of prayer. This was the original aim of the founders of the Jerusalem Chamber Fellowship, of whom I am one. We were convinced that the real failures, difficulties and weaknesses of the Church are spiritual and can only be remedied by spiritual effort and sacrifice, and that her deepest need is a renewal, first in the clergy and through them the laiety; of the great Christian tradition of the inner life. The Church wants not more consecrated philanthropists, but a disciplined priesthood of theocentric souls who shall be tools and channels of the Spirit of God: and this she cannot have until Communion with God is recognised as the first duty of the priest. But under modern conditions this is so difficult that unless our fathers in God solemnly require it of us, the necessary efforts and readjustments will not be*

made. With the development of that which is now called "The Way of Renewal" more and more emphasis has been placed on the nurture and improvement of the intellect, less and less, on that of the soul. I do not underrate the importance of the intellectual side of religion. But all who do personal religious work know that the real hunger among laiety is not for halting attempts to reconcile theology and physical science, but for the deep things of the spirit.

Evelyn Underhill was a remarkable person with wide ranging interests. At one stage she was a member of the Order of the Golden Dawn, which had on its membership roll some of the most renowned and also the most notorious occultists of the day as well as some social celebrities. W. B. Yeats and Maude Gonne were also members of this order. One of the members, Aleister Crowley, was described by a magistrate as 'The wickedest man on earth'. The Golden Dawn was based on a miscellany of ideas from various mystical and fraternal organisations and secret societies. Enmities arose between the members resulting in the disintegration and disbanding of the order. Archbishop Dr Cosmo Gordon Lang (1864–1945) was known for his interest in spiritualism. In 1943 he wrote a short one act play "The Mongolian Master and his Disciple in Quest of the Little World of White by Frater Om-soc" which was a spoof on occultism and on Aleister Crowley in particular. Om-soc is Cosmo spelled backwards. Let us take a look at another excerpt from Underhill – this time from a talk given on the BBC in 1936. This too is timeless writing and still fully relevant as it was then:

... when we lift our eyes from the crowded by-pass to the eternal hills; then, how much the personal and practical things we have to deal with are enriched. What meaning and coherence comes into our scattered lives. We mostly spend those lives conjugating three verbs: to Want, to Have, to Do. Craving, clutching, and fussing, on the material, political, social, emotional, intellectual - even the religious - plane, we are kept in perpetual unrest; forgetting that none of these verbs have any ultimate significance, except so far as they are transcended by and included in, the fundamental verb, to Be; and that Being, not wanting, having and doing, is the essence of a spiritual life. But now, with this widening of the horizon, our personal ups and downs, desires, cravings, efforts, are seen in scale; as small and transitory spiritual facts, within a vast, abiding spiritual world, and lit by a steady spiritual light. And at once, a new coherence comes into our existence, a new tranquillity, and release. Like a chalet in the Alps, that homely existence gains atmosphere,

dignity, significance from the greatness of the sky above it and the background of
the everlasting hills ...

Evelyn Underhill collaborated with Rabindranath Tagore in the almost impossible task of rendering the poetical works of medieval Indian poet Kabir in English. Let us take a few brief glimpses into Underhill's masterly introduction to the works of Kabir.

The poet Kabir, a selection from whose songs is here for the first time offered to English readers, is one of the most interesting personalities in the history of Indian mysticism. Born in or near Benares, of Mohammedan parents, and probably about the year 1440, he became in early life a disciple of the celebrated Hindu ascetic Ramananda. Ramananda had brought to Nothern India the religious revival which Ramanuja, the great twelfth century reformer of Brahamanism, had initiated in the South ... We may safely assert, however, that in their teachings, two - perhaps three - apparently antagonistic streams of intense spiritual culture met ... and it is one of the outstanding characteristics of Kabir's genius that he was able in his poems to fuse them into one... From the point of view of orthodox sanctity, whether Hindu or Mohammedan, Kabir was plainly a heretic; and his frank dislike of all institutional religion, all external observance completed, so far as ecclesiastical opinion was concerned, his reputation as a dangerous man. God he proclaimed was "neither in Kaaba nor in Kailash." Those who sought Him needed not go far; for He awaited discovery everywhere, more accessible to "the washerwoman and the carpenter" than to the selfrighteous holy man. ... therefore the whole apparatus of piety, Hindu and Moslem alike – the temple and mosque, idol and holy water, scriptures and priests – were denounced by this inconveniently clear-sighted poet as mere substitutes for reality; dead things intervening between the soul and its love.

The images are all lifeless, they cannot speak:
I know, for I have cried aloud to them.
The Purana and Koran are mere words:
lifting up the curtain, I have seen.

This sort of thing cannot be tolerated by any organised church; and it is not surprising that Kabir, having his headquarters in Benares, the very centre of priestly

influence, was subjected to considerable persecution. The well-known legend of
the beautiful courtesan sent by Brahmans to tempt his virtue, and converted,
like the Magdalen, by her sudden encounter with the initiate of a higher love,
pre serves the memory of the fear and dislike with which he was regarded by the
ecclesiastical powers.

In May 2004, a statue of Evelyn Underhill was dedicated in Guildford Cathedral. It should be noted that the name of Underhill's mentor Friedrich Hügel must not be confused with Friedrich Hegel. Baron Friedrich von Hügel (1852–1925) was of Italian nationality despite his aristocratic German surname. He was born in Florence – his father was an Austrian diplomat and his mother was Scottish. Von Hügel wrote extensively on the philosophy of religion. Georg Wilhelm Friedrich Hegel (1770–1831) was a philosopher of the period 'German Idealism' in the decades following Kant. Hegel is well known for his concept of a *Zeitgeist* and collective human consciousness. Philosopher Schopenhauer was however critical of Hegel's work.

In 1931, Kurt Gödel (1906–1978) proposed his Incompleteness Theorem – a major mathematical milestone. This theorem states that within any given branch of mathematics there would always be some propositions that couldn't be proven either true or false using the rules and axioms of that mathematical branch itself. You might be able to prove every conceivable statement about numbers within a system by going outside the system in order to come up with new rules and axioms, but by doing so you will only create a larger system with its own unprovable statements. The implication is that all logical systems of any complexity are, by definition, incomplete; each of them contains, at any given time, more true statements than it can possibly prove according to its own defining set of rules. This can be taken to imply that you will never be entirely able to understand yourself, since your mind, like any other closed system, can only be sure of what it knows about itself by relying on what it knows about itself. This brings us back to the hierarchy of consciousness suggesting that you cannot completely understand yourself without some access from a higher level of consciousness. Einstein had a similar idea when he said:

The problems that exist in the world today cannot be solved by the level of thinking
that created them.

In 1948, Gödel showed that tunnels in space-time may be possible, linking past and present. This idea was later described by physicist John Wheeler as space-time 'wormholes'. Wheeler also coined the term 'Big Bang' describing the origin of the universe. The concept of 'cosmic strings' is also becoming prominent among scientists on a quest for discovering a means of time travel. We may have to wait some time before this emerges from science fiction into reality.

Computer pioneer Alan Mathison Turing (1912–1954) made a bridge between the logical and the physical worlds, thought and action, which crossed conventional boundaries. Turing built on Gödel's work proposing a hypothetical Turing machine which was in fact a computer program at a time when computers did not yet exist. The universal Turing machine describes a computer which could be programmed to perform an unlimited variety of functions and even function like a brain. Turing's contribution to the development of computer science is immense. His work on cracking military encryption had a decisive influence on the outcome of WWII. Turing's breaking of the Enigma code led to the elimination of the threat of U-boat wolf packs to Atlantic shipping. This was the first time in history that the outcome of a major international military conflict was precipitated by brilliant mathematical logic. Turing's Halting Problem stated that it is impossible for a computer program to know within itself if the program will eventually halt or run indefinitely. Another program analysing this program can of course decide if the program will eventually halt. We have here a suggestion of a higher level of consciousness. The probability that the program will halt is taken further by Gregory Chaitin who defined the Omega number – a probability between zero and one. This number is definable but utterly non-computable.

In 1928, at an international conference, mathematician David Hilbert posed three mathematical challenges which have their origins as far back as Gottfried Leibniz. Kurt Gödel managed to disprove the first two of the propositions in 1934 but the third, the *Entscheidungsproblem* (Decision problem), remained. Turing managed to disprove this proposition in terms of his Turing Machine and Halting Problem, a prestigious mathematical achievement.

The unspeakably despicable persecution and consequent tragic death of Dr. Alan Mathison Turing are almost beyond belief. Turing's cyanide-laced apple almost seems to have a symbolic link to a martyrdom mentioned earlier in this chapter. In September 2012 a special edition of the 'Monopoly' board game was announced in honour of Alan Turing. The first production run was paid for by Google and donated to Bletchley Park to sell for fundraising.

We are left with the disconcerting conclusion that the various branches of mathematics are riddled with holes in their axiomatic provability and that there is no overriding philosophy to bind the various branches of mathematics together.

Let us close this chapter with a quotation from the rich legacy of one of the greatest philosopher scientists of the early twentieth century – Sir Arthur Stanley Eddington (1882–1944):

> 'The substratum of everything is of mental character which in some parts rises to the level of consciousness ... and the universe is of the nature of a thought.'

CONCLUSION

The central focus of this book has been to present an appreciation of the universe between the ultimate extremes of small and large and also to take a look at the interaction of human consciousness with physical matter, and finally, to consider the ultimate purpose of the universe.

After considering the words of some of the most brilliant physicists, mathematicians and philosophers of the twentieth century and earlier, we are left with the idea that the universe is hyper dimensional in a hierarchy of dimensionality, with the space-time dimensionality of the objective universe representing only a sub-set of the whole. We are also left with the concept of a universal consciousness as well as a hierarchical structure of levels of consciousness and that the ultimate nature of the objective universe is that of a thought! This leaves us with the idea that the universal mathematical substrate and universal consciousness is the only continuous structure in the universe.

We present a theory of time-vector gravity which dispenses with the notion of a 'gravitational field'. As the theory involves all eleven dimensions of the space-time continuum we have had to take a close look at the structure of space, Kaluza-Klein dimensions, the ultimate structure of matter energy and also consider a digital time vector. In order to construct a working model it has been necessary to consider up-to-date theoretical physics as well as sources with their roots in ancient Sanskrit writings. In proposing a theory of gravitation it has been necessary to take an iconoclastic stance with the notions of 'gravitational field', 'gravitons' and even 'mass'.

A sobering thought at the level of the objective universe is the insubstantial nature of matter, indeed, the ultimate nature of physical matter is nothing more than a mathematical concept sprinkled with energy!

Stand beside a five ton ingot of steel. This is surely a substantial solid 'thing'. You cannot kick it out of the way; you would need a heavy-duty crane to lift it. Strike it with a hammer and all that you would produce are some sound waves.

From a particle point of view what do we have here? This heavy solid thing is less substantial than the fragrance wafting from a flower assuming of course that the fragrance molecules are 'solid'.

What then is real and what is illusion?

Edgar Cayce hits the nail squarely on the head with:

We are not physical beings having a spiritual experience,
We are spiritual beings having a physical experience.

A sobering thought at the other end of the scale is the extent of the universe and the question of whether it is populated with human life. If we assume that the occurrence of stars with planets is very unusual and in turn that the occurrence of planets with life is very unusual, and again that the occurrence of planets with life and having intelligent life is also very small then there could still be thousands of planets with intelligent life in our own Milky Way galaxy. What then of the thousands of millions of other galaxies? Wondering if there is any life 'out there' is much like picking up a sea shell on the beach and wondering if there are more shells in the sea. The extent of the universe is vast beyond any attempts of imagination.

Who better than Beethoven to musically exult in the awesome glory of the firmament with Schiller's words:

Brüder, überm Sternenzelt	Brethren, above the starry canopy of Heaven
muß ein lieber Vater wohnen.	A loving Father must surely dwell.

What then is the purpose of all this?
This is not an impertinent or difficult question.

Surely the ultimate purpose of the universe is the development of human consciousness through every conceivable experience in a trial and error process much like the evolutionary processes in the physical world. Whether this takes place in one lifetime or many is beside the point. What then of the vast numbers of humans involved – do large numbers eventually appear as aspects of a single Being? This idea is seriously proposed by some of the greatest thinkers of our time: that individual consciousnesses must gradually strive to become one with the universal consciousness, or that individual consciousness has actually always been part of the universal consciousness but must go through the evolutionary process of development.

This idea is gloriously stated by Beethoven, with Schiller's words, in his ninth symphony:

Froh, wie seine Sonnen fliegen	Cheerfully, as suns fly
durch des Himmels prächt'gen Plan,	through heaven's glorious firmament,
laufet, Brüder, eure Bahn,	run your courses brethren,
freudig wie ein Held zum Siegen.	Joyously as a hero to victory.

These words are sung *alla marcia* by a tenor voice; Beethoven does not treat this with awe but with confident cheerfulness.

Artist and poet William Blake (1757–1827) also takes the idea of a heroic and triumphant course through the heavens, at a more parochial level, in his hymn:

'And did those feet in ancient times ... '

Bring me my bow of burning gold!
Bring me my arrows of desire!
Bring me my spear! O clouds unfold!
Bring me my chariot of fire!

In the finale of Beethoven's Op. 135 quartet he takes the theme '*Muß es sein? Es muß sein! Es muß sein!* (Must it be? It must be! It must be!) At musical level this work can be seen as a final and inevitable break with classicism. From a classic point of view, this music is shocking, revolutionary and exhilarating. The ominous '*Es muß sein!*' is not written with gloom but with gaiety. This is not a funeral but a celebration. On a larger scale this work can be seen to represent the development and destiny of Beethoven's life; on the cosmic scale it can represent the evolution and ultimate fulfillment of the conscious universe.

The idea that ultimate reality lies in hyperdimensionality, beyond the space-time continuum, from where the objective universe is but an illusion, is an awesome prospect to contemplate. The time vector is so much an essential part of movement and experience in the objective universe that it is practically impossible for us to imagine any type of existence without time.

We can sympathise with Bernard of Cluny when he wrote in his 12th century Latin hymn: *Urbs Sion aurea, patria lactea:*

I know not, O I know not
What social joys are there,
What radiancy of glory,
What light beyond compare.

Do we really need to pursue such overwhelmingly large issues, or should we rather heed the words of Cardinal John Henry Newman in his famous hymn 'Lead kindly Light':

Keep Thou my feet; I do not ask to see
The distant scene,–
One step enough for me.

With the publication of Newton's Principia Mathematica in 1687 the educated public could have felt quite comfortable with the idea that science and mathematics had finally been fully sorted out. This was unquestionably a work of towering genius, one of the great books of all time, and today still fully relevant to the requirements of civil and mechanical engineering. Paradoxically what are we to make of Newton's bizarre 'bodkin in the eye' experiment and the years spent in fruitless dabbling in alchemy? Science's state of fulfillment was doomed not to last. A shocking awakening came with Einstein's theories of relativity followed by uncertainty theory even too shocking for Einstein. Since then research into particle physics has developed into huge industries in the industrialised countries. Have we now attained a position where the end is in sight? Hardly; even with our astonishing advances in particle physics and mathematics, can anyone accurately describe the structure of electrons, photons or even atomic nuclei? We are still at an embryonic stage! And what of the development of the human consciousness – there seems to be little visible progress in the past few thousand years.

We have a long way to go.

Let us end with what are possibly some of the most comforting words ever phrased:

His light shineth even in our darkness.

APPENDIX

THE SI PREFIXES

The SI (Le Systeme International d'Unités) prefixes provide a most convenient means of referring to the wide range of numeric values used in science and engineering. It will be seen that the prefixes of high values have the Italian female 'a' ending and those of the low values the male 'o' and that the names of the prefixes are loosely derived from Greek numbers for the groupings of three. The original small range metric prefixes however are not in accordance with this usage and are indented in the table. The range of the prefixes has been steadily growing over the years, the latest, yotta and yocto being added as recently as 1991. Does it need to extend further? Probably not – the ordinary decimal exponent notation is well suited to handle any value that may be needed.

The verbal terms billion, trillion, quadrillion etc. have the confusion of different values on opposite sides of the Atlantic and the larger values do not see sufficient use to provide an easily grasped meaning. The terms milliard and milliardth are not too bad but do not seem to enjoy much usage outside Europe. There is a way out of this problem – simply use an SI prefix with million. Million is an easily grasped value and the prefixes are easily remembered. This will give us kilomillion, megamillion, gigamillion, teramillion etc. – no problem! NB. This is only a suggestion and definitely not part of the standard!

The standard does not permit the combination of prefixes. A micro-kilogram must therefore be simplified to milligram. The prefixes must be joined to the unit without space or hyphen.

One frequently comes across curious misusages of the prefixes. A millipede could rather be called a kilopede and a centipede a hectopede as no creature could possibly locomote on a thousandth or a hundredth of a leg. This is not actually a prefix usage error as the creepy crawlies were named before the definition of the metric prefixes. Sometimes the prefixes are used on their own without mention of the quantity being referred to – this is frequently the case with groceries, travelling

				Long Scale	Short Scale
yotta	Y	10^{24}		Quadrillion	Septillion
zetta	Z	10^{21}			Sextillion
exa	E	10^{18}		Trillion	Pentillion
peta	P	10^{15}			Quadrillion
tera	T	10^{12}		Billion	Trillion
giga	G	10^{9}			Billion
mega	M	10^{6}	Million		
kilo	k	10^{3}	Thousand		
hecto	h	10^{2}	Hundred		
deca	da	10^{1}	Ten		
		10^{0}	One		
deci	d	10^{-1}	Tenth		
centi	c	10^{-2}	Hundredth		
milli	m	10^{-3}	Thousandth		
micro	μ	10^{-6}	Millionth		
nano	n	10^{-9}			
pico	p	10^{-12}			
femto	f	10^{-15}			
atto	a	10^{-18}			
zepto	z	10^{-21}			
yocto	y	10^{-24}			

distances, money and computer memory. Such are the delightful inconsistencies of language. The decimal exponent notation is most convenient for handling very large or small numbers – perhaps too convenient for one to grasp the enormity of the extreme ranges used in cosmology.

How much can one readily comprehend? A million, 10^6, or Mega prefix is quite easy. Take a square metre of graph paper ruled in one mm squares and you can see a million quite easily at a glance. Does anyone use graph paper anymore? Before everyone had their own computer, graph paper was obtainable in a variety of scales and coordinate systems. Now make the square metre into a cubic metre of one mm cubes - this takes us up to 10^9, a thousand million or Giga prefix. It is not possible to see all the little cubes but one can still have some appreciation for the number - this is possibly the limit to getting a feel for large numbers. We now take our cubic metre and expand it to a kilometre in all directions. This takes the little 1 mm cubes to 10^{18} or the Exa prefix. This number is humanly incomprehensible. Shall we go further? Take the cubic kilometre and expand it to a cube with 1 000 km sides. This takes us to 10^{27} but this is nowhere near the extreme values used in cosmology. Let's go one further and make our cube with sides of one million km. If you tried to put the Sun into a box of this size the sides would bulge and the corners of the cube would protrude. The number of little cubes would now be 10^{36}, still a modest number in cosmological terms!

Considerable confusion has arisen with the use of SI prefixes in the Data Processing and Data Transmission industries. The term kilobyte is generally considered to be 1024 bytes and not 1000 bytes; similarly a megabyte is regarded as 1 048 576 bytes and not 1 000 000 bytes as would be required by the SI.

A new system of binary multiples has been developed by the International Electrotechnical Commission with the strong support of the International Committee for Weights and Measures. The binary prefixes are as follows:

2^{10}	kibi	ki
2^{20}	mebi	Mi
2^{30}	gibi	Gi
2^{40}	tebi	Ti
2^{50}	pebi	Pi
2^{60}	exbi	Ei

A few examples should clarify the matter:

One kibibit	1 kibit	$= 2^{10}$ bit	$= 1\ 024$ bit
One kilobit	1 kbit	$= 10^3$ bit	$= 1\ 000$ bits
One mebibyte	1 MiB	$= 2^{20}$ byte	$= 1\ 048\ 576$ bytes
One megabyte	1 MB	$= 10^6$ byte	$= 1\ 000\ 000$ bytes
One gibibyte	1 GiB	$= 2^{30}$ byte	$= 1\ 073\ 741\ 824$ bytes
One gigabyte	1 GB	$= 10^9$ byte	$= 1\ 000\ 000\ 000$ bytes

The system of binary prefixes has been provided for use by the data processing and data transmission industries and does not form part of the SI.

THE MEASURES OF THE UNIVERSE

The weights and measures used by the people of our planet over the millennia is a huge subject which could probably fill several books, but fortunately, we live in enlightened times and the basic definitions of physical units have been reduced to a Magnificent Seven.

Quantity	Name	Symbol	CGPM Definition (Conférence Générale des Poids et Mesures)
length	metre	m	The length of path travelled by light in a vacuum during a time interval of 1/299 792 458 of a second.
mass	kilogram	kg	The mass of the international prototype of the kilogram recognized by the CGPM and in the custody of the Bureau International des Poids et Mesures, Sevres, France.
time	second	s	The duration of 9 192 631 770 periods of the radiation corresponding to the transition between the two hyperfine levels of the ground state of the caesium 133 atom.
electric current	ampere	A	That constant current which, if maintained in two straight parallel conductors of infinite length, of negligible circular cross-section, and placed one metre apart in vacuum, would produce between these conductors a force of 2×10^{-7} newton per metre of length.
thermodynamic temperature	kelvin	K	The fraction 1/273,16 of the thermodynamic temperature of the triple point of water.

| amount of substance | mole | mol | The amount of substance of a system which contains as many elementary entities as there are atoms in 0,012 kg of carbon 12. |
| luminous intensity | candela | cd | The luminous intensity, in a given direction, of a source that emits monochromatic radiation of frequency 540 x 10^{12} hertz and that has a radiant intensity in that direction of 1/683 watt per steradian. |

It should be noted that the symbol for the thermodynamic temperature unit Kelvin is written simply as 'K' and not with a degree symbol as in the Celsius scale. Kelvin units and Celsius degrees are numerically equivalent. Simply add 273,16 to Celsius to get Kelvins.

The definition of the length unit has come a long way since the platinum-iridium bar prototype. From 1960 to 1983 the metre was defined in terms of light wavelengths of krypton 86 but astonishing improvements in the measurement of time have resulted in the present definition based on the speed of light. Fifty years ago this would have seemed ludicrous. The metre is defined to nine significant digits – this gives a precision to about 3 microns per kilometre – something which would be quite impossible to attain from a prototype. The original definition of the metre was 1/10 millionth of the circumference of the earth from the north pole to the equator and passing through Paris. This impractical standard was later defined in terms of the length of a pendulum and then into metal bar prototypes. We now have a most favourable situation with the value of the speed of light. Since the 19th century a great many measurements of the speed of light have been made with varying degrees of accuracy. Now that the metre is defined in terms of the speed of light we have an exact value for this most fundamental constant of the universe.

The time unit has also seen a move away from a prototype – in this case the Solar System. Transit times in a multiple body orbiting system, where everything influences everything else, are hardly suited for setting an absolute time standard. The ephemeris second is still in use for astronomical purposes.

SI SUPPLEMENTARY UNITS

Plane angle	radian	rad	The plane angle between two radii of a circle which cut off on the circumference an arc equal to the length of the radius.
Solid angle	steradian	sr	The solid angle which, having its vertex in the centre of a sphere, cuts off an area of the surface of a sphere equal to that of a square with sides equal to the radius of the sphere.

These supplementary units are actually geometric definitions but are essential for defining some of the physical units.

The SI system has been a major development in the simplification of definitions based on the MKS (Metre, kilogram, second) system of units. A curious anomaly is that the mass unit has a built-in kilo prefix and is the only remaining definition to be based on a securely preserved prototype. It would of course be possible to define mass in terms of the rest mass of the proton, eliminating prototypes altogether, but this would leave us with a totally impractical standard. The CGS (Centimetre, gram, second) standard of units has mercifully faded into history. This had a curious centi prefix for the length unit. The Giorgi system which later was to become the MKS was introduced by Giovanni Giorgi to overcome anomalies in the CGS electrical and magnetic units and also set the value of the magnetic permeability of space to fit in with the definition of the ampere. The unit of force, the newton, can now be simply expressed in terms of the acceleration of a kilogram of mass or alternatively in terms of the ampere and metre. This is a most satisfactory state of affairs.

A great many units are derived from the seven base units – a selection of named units follows:

Quantity	Name	Symbol	Expression in terms of other SI units	Expression in terms of base units
admittance	Siemens	S	Ω^{-1}	$m^{-2}.kg^{-1}.s^3.A^2$
capacitance	farad	F	C/V	$m^{-2}.kg^{-1}.s^4.A^2$
conductance	Siemens	S	Ω^{-1}	$m^{-2}.kg^{-1}.s^3.A^2$
electrical resistance	ohm	Ω	V/A	$m^2.kg.s^{-3}.A^{-2}$
electric charge	coulomb	C	A.s	A.s
electric potential	volt	V	W/A	$m^2.kg.s^{-3}.A^{-1}$
energy	joule	J	N.m or W.s	$m^2.kg.s^{-2}$
force	newton	N	$kg.m /s^2$	$kg.m.s^{-2}$
frequency	hertz	Hz	s^{-1}	s^{-1}
illuminance	lux	lx	lm/m^2	$m^{-2}.cd.sr$
impedance	ohm	Ω	V/A	$m^2.kg.s^{-3}.A^2$
inductance	henry	H	Wb/A	$m^2.kg.s^{-2}.A^{-2}$
luminous flux	lumen	lm	cd.sr	cd.sr
magnetic flux	weber	Wb	V.s	$m^2.kg.s^{-2}.A^{-1}$
magnetic induction	tesla	T	Wb/m^2	$kg.s^{-2}.A^{-1}$
power	watt	W	J/s	$m^3.kg.s^{-3}$
pressure	pascal	Pa	N/m^2	$m^{-1}.kg.s^{-2}$
reactance	ohm	Ω	V/A	$m^2.kg.s^{-3}.A^2$
stress	pascal	Pa	N/m^2	$m^{-1}.kg.s^{-2}$
susceptance	siemens	S	Ω^{-1}	$m^{-2}.kg^{-1}.s^3.A^2$
weight	newton	N	$kg.m/s^2$	$m.kg.s^{-2}$
work	joule	J	N.m	$m^2.kg.s^{-2}$

Two commonly used units, the litre and the metric ton, are classified as commercial units and are not intended for use in scientific calculations.

The litre is exactly one cubic decimetre and the metric ton one thousand kilograms.

The unit of magnetic induction or magnetic flux density, the tesla, is a very large unit so that the gauss is often used instead. (One tesla = 10 000 gauss). The gauss is a CGS unit and not part of the SI standard.

The unit of capacitance, the farad, is also very large so that the values of capacitors are commonly given in microfarads or picofarads.

Names of units which are derived from the name of a person are not written with a capital letter except at the beginning of a sentence or if the entire sentence is written in capital letters. Symbols which are derived from the name of a person are always written with a capital letter. In electrical calculations the symbol "I" is used for ampere current. This is derived from the French Intensité de courant.

There are some units which are not part of the SI standard but which are in use for specific purposes.

Unit	Symbol	Value	Usage
Ångstrom unit	Å	1×10^{-10} m	Optics
Astronomical unit	AU	149,6 Gm	Astronomy
Atomic mass unit	amu	$1,660\ 538 \times 10^{-27}$ kg	Atomic & nuclear sciences
Electron volt	eV	0,160 217 aJ	Atomic & nuclear sciences
Light year	LY	9,460 55 Pm	Astronomy
Parsec	pc	30,857 Pm	Astronomy
Jansky	Ja	10^{-26} W m^{-2} Hz^{-1}	Radio astronomy

The AU is the length of the radius of the unperturbed circular orbit of a body of negligible mass moving around the Sun with a sidereal angular velocity of 17,202 098 950 milliradians per day of 86 400 ephemeris seconds. The AU and LY are actually abbreviations, not symbols, and depend on the language being used. This definition of the AU does not take relativistic effects into account which would cause the value to be slightly different when observed from different locations. Members of the International Astronomical Union have therefore proposed a new definition: 149 597 870 700 metres.

amu is equal to the fraction 1/12 of the mass of an atom of the nuclide carbon-12.

A pc is the distance at which 1 AU subtends an angle of one second of arc.

The eV is the energy acquired by an electron in passing through a potential difference of 1 Volt in vacuum.

There are many units which are not recognised by the SI and should not be used, and certainly not used for the calibration of instruments. Electromagnetic units of the 3-dimensional CGS system cannot strictly speaking be compared to the corresponding units of the SI which are 4-dimensional as far as electrical units are concerned.

Use of the following units should be avoided:

Poise (dynamic viscosity), erg (energy), calorie (energy), dyne (force), phot (illuminance), stokes (kinematic viscosity), fermi (length), micron (length), stilb (luminance), gauss (magnetic flux density), oersted (magnetic field strength), maxwell (magnetic flux), torr (pressure), pieze (pressure), stere (volume).

The Greek letter μ should only be used as the 'micro' prefix. This symbol is still in common use as a unit of length, the 'micron' which represents one millionth of a metre. The micron is in such widespread use that it is most unlikely that it will be phased out. A micron should be given μm. A micrometer is not a unit of length - this is an instrument for measuring fine tolerance machine parts and is commonly calibrated in hundredths of a millimetre or thousandths of an inch. A selection of micrometers will be found in every precision engineering machine shop.

THE BEAUTIFUL NATURAL PLANCK UNITS OF THE UNIVERSE

In setting units of measure it is quite natural to use objects which come readily to hand, hence such units as foot, cubit, hair's breadth, hand, stone, bushel, cup, teaspoon, pennyweight, day, horsepower and many more but none of these is suited to scientific work. The SI has provided us with a wonderful set of measures for any conceivable scientific or engineering purpose. In 1899, Max Planck conceived the idea of a system of natural units where physical constants (including the famous Planck's constant) would be reduced to unity. The units defined were:

Time T, Length L, Mass M, Electric Charge Q, and Temperature Θ. These units have also been described as 'God's units'. Planck said of his units:

> ... these necessarily retain their meaning for all times and for all civilisations, even extraterrestrial and non-human ones, and can therefore be designated as "natural units" ...

The following five universal constants become unity when expressed in terms of Planck units:

Constant	Dimensions
Speed of light in vacuum	LT^{-1}
Gravitation constant	$M^{-1}L^3 T^{-2}$
Dirac's constant (Planck's constant divided by 2π)	$M L^2 T^{-1}$
Coulomb force constant	$Q^{-2} M L^3 T^{-2}$
Boltzmann constant	$M L^2 T^{-2} \Theta^{-1}$

Using the constants set to unity we get the following values for the five Planck units:

Unit	SI equivalent value	
Planck time	$5{,}39072 \times 10^{-44}$	s
Planck length	$1{,}61624 \times 10^{-35}$	m
Planck mass	$2{,}17665 \times 10^{-8}$	kg
Planck charge	$1{,}8755459 \times 10^{-18}$	C
Planck temperature	$1{,}41679 \times 10^{32}$	K

Many other units can of course be derived from the basic five units:

Planck energy	$1{,}9561 \times 10^{9}$	J
Planck area	$2{,}61177 \times 10^{-70}$	m^2
Planck force	$1{,}21027 \times 10^{44}$	N
Planck power	$3{,}62831 \times 10^{52}$	W
Planck density	$5{,}15500 \times 10^{96}$	kg m^{-3}
Planck pressure	$4{,}63309 \times 10^{113}$	Pa
Planck angular frequency	$1{,}85487 \times 10^{43}$	s^{-1}
Planck current	$3{,}4789 \times 10^{25}$	A
Planck voltage	$1{,}04295 \times 10^{27}$	V
Planck impedance	$2{,}99792458 \times 10^{1}$	Ω

The Planck quantum of action (Planck's constant) is a fundamental value in quantum theory. The quantity 'action' is the product of momentum and length and is equal for all photon quanta. Multiplied by frequency this gives the energy value of a photon. From here it is only one or two steps to Einstein's $E = mc^2$. Interestingly, photons do have the quantity 'momentum' which is the product of mass and velocity, even though they do not have mass. Photons do however have energy and momentum can be expressed in terms of energy and velocity.

From the values given here it will be seen that the extremely large and small equivalents in the SI make the Planck units totally unsuited for use in the engineering situation but they are useful in quantum gravity research. The famous equations of physics, the Schrödinger wave equation, Maxwell-Heaviside equations, Einstein energy equation etc. all take on a 'clean shaven' look when expressed in Planck units. It remains necessary however to retain the unity-constants when dimensional consistency is required.

The length unit is smaller than the smallest subatomic particle, the temperature unit is hotter than the centre of a star, the velocity unit is the velocity of light. If you would wish to pass an electrical current of one Planck unit through a wire with the same current density used in ordinary house wiring, what thickness of wire would you need? – You would need a wire with a diameter larger than the diameter of the Sun! How about designing an electrical generator with an output of one Planck unit of power? – It would have a size of nearly a light year across! It remains eminently sensible to measure the height of your horse's withers in hands rather than in Planck lengths.

The Planck unit of electric charge and the electron charge are both fundamental quantities of the universe yet the Planck charge is nearly twelve times larger than the electron charge. This is a puzzle. The Planck charge does not correspond to any physical particle in whole, fraction or multiple, nevertheless the Planck charge leads us to another fundamental constant of the universe. The electron charge/Planck charge ratio squared gives us the Sommerfeld fine structure constant which deals with the behaviour of electrons on collision course, with photon ejection and absorption and other matters in particle physics. The value of the fine structure constant is important. If it were different by 4%, carbon would not be produced in stellar fusion, and if it were greater than 0,1 stellar fusion would not occur at all leaving a universe consisting of nothing but hydrogen. Further detail of the fine structure constant and comments by physicist Richard Feynman are given under the next heading "The Numeric Values of the Universe". Richard Feynman is known for his 'Feynman diagrams' which illustrate the reactions between fundamental particles showing the annihilation and creation of particles and photons. An interesting aspect of these processes is that parts can occur in negative time, provided that this is less time than can be noticed.

THE NUMERIC VALUES OF THE UNIVERSE

Size of Universe	$1{,}296 \times 10^{26}$ m
Age of Universe	432×10^{15} s ($13{,}7 \times 10^{9}$ years)
Number of galaxies in universe	350×10^{9} (rough estimate)
Number of dwarf galaxies	7×10^{12}
Number of stars in universe	3×10^{22} (Thirty billion trillion)
Number of stars in our galaxy	400×10^{9} (rough estimate)
Proportion of universe which is ordinary matter	4,6%
Proportion that is dark matter	23,3%
Proportion that is dark energy	72,1%
Number of atoms in universe	10^{78}
Number of photons in background radiation of universe	10^{88}

It is not possible to obtain an accurate figure for the number of stars in our galaxy as many stars are not visible at all due to distance and intervening dust. As for other galaxies, individual stars are hardly discernible, except for some exceptionally bright stars in close by galaxies or the occasional supernova. The number of galaxies in the universe can likewise be only roughly estimated.

c	Velocity of light in space	$2{,}997\ 924\ 58 \times 10^{8}$ m/s (exact)	$m.s^{-1}$
ε_0	Electrical permittivity of space	$8{,}854\ 187\ 817 \times 10^{-12}$ F/m	$s^{4}.A^{2}.m^{-3}.kg^{-1}$
μ_0	Magnetic permeability of space	$4\pi \times 10^{-7}$ H/m	$m.kg.s^{-2}.A^{-2}$

These three values are now exactly defined. The speed of light is exact as the metre is now defined by this value. Magnetic permeability μ_0 is taken as $4\pi \times 10^{-7}$ as dictated by the definition of the ampere.

The three are related by the Maxwellian: $\varepsilon_0 \mu_0 c^2 = 1$

This is a beautiful little equation linking the most fundamental properties of the universe. If you like, why not substitute the values and SI base units in the equation and see everything reduce to unity. This is a profound equation that made Maxwell famous.

Another fundamental equation of the universe is the Newtonian: $f = ma$

which links force, mass and acceleration. This equation, which finds widespread use in engineering and physics, needs a slight addition to fit in with Einstein's relativity. In the ordinary engineering situation the difference is vanishingly small.

Newtonian gravitational constant	$6{,}674\ 28(67) \times 10^{-11}$ m³.kg⁻¹.s⁻²	
Proton mass	$1{,}672621\ 637(83) \times 10^{-27}$ kg	938.272029(80) MeV
Neutron mass	$1{,}674927211(84) \times 10^{-27}$ kg	
Electron mass	$9{,}109382\ 15(45) \times 10^{-31}$ kg	0.510976 MeV
Electron charge	$1{,}602\ 176487(40) \times 10^{-19}$ C	
Planck's constant	$6{,}6260693(11) \times 10^{-34}$ J.s	
Avogadro's number (mole)	$6{,}022\ 141\ 79(30) \times 10^{23}$	
Faraday constant F	$96\ 485{,}3399(24)$ C mol⁻¹	
Magnetic flux quantum	$2{,}067\ 833\ 667(52) \times 10^{-15}$ Wb	
Boltzmann constant	$1{,}380\ 650\ 4(24) \times 10^{-23}$ J K⁻¹	
Loschmidt constant	$2{,}686\ 777\ 4(47) \times 10^{25}$ m⁻³	

The mass of sub atomic particles is often given in energy units, joules or electron volts, however we have given all mass values in kg for simplicity. The numbers in brackets denote the plus/minus uncertainty on the final digits.

A few energy conversions:

Joule per kilogram	$8,987\ 551\ 787 \times 10^{16}$
Kilogram per joule	$1,112\ 650\ 056 \times 10^{-17}$
Joule per electron volt	$1,602\ 17653(14) \times 10^{-19}$
Electron volt per joule	$6,241\ 50947 \times 10^{18}$
Kilogram per electron volt	$1,782661\ 81(15) \times 10^{-36}$
Electron volt per kilogram	$5,609\ 588\ 96 \times 10^{35}$

In any table of constants, the ubiquitous irrational pair of natural mathematical constants π and e cannot be omitted. We give them here to 24 significant digits which should be more than enough for most purposes.

π	$3,141\ 592\ 653\ 589\ 793\ 238\ 462\ 64 \ldots$
e	$2,718\ 281\ 828\ 459\ 045\ 235\ 360\ 28 \ldots$

With this value of π you can calculate the circumference of a circle the size of the Sun to an accuracy of much less than the size of an atom.

It is actually quite easy to remember this value of π by means of a mnemonic. The number of characters in each word is the corresponding value in π. Ignore the commas.

Now I need a drink, alcoholic of course, after the heavy lectures involving quantum mechanics. All of thy geometry, Herr Planck, is fairly hard.

Complex number theory can bring π and e, the base of natural logarithms, together with the beautiful little Euler equation $e^{i\pi} + 1 = 0$ (Where $i^2 = -1$)

This is the straight line real number case of: $e^{ix} = \cos x + i \sin x$

Another fundamental constant of the universe is the Fibonacci number Φ. This is a natural number to be found in snail shells, seed patterns of flowers and even Greek temples. Φ is easily demonstrated graphically – simply take a square and extend one side so that the length to width ratio of the rectangle formed is the same as for the small rectangle added. You can now cut a square off the small rectangle and the piece left will still have the same length to width ratio. This process can carry on ad infinitum showing the recursive fractal nature of the number. Φ is easily calculated. From the square example we get $\Phi = 1/(\Phi - 1)$ giving $\Phi = (1 + \sqrt{5})/2$

$\Phi = 1{,}618\ 033\ 988\ 749\ 894\ 848\ 204\ 586 \ldots$

The irrational numbers π and e belong to a class of numbers called transcendental numbers. Φ, π and e can each be computed from an infinite convergent series of terms.

The value of π has been in use since ancient times. A simple practical way of determining the value would be to take a wheel, make a mark on the circumference, and then roll it for precisely one revolution and divide this length by the diameter. The value has steadily been improved upon through the centuries but it can now be computed to millions of decimal places. The only useful purpose in doing this is to have a measure for comparing the number processing speed of computers. The ancient Egyptians must have had some or other method of obtaining π as the Giza pyramids have the same base perimeter and height as hemispheres. The use of the Greek letter π as the circumference to diameter ratio dates from the time of mathematician Euler. It is interesting to note that Euler did not define π in terms of a circle. He used π as a summation of an infinite convergent series for use in advanced mathematics - the fact that it was also the circumference to diameter ratio of a circle was a coincidence. There is a case on record when an actuarial formula of Euler containing π was explained to an actuary, the puzzled man exclaimed: 'My dear friend, that must be a delusion: what can a circle have to do with the number of people alive at the end of a given time?'

If we take the complex number Euler equation $e^{i\pi} = -1$ and describe it in terms of ordinary real numbers we get a curious result: the logarithm of -1 (an impossible number) divided by the square root of -1 (another impossible number) gives the circumference of a circle divided by its diameter. It is quite amazing what one can do with numbers that work sideways!

The two dimensional complex number identity $i^2 = -1$ was extended by mathematician Sir William Rowan Hamilton to four dimensional quaternions with the fundamental multiplication identity $i^2 = j^2 = k^2 = ijk = -1$.

There is a rather unusual constant of the universe proposed by Professor Arnold Sommerfeld in 1916 – The Fine Structure Constant. There has been much confusion and mystery surrounding this dimensionless number and was beautifully described by Richard P. Feynman (1918–1988) some 40 years ago:

> *There is a most profound and beautiful question associated with the observed coupling constant e, the amplitude for a real electron to emit or absorb a real photon. It is a simple number that has been experimentally determined to be close to 0.08542455. (My physicist friends won't recognize this number, because they like to remember it as the inverse of its square: about 137,03597 with an uncertainty of about 2 in the last decimal place. It has been a mystery ever since it was discovered more than fifty years ago, and all good theoretical physicists put this number up on their wall and worry about it.) Immediately you would like to know where this number for a coupling comes from: is it related to π or perhaps to the base of natural logarithms? Nobody knows. It's one of the greatest damn mysteries of physics: a magic number that comes to us with no understanding by man. You might say the "hand of God" wrote that number, and "we don't know how He pushed his pencil." We know what kind of a dance to do experimentally to measure this number very accurately, but we don't know what kind of dance to do on the computer to make this number come out, without putting it in secretly!*

It has subsequently been found that the fine structure constant α can be computed from:

$$\alpha = e^2 / 2\varepsilon_0 hc \qquad \text{where} \quad e = \text{electron charge}$$

where e = electron charge

ε_0 = electric permittivity of space

h = Planck's constant

c = velocity of light in space

This gives a value of 7,297 352 537 6 x 10^{-3} with an uncertainty of 50 on the last two digits. The constant can also be formulated in terms of the magnetic permeability of space by expressing electric permittivity in terms of magnetic permeability and the velocity of light.

If you like, why not substitute the SI base units in the equation and see them all cancel out. The constant can also be computed by dividing the electron charge by the Planck charge and squaring the result.

The CODATA 2006 values are:

$\alpha = 7{,}297\ 352\ 537\ 6(50) \times 10^{-3}$ $\qquad\qquad \sqrt{\alpha} = 0.085\ 424\ 5429$

$1/\alpha = 137{,}035999679(94)$

There is speculation that the fundamental constants of the universe may actually change very slightly over the passage of millions of years but there is of course no experimental possibility of measuring this. Few people should see this as a problem.

There is also speculation that the fundamental laws of physics could be quite different in various remote parts of the universe. Let us steer well clear of this matter! Another profound problem in theoretical physics is 'The Information Paradox' which we will not look into here – read about this elsewhere at your peril!

THE VIBES OF THE UNIVERSE

The universe is flooded with electromagnetic radiation, so much of it that the energy of this radiation exceeds that of all physical matter by about three to one. The photon/wave duality of electromagnetic radiation has been looked into in the chapter on photons. Here we take a look at the range of frequencies and wavelengths in which this energy manifests itself. The boundaries are approximate as there is some overlap of the various names of radiation. There is no difference between a short X-ray and a long gamma ray. Also, a 'soft' ray is one with a lower frequency and energy than a 'hard' ray.

Name of radiation	Wavelength metres	Frequency Hz	Photon energy yoctojoules
Zero-point energy	1.616×10^{-35}	1.85×10^{45}	1.229×10^{34}
Gamma rays	$< 10 \times 10^{-12}$	$> 30 \times 10^{18}$	$> 20 \times 10^{9}$
X-rays	$10 \times 10^{-12} - 10 \times 10^{-9}$	$30 \times 10^{15} - 30 \times 10^{18}$	$20 \times 10^{6} - 20 \times 10^{9}$
Ultra violet light	$10 \times 10^{-9} - 400 \times 10^{-9}$	$750 \times 10^{12} - 30 \times 10^{15}$	$500 \times 10^{3} - 20 \times 10^{6}$
Visible light	$400 \times 10^{-9} - 700 \times 10^{-9}$	$400 \times 10^{12} - 750 \times 10^{12}$	$300 \times 10^{3} - 500 \times 10^{3}$
Infra red, heat	$700 \times 10^{-9} - 1 \times 10^{-3}$	$300 \times 10^{9} - 400 \times 10^{12}$	$200 - 300 \times 10^{3}$
Microwaves	$1 \times 10^{-3} - 100 \times 10^{-3}$	$3 \times 10^{9} - 300 \times 10^{9}$	$2 - 200$
Microwave oven	0,12	$2,45 \times 10^{9}$	1,6
Satellite bands	$10 \times 10^{-3} - 200 \times 10^{-3}$	$1,5 \times 10^{9} - 31 \times 10^{9}$	$1 - 20$
RADAR short rge.	$7 \times 10^{-3} - 300 \times 10^{-3}$	$1 \times 10^{9} - 40 \times 10^{9}$	$0,6 - 26$
RADAR UHF	$0,3 - 1$	$300 \times 10^{6} - 1000 \times 10^{6}$	$0,2 - 0,6$
RADAR VHF	$1 - 6$	$50 \times 10^{6} - 330 \times 10^{6}$	$33 \times 10^{-3} - 218 \times 10^{-3}$
TV broadcasting	$0,4 - 6,0$	$50 \times 10^{6} - 800 \times 10^{6}$	$33 \times 10^{-3} - 530 \times 10^{-3}$
FM broadcasting	$3 - 3,5$	$87 \times 10^{6} - 110 \times 10^{6}$	$57 \times 10^{-3} - 73 \times 10^{-3}$
SW broadcasting	$11 - 120$	$3 \times 10^{6} - 30 \times 10^{6}$	$2 \times 10^{-3} - 20 \times 10^{-3}$
AM broadcasting	$170 - 600$	$500 \times 10^{3} - 1700 \times 10^{3}$	$0,3 \times 10^{-3} - 1,1 \times 10^{-3}$

Frequencies lower than the kilohertz range are regarded as em-disturbances.

The values given in this table are very simply related by:

$$c = \lambda f \qquad \text{and} \qquad e = h f \qquad \text{where} \quad c = \text{velocity of light}$$

$$\lambda = \text{ wavelength}$$

$$f = \text{frequency}$$

$$e = \text{photon energy}$$

$$h = \text{Planck's constant}$$

The symbols 'c' and 'e' are heavily overworked in science and mathematics and used for a variety of purposes.

If you prefer photon energy expressed in electron volts:

one MeV = 160,217 653 femtojoules.

When giving the colour of starlight or other light sources this is usually given as a temperature. This is the colour of the output peak on the black body radiation curve.

The energy of the photons is proportional to the frequency which is very high in the case of gamma rays. The figures for zero-point energy are crazy beyond human comprehension; perhaps new theory will shed fresh light upon this matter! Gamma rays are at best hazardous and at worst deadly. The band of visible light is one of the greatest blessings of the universe but all em-radiation can be dangerous in extreme doses. Even visible light can be hazardous – the light from an arc welder or a laser beam are examples. A mild dose of ultra violet light can be quite nice for a sun tan but there is a painful price to pay for too much exposure. The danger from ultra violet light and higher frequencies is due to this radiation being ionising, which can break chemical bonds. The danger from microwaves and longer wavelengths is from heating caused by induced eddy currents. Birds foolish enough to perch on high power RF antennae can be zapped to oblivion. There is current speculation that microwave radiation resulting from the ubiquitous use of cellular phones is placing the insect life of this planet in jeopardy. The possible risks to persons using cellular phones remain undecided; the chief concerns here being risk of brain cancer and impairment of fertility. Microwave ovens have become ubiquitous, being used in domestic and restaurant kitchens worldwide. The heat is produced mostly by the violent agitation of water molecules in the foodstuffs to be heated. Water molecules are electrically bipolar making them susceptible to high frequency radiation. There is much controversy surrounding the use of microwave ovens. It is claimed that microwave ovens can partially destroy the nutritional value of the food as well as generate toxic and carcinogenic substances. It is also claimed that there are cover-up conspiracies in place to protect the industries benefiting from the manufacture and distribution of microwave ovens.

There is some controversy over potential hazards from heavy current carrying power lines. Power lines can and do induce eddy currents in their vicinity. Power line eddy currents induced in humans do not produce any significant heating but the possible harmful effects of long exposure to these to the central nervous system and electrolysis in the metabolism remain to be determined. Electrical distribution authorities will of course insist that powerline radiation is harmless but there are cases where people regularly sleeping in proximity to heavy current carrying conductors have become gravely ill from undetermined causes. No one is suggesting that house wiring may be harmful – this is a problem concerning kiloamps.

There are two other 'rays' which must not be confused with em-radiation. Cosmic Rays consist of high energy extraterrestrial particles streaming down to Earth, often at a large fraction of the speed of light. Many of these come from the Sun and others from interstellar space originating from other stars and supernovas. The cosmic ray particles are of atomic nuclei of a wide assortment of elements and other particles but mostly hydrogen nuclei and also some electrons. All of these particles carry charges and are therefore greatly affected by the magnetic field of the Earth. The cosmic rays collide with atoms in the atmosphere resulting in showers and bursts of secondary radiation. Needless to say, cosmic rays are a huge and complex study of their own. The Earth also receives much em-radiation from the cosmos covering practically the entire electromagnetic spectrum. An alarming component of this is gamma ray bursts from supernovas.

Cathode rays are streams of electrons. In a TV tube the rays are produced at the cathode or electron gun at the back of the tube from where they stream forward to produce a picture on the screen which is the anode. A similar arrangement is found in oscilloscopes and many computer screen monitors.

In the pioneering days of radioactivity study the three types of radiation detected were simply named alpha, beta and gamma rays. The alpha rays are helium nuclei, the beta rays electrons, and gamma rays, the most penetrating, are high energy em-radiation. Alpha and beta radiation is highly ionising posing an extreme health hazard if particles emitting this radiation are inhaled or ingested. Ionising radiation targeting cellular DNA can cause cancers and mutations. Gamma rays are similar to X-rays but of higher frequency and energy and consequently more dangerous. When you have X-ray plates taken the radiographer or dentist will always get well out of the way, and with good reason. Having a few plates taken will do no harm but repeated exposure to this radiation on a daily basis would be extremely harmful.

In current medical practice the risk of X-rays has been greatly reduced by the use of low-dosage radiation and electronic image capturing. To complicate matters greatly, solid particles in flight also have an effective wavelength and frequency. Particles with mass cannot move at the speed of light so this is definitely not em-radiation. This concept was first proposed by Louis de Broglie in 1924. A practical result of this was the development of the electron microscope which uses electrons instead of photons and magnetic lenses instead of glass optics. Matter waves are mathematically described by the famous Schrödinger wave equation.

Besides visible light, the vibes that we are constantly aware of are of course sound waves. For us to hear them they must be in the air but they can travel much faster in solids and liquids. Sound waves are quite different to em-radiation as they require a medium through which to travel and quickly dissipate as they spread out. Sound waves are extremely slow compared to em-radiation. A sound wave in the air at sea level travels at only about 332 metres per second. Let us see what happens if we transpose the em-radiation frequencies to the sound waves of a musical keyboard. Accompanists often transpose a few semitones to suit the voice of a singer. Let's transpose visible light to start at middle C on a piano keyboard. All the colours will occupy less than the notes of one octave. An increase of one octave represents a doubling of frequency. A piano has three octaves higher in the treble and also three lower in the bass. To accommodate the whole em-spectrum we will need fifteen octaves to reach gamma rays in the treble and 29 octaves in the bass to reach AM broadcasting.

Electromagnetic radiation can be good, bad, ugly or even downright horrible. An obvious candidate for the last category is electromagnetic pulse which is commonly known as EMP. The first known source was from thermonuclear explosions where a vast surge of gamma rays interacts with atoms in the atmosphere and through a somewhat complex process becomes a continuous very broad spectrum of em-radiation. The duration of the pulse is extremely short, a few nanoseconds, about the time taken for light to travel one metre but the energy involved is immense. There is no particular frequency for the pulse – some of the energy can actually be of very low frequency. So where is the problem? The pulse can be picked up by almost any metallic object – power lines, electric wiring, fences, telephone wires, railway tracks or whatever. The induced electrical surges can cause damage on a mind boggling scale. Particularly at risk are any instruments or appliances using microelectronics – which obviously means all computers and communication systems, and more frighteningly, aircraft in flight. How bad is the danger? A large thermonuclear explosion occurring a few hundred kilometers above the Earth could disable the electronic circuits of an entire continent in a single flash! The damage would be terminal, requiring replacement of the destroyed components. Even components which have received a mild exposure could be sufficiently weakened to self destruct from their own power supply. Needless to say this has not yet happened. Deploying a nuclear weapon vertically downwards from deep space to detonate hundreds of kilometres above the Earth would be a military planner's worst nightmare. The attack would be completely undetectable until after detonation and would disable the entire military, industrial and economic infrastructure of a continent in milliseconds. The use of nuclear weapons is to be discouraged! A nuclear explosion in space would appear as an intense flash many times brighter than the sun. It would only be an instantaneous point source of light, heat and gamma rays; no sound, no fireball and no cloud – heat radiation striking the atmosphere may cause a thunder-like rumble. Hopefully no one will ever have to experience such an event. EMP is not only produced by nuclear explosions – on a smaller scale new military applications have been developed. EMP weapons can disable aircraft and missiles in flight, and even detonate artillery shells before they reach their targets. Is this not the 'death ray' weapon that Nikola Tesla wrote about nearly a century ago?

We should also look at another fairly recent entry to the vibe scene – the LASER which is an acronym for Light Amplification by Stimulated Emission of Radiation. This has absolutely nothing to do with 'lazar' which is derived from 'Lazarus' meaning death and decay. A laser is a very intense pencil of coherent light. Lasers were first produced in 1960 and soon became ubiquitous; they read our CDs and barcodes, select the pixels on our laser printers, are widely used in land surveying and even create reference 'stars' for use by astronomers. One of the most important uses of lasers is for fibre-optic communications. Lasers are also used in industry for boring ultra small holes and are also widely used in medicine for precision surgery. Lasers have also found use in the military situation. Laser weapons can zap aircraft and other hostile objects in flight.

There are also other vibes of as yet scientifically undetermined nature such as those investigated by Nikola Tesla, but these subtle energy life force vortices and feng shui effects do not lend themselves to discussion in the quantitative engineering context.

ACRONYMS AND ABBREVIATIONS

2MASS	Two Micron All Sky Survey
AAS	American Astronomical Society
ACE	Advanced Composition Explorer
ADEPT	Advanced Dark Energy Physics Telescope (Proposed mission)
AGILE	Astro-rivelatore Gamma a Immagini L'Eggero
AGN	Active Galactic Nucleus
AKARI	Infrared Imaging Surveyor - Japan (Previously ASTRO-F or IRIS)
ALICE	Ultraviolet imaging spectrometer
ALMA	Atacama Large Millimetre Array - radio observatory
ALPACA	Advanced Liquid mirror Probe for Astrophyics, Cosmology and Asteroids
ANTS	Autonomous Nano Technology Swarm Swarming technology for investigating asteroids
AOS	Acousto Optical Spectrometer
APO	Apache Point Observatory
ARC	Astrophysical Research Consortium
ARCSAT	ARC Small Aperture Telescope
ARECIBO	Giant radio telescope near town of Arecibo, Puerto Rico
ASC	Agence Spatiale Canadienne (CSA)
ASD	Astrophysics Science Division
ASI	Agenzia Spaziale Italiana (Italian space agency)
ASKAP	Australian SKA Pathfinder
ASPERA	Analyser of Space Plasmas and Energetic Atoms
ATA	Allen Telescope Array – SETI radio observatory

Baade	Magellan-1 6,5m optical telescope named in honour of astronomer Walter Baade
BAT	Burst Alert Telescope
BepiColombo	Mercury mission named in honour of Giuseppe Colombo ESA Cornerstone mission
BiSON	Birmingham Solar Oscillation Network
BLAZAR	Active galactic nucleus with highly variable radiation
BMDO	Ballistic Missile Defense Organisation
CASSEGRAIN	Telescope optical design by Laurent Cassegrain
CASSINI	Space mission to Saturn named in honour of Giovanni Domenica Cassini
CAUP	Centro de Astrofisica da Universidade do Porto
CCD	Charge Coupled Device
CGRO	Compton Gamma Ray Observatory
CHANDRA	X-ray space telescope named in honour of Subrahmanyan Chandrasekhar
Chandra	'Moon' (Sanskrit)
Chandrayaan	'Moon craft' (Sanskrit). ISRO Moon missions
CHARA	Centre for High Angular Resolution Astronomy
CICLOPS	Cassini Imaging Central Laboratory for OPerationS
Clay	Magellan-2 6,5m optical telescope named in honour of philanthropist Landon Clay
Clementine	Nickname for DSPSE
CME	Coronal Mass Ejection
CNSR	Comet Nucleus Sample Return
CNRS	Centre National de la Recherche Scientifique
COBE	COsmic Background Explorer

COBRAS/SAMBA	Microwave background mission renamed PLANCK Mission
CONSERT	Comet Nucleus Sounding Experiment by Radiowave Transmission
COROT	COnvection ROtation and planetary Transits
COS	Cosmic Origins Spectrograph
COSIMA	Cometary Secondary Ion Mass Analyser
CRAF	Comet Rendesvous Asteroid Flyby
CSA	Canadian Space Agency (ASC)
CTIO	Cerro Tololo Inter-American Observatory
CXO	Chandra X-Ray Observatory
DAMTP	Department of Applied Mathematics and Theoretical Physics
DARWIN	Space mission for detecting extraterrestrial life named in honour of Charles Darwin
DAWN	Space Mission to orbit two asteroids
DIRBE	Diffuse Infrared Background Experiment
DLR	Deutschen Zentrums für Luft- und Raumfahrt
DMR	Differential Microwave Radiometer
DSPSE	Deep Space Program Science Experiment (Clementine)
DTF	Digital Fourier Transform spectrometer
EDLS	ExoMars Entry, Descent and Landing System
E-ELT	European Extremely Large telescope
EHT	Event Horizon Telescope
ELODIE	High resolution spectrograph installed at the Observatoire de Haute-Provence
EPIC	European Photon Imaging Camera
ESA	European Space Agency

ESO	European Southern Observatory
ESTEC	European Space Research and Technology Centre
EVN	European VLBI Network
FAST	Five hundred metre Aperture Spherical Telescope (China)
Fermi	Fermi Gamma-ray Space Telescope (Formerly GLAST) Named in honour of Enrico Fermi
FINESSE	Frequency domain INterfErometry Simulation SoftwarE
FIPS	Fast Imaging Plasma Spectrometer
FIRAS	Far InfraRed Absolute Spectrophotometer
FIRST	Far InfraRed Space Telescope (renamed Herschel)
FPMI	Fabry-Perot Michelson interferometer
FUSE	Far Ultraviolet Spectroscopic Explorer
GALEX	GALaxy Evolution eXplorer. NASA small explorer class mission to study structure and evolution of universe
GBT	Green Bank Telescope
GBM	GLAST Burst Monitor
GEMS	Gravity and Extreme Magnetism SMEX mission
GEO600	European Gravitational Observatory with 600 metre interferometers
GEODSS	Ground-based Electro-Optical Deep Space Surveillance telescope
GIOTTO	ESA mission launched in 1986 to study Halley's comet
GLAST	Gamma ray Large Area Space Telescope (Renamed Fermi)
GMC	Giant Molecular Cloud (Stellar nursery)
GMT	Great Magellan Telescope
GRACE	Gravity Recovery And Climate Experiment

GREGORIAN	Telescope optical design by James Gregory
GSFC	NASA's Goddard Space Flight Centre
HARPS	High Accuracy Radial velocity Planet Searcher
HERSCHEL	Infrared space mission named in honour of William Herschel
HET	Hobby-Eberly Telescope
HETE	High Energy Transient Explorer
HEX	High Energy X-ray/gamma ray spectrometer
HIFI	Heterodyne Instrument for the Far Infrared
HiRISE	High Resolution Imaging Science Experiment
HRSC	High Resolution Stereo Camera
HST	Hubble Space Telescope
HUBBLE	Earth orbiting telescope (HST) named in honour of Edwin Hubble
HySI	Hyper Spectral Imager
IAU	International Astronomical Union
IMPACT	In-situ Measurement of Particles And cmE Transients.
INFN	Instituto Nazionale de Fisica Nucleare
INTEGRAL	INTErnational Gamma Ray Astrophysics Laboratory
IRIS ISAS	InfraRed Imaging Surveyor (Renamed AKARI) Institute of Space and AStronomical research - Japan
ISRO	Indian Space Research Organisation
JAXA	Japan Aerospace eXploration Agency
JCMT	James Clerk Maxwell Telescope
JDEM	Joint Dark Energy Mission (Proposed mission)
JPL	Jet Propulsion Laboratory – a division of the California Institute of Technology in Pasedena
JWST	James Webb Space Telescope (Formerly NGST)

KAGUYA	Nickname for Selene mission named after Japanese princess
KAT-7	Seven dish testbed for meerKAT precursor to SKA
KBO	Kuiper Belt Object
KECK	Very large telescopes named in honour of William Keck
KEPLER	Space mission for detecting extraterrestrial planets named in honour of Johannes Kepler
LAGEOS	LAser GEOdynamics Satellite
LAMA	Large Aperture liquid Mirror Array
LBT	Large Binocular Telescope
LCROSS	Lunar Crater Observation and Sensing Satellite
LIBS	Laser-induced Breakdown Spectroscopy
LIGO	Laser Interferometer Gravitational wave Observatory
LINEAR	LIncoln Near Earth Asteroid Research
LISA	Laser Interferometer Space Antennae – gravitational mission
LLNL	Lawrence Livermore National Laboratory
LLRI	Lunar Laser Ranging Instrument
LMT	Liquid Mirror Telescope
LRO	Lunar Reconnaissance Orbiter
LSST	Large (8m) Synoptic Survey Telescope (Wide angle)
PLRP	Lunar Precursor Robotic Program
MARCI	Mars Colour Imager
MaRS	Mars Radio Science Experiment
MARVELS	Multi-object APO Radial Velocity Exoplanet Large-area Survey
MASER	Microwave Amplification by Stimulated Emission of Radiation

meerKAT	Karoo Array Telescope (precursor to SKA)
MERLIN	Multi Element Radio Linked Interferometric Network
MESSENGER	MErcury Surface, Space ENvironment, GEochemistry and Ranging
MIDAS	Micro-Imaging Dust Analysis System
MIDEX	Medium class Explorer missions
M3	Moon Mineralogy Mapper
MMO	Mercury Magnetospheric Orbiter
MMT	Multiple Magnum Mirror Telescope
MMX	Michelson Morley Experiment
MOA	Microlensing Observations in Astrophysics
MPO	Mercury Planetary Orbiter
MRO	Mars Reconnaissance orbiter
MSL	Mars Science Laboratory
NAOJ	National Astronomical Observatory of Japan
NASA	National Aeronautics and Space Administration
NCRA	National Centre for Radio Astrophysics
NEAT	Near Earth Asteroid Tracking
NEO	Near Earth Object
NGST	Next Generation Space Telescope (Now named JWST)
NIAC	NASA Institute for Advanced Concepts
NICMOS	Near Infrared Camera and Multi-Object Spectrometer
NOAO	National Optical Astronomy Observatory
NRAO	National Radio Astronomy Observatory
NRL	Naval Research Laboratory
OPPORTUNITY	NASA Mars exploration rover – Name chosen by competition

OSIRIS	Optical, Spectroscopic, and Infrared Remote Imaging System
PACS	Infrared Photoconductor Array Camera and Spectrometer.
PARANAL	ESO very large telescopes atop Cerro Paranal, Chile
PAM	Prospecting Asteroid Mission
Pan-STARRS	Panoramic Survey Telescope And Rapid Response System
PFS	Planetary Fourier Spectrometer
PHARO	Palomar High Angular Resolution Observer
PLANCK	Cosmic background radiation mission named in honour of Max Planck
PLASTIC	PLAsma and SupraThermal Ion Composition.
PPARC	Particle Physics and Astronomy Research Council
PFS	Planetary Fourier Spectrometer
RADOM	RAdiation DOse Monitor
RGO	Royal Greenwich Observatory
RHESSI	Ramaty High-Energy Solar Spectroscopic Imager
RKA	Russian Space Agency
RTG	Radioisotope Thermoelectric Generator
ROSETTA	ESA Robotic space mission launched in 2004 to study comet 67P/Churyumov-Gerasimenko
RXTE	Rossi X-ray Timing Explorer
SAFIR	Single Aperture Far-InfraRed observatory
SALT	South African Large Telescope
SAO	Smithsonian Astrophysical Observatory
SAR	Specific Absorption Rate
SAR	Synthetic Aperture Radar

SARA	Sub-keV Atom Reflecting Analyser
SCUBA	Submillimetre Common Users Bolometer Array
SDO	Scattered Disk Object
SDO	Solar Dynamics Observatory
SECCHI	Sun Earth Connection Coronal and Heliospheric Investigation
SELENE	SELenological and ENgineering Explorer - Lunar orbiting mission
SETI	Search for ExtraTerrestrial Intelligence
SIM	Space Interferometer Mission
SKA	Square Kilometre Array radio telescope
SNAP	SuperNova Acceleration Probe (Proposed mission)
SMART	Small Missions for Advanced Research in Technology - ESA missions
SOAR	SOuthern Astrophysical Research telescope
SOFIA	Stratospheric Observatory For Infrared Astronomy
SOHO	SOlar and Heliospheric Observatory (ESA and NASA)
SPDS	ExoMars Drill and Sample Preparation and Distribution System
SPICAV	Spectroscopy for Investigation of Characteristics of the Atmosphere of Venus
SPIRIT	NASA Mars exploration rover – Name chosen by competition
SPT	South Pole Telescope
SQUID	Superconducting QUantum Interference Device
SST	Spitzer Space Telescope named in honour of Lyman Spitzer
STARDUST	NASA mission to collect samples from comet Wild 2
STEREO	Solar TErrestrial RElations Observatory

SUSAKU	Gamma ray mission – formerly Astro-E2
SWAS	Submillimetre Wave Astronomy Satellite
SWEEPS	Saggitarius Window Eclipsing Extrasolar Planet Search
SWIFT	Space mission for swiftly observing gamma ray bursts
TAMA300	300m Laser interferometer gravitational wave antenna
THEMIS	Time History of Events and Macroscale Interactions during Substorms
TMC	Terrain Mapping Camera
TNO	Trans-Neptunian Object
TIFR	TATA Institute of Fundamental Research
TPF	Terrestrial Planet Finder
UCLA	University of California, Los Angeles
UVOT	Ultra Violet and Optical Telescope
VIRTIS	Visible and Infrared Thermal Imaging Spectrometer
VISTA	Visible and Infrared Survey Telescope for Astronomy
VLA	Very Large Array
VLBI	Very Large Baseline Interferometry (Radio astronomy)
VSOP-2	Very long baseline interferometer Space Observatory, Program-2
WISDOM	ExoMars Water Ice and Subsurface Deposit Observations on Mars
WISE	Wide-field Infrared Survey Explorer
WMAP	Wilkinson Microwave Anisotropy Probe
XEUS	X-Ray Evolving Universe Spectrometer
XMM	X-Ray Multi Mirror mission (ESA Cornerstone)
XSM	Solar X-ray Monitor

Index

H

I

J

K

www.ingramcontent.com/pod-product-compliance
Lightning Source LLC
Chambersburg PA
CBHW051203200326
41519CB00025B/6986